T0205699

A Course in Modern Analysis and Its Applications

AUSTRALIAN MATHEMATICAL SOCIETY LECTURE SERIES

Editor-in-chief: Associate Professor Michael Murray, University of Adelaide

Editors:
Professor P. Broadbridge,
University of Delaware, USA

Professor C. C. Heyde,
Australian National University, Canberra, Australia

Associate Professor C. E. M. Pearce,
University of Adelaide, Australia

Professor C. Praeger,
University of Western Australia, Australia

A Course in Modern Analysis
and Its Applications

Graeme L. Cohen
University of Technology, Sydney

CAMBRIDGE
UNIVERSITY PRESS

CAMBRIDGE
UNIVERSITY PRESS

University Printing House, Cambridge CB2 8BS, United Kingdom

One Liberty Plaza, 20th Floor, New York, NY 10006, USA

477 Williamstown Road, Port Melbourne, VIC 3207, Australia

4843/24, 2nd Floor, Ansari Road, Daryaganj, Delhi - 110002, India

79 Anson Road, #06-04/06, Singapore 079906

Cambridge University Press is part of the University of Cambridge.

It furthers the University's mission by disseminating knowledge in the pursuit of education, learning and research at the highest international levels of excellence.

www.cambridge.org
Information on this title: www.cambridge.org/9780521526272

© Cambridge University Press 2003

First published 2003

A catalogue record for this publication is available from the British Library

ISBN   978-0-521-81996-1   Hardback
ISBN   978-0-521-52627-2   Paperback

Contents

Preface

This book offers an introduction to some basic aspects of modern analysis. It is designed for students who are majoring in some area of mathematics but who do not necessarily intend to continue their studies at a graduate level.

The choice of material and the method of presentation are both aimed at as wide a readership as possible. Future teachers of high school mathematics should be given an introduction to the mathematical future as much as they must be given some knowledge of the mathematical past; students of mathematical engineering, biology or finance may need to read current literature without desiring to contribute to it. These are perhaps the extremes in the type of student to whom this book is directed. At the same time, students who do need to go on to courses in measure theory and functional analysis will find this book an easy introduction to the initial concepts in those areas.

Syllabus requirements would rarely allow more than one semester to be available for work of this nature in an undergraduate course and this imposes restrictions on topics and the depth of their presentation. In line with the above thoughts, I have tried throughout to merge the nominal divisions of pure and applied mathematics, leaving enough for students of either inclination to have a feeling for what further developments might look like. After a somewhat objective choice of topics, the guiding rule in the end was to carry those topics just far enough that their applications might be appreciated. Applications have been included from such fields as differential and integral equations, systems of linear algebraic equations, approximation theory, numerical analysis and quantum mechanics.

The better the reader's knowledge of real variable analysis, the easier this book will be to work through. In particular, a thorough understanding of convergence of sequences and series is desirable. However, it should be possible to manage with little more than the quite detailed summary of these notions in Chapter 1. This is a lengthy chapter and the reasons for its length must be explained. It aims essentially to review or at least mention all topics required in the following chapters. But considerable attention has been given to maintaining from beginning to end a stream of thought which justifies and anticipates the generalisations that follow. The central and recurring theme is the completeness of the real number system. It is not advised to take the chapter too seriously at its first reading. Read as much as possible at a sitting, skipping proofs and difficult passages, and just retaining sufficient to be able to follow the development. Return later to the less understood pieces. Review exercises that imply a suitable level of understanding have been included throughout this chapter.

Nothing is used in this book from the theory of functions of a complex variable, from theories of measure and integration, such as Lebesgue integration, or from modern algebra. Topics like completeness and compactness are approached initially through convergence of sequences in metric space, and the emphasis remains on this approach. However, the alternative topological approach is described in a separate chapter. This chapter, Chapter 5, gives the book more flexibility as an introductory text for subsequent courses, but there are are only a few later references to it and it may be omitted if desired.

Except for the exercises in Chapter 1, each exercise set is split in two by a dotted line. Those exercises before the line are essential for an understanding of the concepts that precede them and in some cases are referred to subsequently; those after the line are either harder practice exercises or introduce theoretical ideas not later required. The book includes a large number of solved problems which should be considered as an integral part of the text. Furthermore, many of the exercises before the line in each set have complete solutions given at the end of the book.

This edition is a completely revised and extended version of notes I produced in 1978 and have been using ever since. Many colleagues, of whom I mention Dr Gordon McLelland in particular, read sections from that earlier manuscript and I am grateful for their comments. Dr Xuan Tran, as an undergraduate, solved all the exercises in the book, when no solutions were included, and was therefore of great assistance in compiling the solutions given here.

David Tranah, from Cambridge University Press, has been of great assistance in guiding the preparation of this edition, and I am extremely grateful to him. The copious comments of an unknown referee are also very much appreciated.

A little belatedly, I must also thank Professor John Ward, from whose *Young Mathematician's Guide* I quote overleaf. His subject matter may have differed considerably from mine, but our philosophies in writing seem to coincide remarkably.

Graeme L. Cohen
University of Technology, Sydney

October 2002

This I may (without vanity) presume to say, that whoever Reads it over, will find more in it than the Title doth promise, or perhaps he expects. 'Tis true indeed, the Dress is but Plain and Homely, it being wholly intended to Instruct, and not to Amuse or Puzzle the young Learner with hard Words; nor is it my Ambitious Desire of being thought more Learned or Knowing than really I am ...; However in this I shall always have the Satisfaction, That I've sincerely Aim'd at what is Useful, altho' in one of the meanest Ways; 'Tis Honour enough for me to be accounted as one of the under Labourers in Clearing the Ground a little, and Removing some of the Rubbish that lies in the way to Knowledge. How well I have performed that, must be left to proper Judges.

From the Preface of *The Young Mathematicians's Guide*, by John Ward, third edition, 1719.

1

Prelude to Modern Analysis

1.1 Introduction

The primary purpose of this chapter is to review a number of topics
from analysis, and some from algebra, that will be called upon in the
following chapters. These are topics of a classical nature, such as appear
in books on advanced calculus and linear algebra. For our treatment of
modern analysis, we can distinguish four fundamental notions which will
be particularly stressed in this chapter. These are

(a) set theory, of an elementary nature;
(b) the concept of a function;
(c) convergence of sequences; and
(d) some theory of vector spaces.

On a number of occasions in this chapter, we will also take the time
to discuss the relationship of modern analysis to classical analysis. We
begin this now, assuming some knowledge of the points (a) to (d) just
mentioned.

Modern analysis is not a new brand of mathematics that replaces the
old brand. It is totally dependent on the time-honoured concepts of
classical analysis, although in parts it can be given without reference to
the specifics of classical analysis. For example, whereas classical analysis
is largely concerned with functions of a real or complex variable, modern
analysis is concerned with functions whose domains and ranges are far
more general than just sets of real or complex numbers. In fact, these
functions can have domains and ranges which are themselves sets of
functions. A function of this more general type will be called an operator
or mapping. Importantly, very often any set will do as the domain of a
mapping, with no specific reference to the nature of its elements.

1

This illustrates how modern analysis generalises the ideas of classical analysis. At the same time, in many ways modern analysis simplifies classical analysis because it uses a basic notation which is not cluttered with the symbolism that characterises many topics of a classical nature. Through this, the unifying aspect of modern analysis appears because when the symbolism of those classical topics is removed a surprising similarity becomes apparent in the treatments formerly thought to be peculiar to those topics.

Here is an example:

$$\int_a^b k(s,t)x(t)\,dt = f(s), \quad a \leqslant s \leqslant b,$$

is an *integral equation*; f and k are continuous functions and we want to solve this to find the continuous function x. The left-hand side shows that we have operated on the function x to give the function f, on the right. We can write the whole thing as

$$Kx = f,$$

where K is an operator of the type we just mentioned. Now the essence of the problem is clear. It has the same form as a matrix equation $A\mathbf{x} = \mathbf{b}$, for which the solution (sometimes) is $\mathbf{x} = A^{-1}\mathbf{b}$. In the same way, we would like the solution of the integral equation to be given simply as $x = K^{-1}f$. The two problems, stripped of their classical symbolism, appear to be two aspects of a more general study.

The process can be reversed, showing the strong applicability of modern analysis: when the symbolism of a particular branch of classical analysis is restored to results often obtained only because of the manipulative ease of the simplified notation, there arise results not formerly obtained in the earlier theory. In other cases, this procedure gives rise to results in one field which had not been recognised as essentially the same as well-known results in another field. The notations of the two branches had fully disguised the similarity of the results.

When this occurs, it can only be because there is some underlying structure which makes the two (or more) branches of classical analysis appear just as examples of some work in modern analysis. The basic entities in these branches, when extracted, are apparently combined together in a precisely corresponding manner in the several branches. This takes us back to our first point of the generalising nature of modern analysis and of the benefit of working with quite arbitrary sets. To combine the elements of these sets together requires some basic ground rules and this is why, very often and predominantly in this book, the sets are assumed to be vector spaces: simply because vector spaces are sets with certain rules attached allowing their elements to be combined together in a particular fashion.

We have indicated the relevance of set theory, functions and vector spaces in our work. The other point, of the four given above, is the springboard that takes us from algebra into analysis. In this book, we use in a very direct fashion the notion of a convergent sequence to generate virtually every result.

We might mention now, since we have been comparing classical and modern analysis, that another area of study, called functional analysis, may today be taken as identical with modern analysis. A functional is a mapping whose range is a set of real or complex numbers and functional analysis had a fairly specific meaning (the analysis of functionals) when the term was first introduced early in the 20th century. Other writers may make technical distinctions between the two terms but we will not.

In the review which follows, it is the aim at least to mention all topics required for an understanding of the subsequent chapters. Some topics, notably those connected with the points (a) to (d) above, are discussed in considerable detail, while others might receive little more than a definition and a few relevant properties.

1.2 Sets and numbers

A *set* is a concept so basic to modern mathematics that it is not possible to give it a precise definition without going deeply into the study of mathematical logic. Commonly, a set is described as any collection of objects but no attempt is made to say what a 'collection' is or what an 'object' is. We are forced in books of this type to accept sets as fundamental entities and to rely on an intuitive feeling for what a set is.

The objects that together make up a particular set are called *elements* or *members* of that set. The list of possible sets is as long as the imagination is vivid, or even longer (we are hardly being precise here) since, importantly, the elements of a set may themselves be sets.

Later in this chapter we will be looking with some detail into the properties of certain sets of numbers. We are going to rely on the reader's experience with numbers and not spend a great deal of time on the development of the real number system. In particular, we assume familiarity with

(a) the *integers*, or whole numbers, such as -79, -3, 0, 12, $4{,}063{,}180$;

(b) the *rational numbers*, such as $-\frac{5}{3}$, $\frac{11}{17}$, which are numbers expressible as a ratio of integers (the integers themselves also being examples);

(c) those numbers which are not rational, known as *irrational numbers*, such as $\sqrt{2}$, $\sqrt[3]{15}$, π;

(d) the *real numbers*, which are numbers that are either rational or irrational;

(e) the ordering of the real numbers, using the inequality signs $<$ and $>$ (and the use of the signs $\leqslant$ and $\geqslant$);

(f) the representation of the real numbers as points along a line; and

(g) the fact, in (f), that the real numbers fill the line, leaving no holes: to every point on the line there corresponds a real number.

The final point is a crucial one and may not appear to be so familiar. On reflection however, it will be seen to accord with experience, even when expressed in such a vague way. This is a crude formulation of what is known as the *completeness* of the real number system, and will be referred to again in some detail subsequently.

By way of review, we remark that we assume the ability to perform simple manipulations with inequalities. In particular, the following should be known. If a and b are real numbers and $a < b$, then

$$-a > -b;$$

$$\frac{1}{a} > \frac{1}{b}, \text{ if also } a > 0 \text{ or } b < 0;$$

$$\sqrt{a} < \sqrt{b}, \text{ if also } a \geqslant 0.$$

With regard to the third property, we stress that the use of the radical sign ($\sqrt{}$) always implies that the nonnegative root is to be taken. Bearing this comment in mind, we may define the *absolute value* $|a|$ of any real number a by

$$|a| = \sqrt{a^2}.$$

More commonly, and equivalently of course, we say that $|a|$ is a whenever $a > 0$ and $|a|$ is $-a$ whenever $a < 0$, while $|0| = 0$. For any real numbers a and b, we have

$$|a + b| \leqslant |a| + |b|, \qquad |ab| = |a|\,|b|.$$

These may be proved by considering the various combinations of positive and negative values for a and b.

We also assume a knowledge of *complex numbers*: numbers of the form $a + ib$ where a and b are real numbers and i is an imaginary unit, satisfying $i^2 = -1$.

This is a good place to review a number of definitions and properties connected with complex numbers. If $z = a+ib$ is a complex number, then we call the numbers a, b, $a - ib$ and $\sqrt{a^2 + b^2}$ the *real part, imaginary part, conjugate* and *modulus*, respectively, of z, and denote these by $\operatorname{Re} z$, $\operatorname{Im} z$, $\bar{z}$ and $|z|$, respectively. The following are some of the simple properties of complex numbers that we use. If z, z_1 and z_2 are complex numbers, then

$$\overline{\bar{z}} = z,$$
$$\overline{z_1 + z_2} = \bar{z}_1 + \bar{z}_2,$$
$$\overline{z_1 z_2} = \bar{z}_1 \bar{z}_2,$$
$$|\operatorname{Re} z| \leqslant |z|, \ |\operatorname{Im} z| \leqslant |z|,$$
$$z\bar{z} = |z|^2,$$
$$|z_1 + z_2| \leqslant |z_1| + |z_2|,$$
$$|z_1 z_2| = |z_1| |z_2|.$$

It is essential to remember that, although z is a complex number, the numbers $\operatorname{Re} z$, $\operatorname{Im} z$ and $|z|$ are real. The final two properties in the above list are important generalisations of the corresponding properties just given for real numbers. They can be generalised further, in the natural way, to the sum or product of three or four or more complex numbers.

Real numbers, complex numbers, and other sets of numbers, all occur so frequently in our work that it is worth using special symbols to denote them.

Definition 1.2.1 The following symbols denote the stated sets:

N, the set of all positive integers;
Z, the set of all integers (positive, negative and zero);
Q, the set of all rational numbers;
R, the set of all real numbers;
R$_+$, the set of all nonnegative real numbers;
C, the set of all complex numbers.

Other sets will generally be denoted by ordinary capital letters and their elements by lower case letters; the same letter will not always refer to the same set or element. To indicate that an object x is an element of a set X, we will write $x \in X$; if x is not an element of X, we will write $x \notin X$. For example, $\sqrt{2} \in \mathbf{R}$ but $\sqrt{2} \notin \mathbf{Z}$. A statement such as $x, y \in X$ will be used as an abbreviation for the two statements $x \in X$ and $y \in X$. To show the elements of a set we always enclose them in braces and give either a complete listing (for example, $\{1, 2, 3\}$ is the set consisting of the integers 1, 2 and 3), or an indication of a pattern (for example, $\{1, 2, 3, \dots\}$ is the set $\mathbf{N}$), or a description of a rule of formation following a colon (for example, $\{x : x \in \mathbf{R}, \ x \geqslant 0\}$ is the set $\mathbf{R}_+$). Sometimes we use an abbreviated notation (for example, $\{n : n = 2m, \ m \in \mathbf{N}\}$ and $\{2n : n \in \mathbf{N}\}$ both denote the set of all even positive integers).

An important aspect in the understanding of sets is that the order in which their elements are listed is irrelevant. For example, $\{1, 2, 3\}$, $\{3, 1, 2\}$, $\{2, 1, 3\}$ are different ways of writing the same set. However, on many occasions we need to be able to specify the first position, the second position, and so on, and for this we need a new notion. We speak of *ordered pairs* of two elements, *ordered triples* of three elements, and, generally, *ordered n-tuples* of n elements with this property that each requires for its full determination a list of its elements and the order in which they are to be listed. The elements, in their right order, are enclosed in parentheses (rather than braces, as for sets). For example, $(1, 2, 3)$, $(3, 1, 2)$, $(2, 1, 3)$ are different ordered triples. This is not an unfamiliar notion. In ordinary three-dimensional coordinate geometry, the coordinates of a point provide an example of an ordered triple: the three ordered triples just given would refer to three different points in space.

We give now a number of definitions which help us describe various manipulations to be performed with sets.

Definition 1.2.2

(a) A set S is called a *subset* of a set X, and this is denoted by $S \subseteq X$ or $X \supseteq S$, if every element of S is also an element of X.

(b) Two sets X and Y are called *equal*, and this is denoted by $X = Y$, if each is a subset of the other; that is, if both $X \subseteq Y$ and $Y \subseteq X$. Otherwise, we write $X \neq Y$.

(c) A set which is a subset of any other set is called a *null set* or *empty set*, and is denoted by $\varnothing$.

(d) A set S is called a *proper* subset of a set X if $S \subseteq X$, but $S \neq X$.

(e) The *union* of two sets X and Y, denoted by $X \cup Y$, is the set of elements belonging to at least one of X and Y; that is,

$$X \cup Y = \{x : x \in X \text{ or } x \in Y \text{ (or both)}\}.$$

(f) The *intersection* of two sets X and Y, denoted by $X \cap Y$, is the set of elements belonging to both X and Y; that is,

$$X \cap Y = \{x : x \in X \text{ and } x \in Y\}.$$

(g) The *cartesian product* of two sets X and Y, denoted by $X \times Y$, is the set of all ordered pairs, the first elements of which belong to X and the second elements to Y; that is,

$$X \times Y = \{(x,y) : x \in X, \ y \in Y\}.$$

(h) The *complement* of a set X, denoted by $\sim X$, is the set of elements that do not belong to X; that is, $\sim X = \{x : x \notin X\}$. The complement of X relative to a set Y is the set $Y \cap \sim X$; this is denoted by $Y \backslash X$.

For some simple examples illustrating parts of this definition, we let $X = \{1,3,5\}$ and $Y = \{1,4\}$. Then

$$X \cup Y = \{1,3,4,5\}, \qquad X \cap Y = \{1\},$$
$$X \times Y = \{(1,1),(1,4),(3,1),(3,4),(5,1),(5,4)\},$$
$$Y \times X = \{(1,1),(1,3),(1,5),(4,1),(4,3),(4,5)\}.$$

We see that in general $X \times Y \neq Y \times X$. The set $Y \backslash X$ is the set of elements of Y that do not belong to X, so here $Y \backslash X = \{4\}$.

The definitions of union, intersection and cartesian product of sets can be extended to more than two sets. Suppose we have n sets X_1, X_2, ..., X_n. Their union, intersection and cartesian product are defined as

$$X_1 \cup X_2 \cup \cdots \cup X_n = \bigcup_{k=1}^{n} X_k$$
$$= \{x : x \in X_k \text{ for at least one } k = 1, 2, \ldots, n\},$$

$$X_1 \cap X_2 \cap \cdots \cap X_n = \bigcap_{k=1}^{n} X_k$$
$$= \{x : x \in X_k \text{ for all } k = 1, 2, \ldots, n\},$$

$$X_1 \times X_2 \times \cdots \times X_n = \prod_{k=1}^{n} X_k$$
$$= \{(x_1,x_2,\ldots,x_n) : x_k \in X_k \text{ for } k = 1, 2, \ldots, n\},$$

respectively (the cartesian product being a set of ordered n-tuples). The notations in the middle are similar to the familiar sigma notation for addition, where we write

$$x_1 + x_2 + \cdots + x_n = \sum_{k=1}^{n} x_k,$$

when x_1, x_2, ..., x_n are numbers.

For cartesian products only, there is a further simplification of notation when all the sets are equal. If $X_1 = X_2 = \cdots = X_n = X$, then in place of $\prod_{k=1}^{n} X_k$ or $\prod_{k=1}^{n} X$ we write simply X^n, as suggested by the $\times$ notation, but note that there is no suggestion of multiplication: X^n is a set of n-tuples. In particular, it is common to write $\mathbf{R}^n$ for the set of all n-tuples of real numbers and $\mathbf{C}^n$ for the set of all n-tuples of complex numbers.

It is necessary to make some comments regarding the definition of an empty set in Definition 1.2.2(c). These are gathered together as a theorem.

Theorem 1.2.3

   (a) *All empty sets are equal.*
   (b) *The empty set has no elements.*
   (c) *The only set with no elements is the empty set.*

To prove (a), we suppose that $\varnothing_1$ and $\varnothing_2$ are any two empty sets. Since an empty set is a subset of any other set, we must have both $\varnothing_1 \subseteq \varnothing_2$ and $\varnothing_2 \subseteq \varnothing_1$. By the definition of equality of sets, it follows that $\varnothing_1 = \varnothing_2$. This proves (a) and justifies our speaking of 'the' empty set in the remainder of the theorem. We prove (b) by contradiction. Suppose $x \in \varnothing$. Since for any set X we have $\varnothing \subseteq X$ and $\varnothing \subseteq \sim X$, we must have both $x \in X$ and $x \in \sim X$. This surely contradicts the existence of x, proving (b). Finally, we prove (c), again by contradiction. Suppose X is a set with no elements and suppose $X \neq \varnothing$. Since $\varnothing \subseteq X$, this means that X is not a subset of $\varnothing$. Then there must be an element of X which is not in $\varnothing$. But X has no elements so this is the contradiction we need.   □

All this must seem a bit peculiar if it has not been met before. In defence, it may be pointed out that sets were only introduced intuitively in the first place and that the inclusion in the concept of 'a set with no elements' is a necessary addition (possibly beyond intuition) to provide consistency elsewhere. For example, if two sets X and Y have no elements in common and we wish to speak of their intersection, we can now happily say $X \cap Y = \varnothing$. (Two such sets are called *disjoint*.)

Manipulations with sets often make use of the following basic results.

Theorem 1.2.4 *Let X, Y and Z be sets. Then*

(a) $\sim(\sim X) = X$,

(b) $X \cup Y = Y \cup X$ *and* $X \cap Y = Y \cap X$ *(commutative rules)*,

(c) $X \cup (Y \cup Z) = (X \cup Y) \cup Z$ *and* $X \cap (Y \cap Z) = (X \cap Y) \cap Z$
   (associative rules),

(d) $X \cup (Y \cap Z) = (X \cup Y) \cap (X \cup Z)$ *and* $X \cap (Y \cup Z) = (X \cap Y) \cup (X \cap Z)$
   (distributive rules).

We will prove only the second distributive rule. To show that two sets are equal we must make use of the definition of equality in Definition 1.2.2(b).

First, suppose $x \in X \cap (Y \cup Z)$. Then $x \in X$ and $x \in Y \cup Z$. That is, x is a member of X and of either Y or Z (or both). If $x \in Y$ then $x \in X \cap Y$; if $x \in Z$ then $x \in X \cap Z$. At least one of these must be true, so $x \in (X \cap Y) \cup (X \cap Z)$. This proves that $X \cap (Y \cup Z) \subseteq (X \cap Y) \cup (X \cap Z)$. Next, suppose $x \in (X \cap Y) \cup (X \cap Z)$. Then $x \in X \cap Y$ or $x \in X \cap Z$ (or both). In both cases, $x \in X \cap (Y \cup Z)$ since in both cases $x \in X$, and $Y \subseteq Y \cup Z$ and $Z \subseteq Y \cup Z$. Thus $X \cap (Y \cup Z) \supseteq (X \cap Y) \cup (X \cap Z)$.

Then it follows that $X \cap (Y \cup Z) = (X \cap Y) \cup (X \cap Z)$, completing this part of the proof. □

The following theorem gives two of the more important relationships between sets.

Theorem 1.2.5 (De Morgan's Laws) *Let X, Y and Z be sets. Then*

$$Z \backslash (X \cap Y) = Z \backslash X \cup Z \backslash Y \quad and \quad Z \backslash (X \cup Y) = Z \backslash X \cap Z \backslash Y.$$

There is a simpler form of de Morgan's laws for ordinary complements:

$$\sim(X \cap Y) = \sim X \cup \sim Y \quad and \quad \sim(X \cup Y) = \sim X \cap \sim Y.$$

To prove the first of these, suppose $x \in \sim(X \cap Y)$. Then $x \notin X \cap Y$ so either $x \notin X$ or $x \notin Y$. That is, $x \in \sim X$ or $x \in \sim Y$, so $x \in \sim X \cup \sim Y$. This proves that $\sim(X \cap Y) \subseteq \sim X \cup \sim Y$. Suppose next that $x \in \sim X \cup \sim Y$. If $x \in \sim X$ then $x \notin X$ so $x \notin X \cap Y$, since $X \cap Y \subseteq X$. That is, $x \in \sim(X \cap Y)$. The same is true if $x \in \sim Y$. Thus $\sim X \cup \sim Y \subseteq \sim(X \cap Y)$, so we have proved that $\sim(X \cap Y) = \sim X \cup \sim Y$.

We can use this, the definition of relative complement, and a distributive rule from Theorem 1.2.4 to prove the first result of Theorem 1.2.5:

$$Z \backslash (X \cap Y) = Z \cap {\sim}(X \cap Y) = Z \cap ({\sim}X \cup {\sim}Y)$$
$$= (Z \cap {\sim}X) \cup (Z \cap {\sim}Y) = Z \backslash X \cup Z \backslash Y.$$

The second of de Morgan's laws is proved similarly. $\square$

Review exercises 1.2

(1) Let a and b be real numbers. Show that

  (a) $||a| - |b|| \leqslant |a - b|$,

  (b) $|a - b| < \epsilon$ if and only if $b - \epsilon < a < b + \epsilon$,

  (c) if $a < b + \epsilon$ for every $\epsilon > 0$ then $a \leqslant b$.

(2) Suppose $A \cup B = X$. Show that $X \times Y = (A \times Y) \cup (B \times Y)$, for any set Y.

(3) For any sets A and B, show that

  (a) $A \backslash B = A$ if and only if $A \cap B = \varnothing$,

  (b) $A \backslash B = \varnothing$ if and only if $A \subseteq B$.

1.3 Functions or mappings

We indicated in Section 1.1 how fundamental the concept of a function is in modern analysis. (It is equally important in classical analysis but may be given a restricted meaning there, as we remark below.) A function is often described as a rule which associates with an element in one set a unique element in another set; we will give a definition which avoids the undefined term 'rule'. In this definition we will include all associated terms and notations that will be required. Examples and general discussion will follow.

Definition 1.3.1 Let X and Y be any two nonempty sets (which may be equal).

(a) A *function f* from X into Y is a subset of $X \times Y$ with the property that for each $x \in X$ there is precisely one element (x, y) in the subset f. We write $f \colon X \to Y$ to indicate that f is a function from X into Y.

(b) The set X is called the *domain* of the function $f \colon X \to Y$.

(c) If $(x, y) \in f$ for some function $f\colon X \to Y$ and some $x \in X$, then we call y the *image* of x under f, and we write $y = f(x)$.

(d) Let S be a subset of X. The set

$$\{y : y \in Y,\ y = f(x) \text{ for some } x \in S\},$$

which is a subset of Y, is called the *image* of the set S under $f\colon X \to Y$, and is denoted by $f(S)$. The subset $f(X)$ of Y is called the *range* of f.

(e) When $f(X) = Y$, we say that the function f is from X *onto* Y (rather than into Y) and we call f an *onto* function.

(f) If, for $x_1, x_2 \in X$, we have $f(x_1) = f(x_2)$ only when $x_1 = x_2$, then we call the function $f\colon X \to Y$ *one-to-one*.

(g) An onto function is also said to be *surjective*, or a *surjection*. A one-to-one function is also said to be *injective*, or an *injection*. A function that is both injective and surjective is called *bijective*, or a *bijection*.

Enlarging on the definition in (a), we see that a function f from a set X into a set Y is itself a set, namely a set of ordered pairs chosen from $X \times Y$ in such a way that distinct elements of f cannot have distinct second elements with the same first element. In (c), we see that the common method of denoting a function as $y = f(x)$ is no more than an alternative, and more convenient, way of writing $(x, y) \in f$. Notice the different roles played by the sets X and Y. The set X is fully used up in that every $x \in X$ has an image $f(x) \in Y$, but the set Y need not be used up in that there may be a $y \in Y$, or many such, which is not the image of any $x \in X$. Paraphrasing (e), when in fact each $y \in Y$ is the image of some $x \in X$, then the function is called 'onto'. Notice that the same term 'image' is used slightly differently in (c) and (d), but this will not cause any confusion.

It follows from Definition 1.2.2(b) that two functions f and g from X into Y are equal if and only if $f(x) = g(x)$ for all $x \in X$.

In Figure 1, four functions

$$f_k\colon X \to Y_k, \qquad k = 1,\ 2,\ 3,\ 4,$$

are illustrated. Each has domain $X = \{1, 2, 3, 4, 5\}$. The function $f_1\colon X \to Y_1$ has $Y_1 = \{1, 2, 3, 4, 5, 6\}$ and the function is the subset $\{(1, 3), (2, 3), (3, 4), (4, 1), (5, 6)\}$ of $X \times Y_1$, as indicated by arrows giving the images of the elements of X. The range of f_1 is the set $f_1(X) = \{1, 3, 4, 6\}$. The other functions may be similarly described.

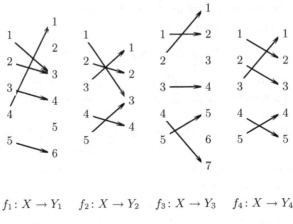

$$f_1\colon X \to Y_1 \quad f_2\colon X \to Y_2 \quad f_3\colon X \to Y_3 \quad f_4\colon X \to Y_4$$

Figure 1

For all four functions, each element of X is the tail of an arrow and of only one arrow, while the elements of the Y's may be at the head of more than one arrow or perhaps not at the head of any arrow. This situation is typical of any function. The elements of Y_2 and Y_4 are all at heads of arrows, so the functions f_2 and f_4 are both onto. Observe that $f_1(1) = 3$ and $f_1(2) = 3$. Also, $f_2(1) = 3$ and $f_2(5) = 3$. This situation does not apply to the functions f_3 and f_4: each element of Y_3 and Y_4 is at the head of at most one arrow, so the functions f_3 and f_4 are both one-to-one.

Only the function f_4 is both one-to-one and onto: it is a bijection. This is a highly desirable situation which we pursue further in Chapters 5 and 7, though we briefly mention the reason now. Only for the function f_4 of the four functions can we simply reverse the directions of the arrows to give another function from a Y into X. We will denote this function temporarily by $g\colon Y_4 \to X$. In full:

$$f_4 = \{(1,2),(2,3),(3,1),(4,5),(5,4)\},$$
$$g = \{(1,3),(2,1),(3,2),(4,5),(5,4)\}.$$

The function g is also a bijection, and has the characteristic properties

$$g(f_4(x)) = x \quad \text{for each } x \in X,$$
$$f_4(g(y)) = y \quad \text{for each } y \in Y_4.$$

We call g the inverse of the function f_4, and denote it by f_4^{-1}. The precise definition of this term follows.

Definition 1.3.2 For any bijection $f\colon X \to Y$, the *inverse function* $f^{-1}\colon Y \to X$ is the function for which

$$f^{-1}(y) = x \text{ whenever } f(x) = y,$$

where $x \in X$ and $y \in Y$.

It follows readily that if f is a function possessing an inverse function, then f^{-1} also has an inverse function and in fact $(f^{-1})^{-1} = f$.

It is sometimes useful in other contexts to speak of the inverse of a function when it is one-to-one but not necessarily onto. This could be applied to the function $f_3\colon X \to Y_3$, above. We can reverse the arrows there to give a function h, but the domain of h would only be $f_3(X)$ and not the whole of Y_3.

The following definition gives us an important method of combining two functions together to give a third function.

Definition 1.3.3 Let $f\colon X \to Y$ and $g\colon Y \to Z$ be two functions. The *composition* of f with g is the function $g \circ f\colon X \to Z$ given by

$$(g \circ f)(x) = g(f(x)), \quad x \in X.$$

Note carefully that the composition $g \circ f$ is only defined when the range of f is a subset of the domain of g. It should be clear that in general the composition of g with f, that is, the function $f \circ g$, does not exist when $g \circ f$ does, and even if it does exist it need not equal $g \circ f$.

For example, consider the functions f_1 and f_4 above. Since $Y_4 = X$, we may form the composition $f_1 \circ f_4$ (but not $f_4 \circ f_1$). We have

$$(f_1 \circ f_4)(1) = f_1(f_4(1)) = f_1(2) = 3,$$

and so on; in full, $f_1 \circ f_4 = \{(1,3),(2,4),(3,3),(4,6),(5,1)\}$.

There are some other terms which require mention. For a function itself, of the general nature given here, we will prefer the terms *map* and *mapping*. The use of the word 'function' will be restricted to the classical sense in which the domain and range are essentially sets of numbers. These are the traditional real-valued or complex-valued functions of one or more real variables. (We do not make use in this book of functions of a complex variable.) The terms *functional* and *operator* will be used later for special types of mappings.

We will generally reserve the usual letters f, g, etc., for the traditional types of functions, and also later for functionals, and we will use letters such as A and B for mappings.

Review exercises 1.3

(1) Let $f = \{(2,2),(3,1),(4,3)\}$, $g = \{(1,6),(2,8),(3,6)\}$. Does f^{-1} exist? Does g^{-1} exist? If so, write out the function in full. Does $f \circ g$ exist? Does $g \circ f$ exist? If so, write out the function in full.

(2) Define a function $f\colon \mathbf{R} \to \mathbf{R}$ by $f(x) = 5x - 2$, for $x \in \mathbf{R}$. Show that f is one-to-one and onto. Find f^{-1}.

(3) For functions $f\colon X \to Y$ and $g\colon Y \to Z$, show that

    (a) $g \circ f\colon X \to Z$ is one-to-one if f and g are both one-to-one,

    (b) $g \circ f\colon X \to Z$ is onto if f and g are both onto.

1.4 Countability

Our aim is to make a basic distinction between finite and infinite sets and then to show how infinite sets can be distinguished into two types, called countable and uncountable. These are very descriptive names: countable sets are those whose elements can be listed and then counted. This has to be made precise of course, but essentially it means that although in an infinite set the counting process would never end, any particular element of the set would eventually be included in the count. The fact that there are uncountable sets will soon be illustrated by an important example.

Two special terms are useful here. Two sets X and Y are called *equivalent* if there exists a one-to-one mapping from X onto Y. Such a mapping is a bijection, but in this context is usually called a *one-to-one correspondence* between X and Y. Notice that these are two-way terms, treating the two sets interchangeably. This is because a bijection has an inverse, so that if $f\colon X \to Y$ is a one-to-one correspondence between X and Y, then so is $f^{-1}\colon Y \to X$, and either serves to show that X and Y are equivalent. Any set is equivalent to itself: the *identity mapping* $I\colon X \to X$, where $I(x) = x$ for each $x \in X$, gives a one-to-one correspondence between X and itself. It is also not difficult to prove, using the notion of composition of mappings, that if X and Y are equivalent sets and Y and Z are equivalent sets, then also X and Z are equivalent sets. See Review Exercise 1.3(3).

We now define a *finite* set as one that is empty or is equivalent to the set $\{1, 2, 3, \ldots, n\}$ for some positive integer n. A set that is not finite is called an *infinite* set. Furthermore:

Definition 1.4.1 *Countable* sets are sets that are finite or that are equivalent to the set **N** of positive integers. Sets that are not countable are called *uncountable*.

It follows that the set **N** itself is countable.

For the remainder of this section, we will be referring only to infinite sets. It will be easy to see that some of the results apply equally to finite sets.

According to the definition, if X is a countable set then there is a one-to-one correspondence between **N** and X, that is, a mapping $f \colon \mathbf{N} \to X$ which is one-to-one and onto. Thus X is the set of images, under f, of elements of **N**:

$$X = \{f(1), f(2), f(3), \ldots\},$$

and no two of these images are equal. This displays the sense in which the elements of X may be counted: each is the image of precisely one positive integer. It is therefore permissible, when speaking of a countable set X, to write $X = \{x_1, x_2, x_3, \ldots\}$, implying that any element of X will eventually be included in the list x_1, x_2, x_3, $\ldots$.

In proving below that a given set is countable, we will generally be satisfied to indicate how the set may be counted or listed, and will not give an actual mapping which confirms the equivalence of the set with **N**. For example, the set **Z** of all integers is countable, since we may write

$$\mathbf{Z} = \{0, -1, 1, -2, 2, -3, 3, \ldots\}$$

and it is clear with this arrangement how the integers may be counted. It now follows that any other set is countable if it can be shown to be equivalent to **Z**. In fact, any countable set may be used in this way to prove that other sets are countable.

The next theorem gives two important results which will cover most of our applications. The second uses a further extension of the notion of a union of sets, this time to a countable number of sets: if X_1, X_2, $\ldots$, are sets, then

$$\bigcup_{k=1}^{\infty} X_k = \{x : x \in X_k \text{ for at least one } k = 1,\, 2,\, 3,\, \ldots\}.$$

Theorem 1.4.2 *If X_1, X_2, ... are countable sets, then*

(a) $\prod_{k=1}^{n} X_k$ *is countable for any integer $n \geqslant 2$,*

(b) $\bigcup_{k=1}^{\infty} X_k$ *is countable.*

Our proof of (a) will require mathematical induction. We show first that $X_1 \times X_2$ is countable. Recall that $X_1 \times X_2$ is the set of all ordered pairs (x_1, x_2), where $x_1 \in X_1$ and $x_2 \in X_2$. Since X_1 and X_2 are countable, we may list their elements and write, using a double subscript notation for convenience,

$$X_1 = \{x_{11}, x_{12}, x_{13}, \dots\}, \qquad X_2 = \{x_{21}, x_{22}, x_{23}, \dots\}.$$

(The first subscript is the set number of any element, the second subscript is the element number in that set.) Writing the elements of $X_1 \times X_2$ down in the following array

$$
\begin{array}{llll}
(x_{11}, x_{21}) & (x_{11}, x_{22}) \rightarrow (x_{11}, x_{23}) & (x_{11}, x_{24}) \rightarrow \cdots \\
\downarrow \quad \nearrow & \swarrow & \nearrow \\
(x_{12}, x_{21}) & (x_{12}, x_{22}) & (x_{12}, x_{23}) & (x_{12}, x_{24}) & \cdots \\
\swarrow & \nearrow \\
(x_{13}, x_{21}) & (x_{13}, x_{22}) & (x_{13}, x_{23}) & (x_{13}, x_{24}) & \cdots \\
\downarrow \quad \nearrow \\
(x_{14}, x_{21}) & (x_{14}, x_{22}) & (x_{14}, x_{23}) & (x_{14}, x_{24}) & \cdots \\
\vdots & \vdots & \vdots & \vdots
\end{array}
$$

and then counting them in the order indicated (those whose subscripts total 5, then those whose subscripts total 6, then those whose subscripts total 7, ...) proves that $X_1 \times X_2$ is countable.

Now assume that $X_1 \times X_2 \times \cdots \times X_{n-1}$ is countable for $n > 2$ and let this set be Y. Then $Y \times X_n$ can be shown to be countable exactly as we showed $X_1 \times X_2$ to be countable. Now, $Y \times X_n$ is the set of ordered pairs $\{((x_1, x_2, \dots, x_{n-1}), x_n) : x_k \in X_k, \ k = 1, \ 2, \ \dots, \ n\}$. The mapping $f : Y \times X_n \to X_1 \times X_2 \times \cdots \times X_n$ given by

$$f(((x_1, x_2, \dots, x_{n-1}), x_n)) = (x_1, x_2, \dots, x_{n-1}, x_n)$$

is clearly a one-to-one correspondence, and this establishes that $X_1 \times X_2 \times \cdots \times X_n$, or $\prod_{k=1}^{n} X_k$, is countable. The induction is complete, and (a) is proved.

The proof of (b) uses a similar method of counting. As before, we write $X_k = \{x_{k1}, x_{k2}, x_{k3}, \dots\}$, for $k \in \mathbf{N}$. We write down the elements of $\bigcup_{k=1}^{\infty} X_k$ in the array

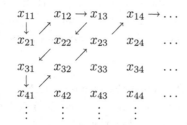

and count them in the order indicated (those whose subscripts total 2, then 3, then 4, ...), this time taking care that any x's belonging to more than one X_k are counted only once. This proves (b), a result which is often expressed by saying: the union of countably many countable sets is itself a countable set. □

It should be clear that the proof of (b) covers the cases where there are only finitely many sets X_k, and where some of these are finite sets. In particular, it implies that the union of two countable sets is countable. We now prove two fundamental results.

Theorem 1.4.3

(a) *The set* **Q** *of rational numbers is countable.*

(b) *The set* **R** *of real numbers is uncountable.*

To prove (a), for each $k \in$ **N** let X_k be the set of all rational numbers that can be expressed as p/k where $p \in$ **Z**. That is,

$$X_k = \left\{ \frac{0}{k}, \frac{-1}{k}, \frac{1}{k}, \frac{-2}{k}, \frac{2}{k}, \dots \right\}.$$

Writing X_k in this way shows that X_k is countable for each k. Any rational number belongs to X_k for some k, so $\bigcup_{k=1}^{\infty} X_k =$ **Q**. Hence, **Q** is countable, by Theorem 1.4.2(b).

We now prove (b), that **R** is uncountable, giving our first example of an uncountable set. The proof relies on the statement that every real number has a decimal expansion. (The following observations are relevant to this. Any real number x has a decimal expansion which, for nonnegative numbers, has the form

$$x = m.n_1 n_2 n_3 \dots = m + \frac{n_1}{10} + \frac{n_2}{10^2} + \frac{n_3}{10^3} + \cdots,$$

where $m, n_1, n_2, n_3, \dots$ are integers with $0 \leqslant n_k \leqslant 9$ for each k. The number is rational if and only if its decimal expansion either terminates or becomes periodic: for example, $\frac{1}{8} = 0.125000\ldots$ terminates and $\frac{1887}{4950} = 0.38121212\ldots$ is periodic, whereas $\sqrt{2} = 1.4142135\ldots$ is neither terminating nor periodic, being irrational. One problem with decimal expansions is that they are not unique for all real numbers. For example, we also have $\frac{1}{8} = 0.124999\ldots$.)

The proof that **R** is uncountable is a proof by contradiction. We suppose that **R** is countable. Then the elements of **R** can be counted, and all will be included in the count. In particular, all real numbers between 0 and 1 will be counted. Let the set $\{x_1, x_2, x_3, \ldots\}$ serve to list all these numbers between 0 and 1 and give these numbers their decimal expansions, say

$$x_1 = 0.n_{11}n_{12}n_{13}\ldots,$$
$$x_2 = 0.n_{21}n_{22}n_{23}\ldots,$$
$$x_3 = 0.n_{31}n_{32}n_{33}\ldots,$$
$$\vdots$$

the double subscript notation again being convenient. Consider the number

$$y = 0.r_1r_2r_3\ldots,$$

where

$$r_k = \begin{cases} 2, & n_{kk} = 1, \\ 1, & n_{kk} \neq 1, \end{cases}$$

for $k \in \mathbf{N}$. This choice of values (which may be replaced by many other choices) ensures that $r_k \neq n_{kk}$ for any k. Hence, $y \neq x_1$ (since these numbers differ in their first decimal place), $y \neq x_2$ (since these numbers differ in their second decimal place), and so on. That is, $y \neq x_j$ for any j. The choice of 1's and 2's in the decimal expansion of y ensures that there is no ambiguity with 0's and 9's. But y is a number between 0 and 1 and the set $\{x_1, x_2, x_3, \ldots\}$ was supposed to include all such numbers. This is the contradiction which proves that **R** is uncountable.   □

We will not prove here the very reasonable statement that a subset of a countable set is itself a countable set, possibly finite. This result was used already in the preceding paragraph and may now be used to prove further that the set **C** of all complex numbers is uncountable: if this were not true then the subset of **C** consisting of all complex numbers with zero imaginary part would be countable, but this subset is **R**.

On the other hand, the set

$$X = \{z : z = x + iy, \ x, y \in \mathbf{Q}\}$$

of all complex numbers with rational real and imaginary parts is countable. This follows using the two theorems above. For $\mathbf{Q}$ is countable, so $\mathbf{Q} \times \mathbf{Q}$ is countable, and there is a natural one-to-one correspondence between X and $\mathbf{Q} \times \mathbf{Q}$, namely the mapping $f : \mathbf{Q} \times \mathbf{Q} \to X$ given by $f((x, y)) = x + iy, \ x, y \in \mathbf{Q}$.

Presumably, uncountable sets are bigger than countable sets, but is $\mathbf{N} \times \mathbf{N}$ bigger than $\mathbf{N}$? To make this notion precise, and thus to be able to compare the sizes of different sets, we introduce cardinality.

Definition 1.4.4 Any set X has an associated symbol called its *cardinal number*, denoted by $|X|$. If X and Y are sets then we write $|X| = |Y|$ if X is equivalent to Y; we write $|X| \leqslant |Y|$ if X is equivalent to a subset of Y; and we write $|X| < |Y|$ if $|X| \leqslant |Y|$ but X is not equivalent to Y. We specify that the cardinal number of a finite set is the number of its elements (so, in particular, $|\varnothing| = 0$), and we write $|\mathbf{N}| = \aleph_0$ and $|\mathbf{R}| = \mathbf{c}$.

There is a lot in this definition. First, it defines how to use the symbols $=$, $<$ and $\leqslant$ in connection with this object called the cardinal number of a set. For finite sets, these turn out to be our usual uses of these symbols. Then, for two specific infinite sets, special symbols are given as their cardinal numbers.

Any infinite countable set is equivalent to $\mathbf{N}$, by definition, so any infinite countable set has cardinal number $\aleph_0$ (pronounced 'aleph null'). So, for example, $|\mathbf{N} \times \mathbf{N}| = |\mathbf{Q}| = \aleph_0$. It is not difficult to see that $n < \aleph_0$ for any $n \in \mathbf{N}$ and that $\aleph_0 < \mathbf{c}$. This is the sense in which uncountable sets are bigger than countable sets.

The arithmetic of cardinal numbers is quite unlike ordinary arithmetic. We will not pursue the details here but will, for interest, list some of the main results. We define addition and multiplication of cardinal numbers by: $|X| + |Y| = |X \cup Y|$ and $|X| \cdot |Y| = |X \times Y|$, where X and Y are any sets, and we define $|Y|^{|X|}$ to be the cardinal number of the *power set* Y^X, which is the set of all functions from X into Y. Then:

$$1 + \aleph_0 = \aleph_0, \quad \aleph_0 + \aleph_0 = \aleph_0, \quad \aleph_0 \cdot \aleph_0 = \aleph_0,$$
$$\mathbf{c} + \mathbf{c} = \mathbf{c}, \quad \mathbf{c} \cdot \mathbf{c} = \mathbf{c}, \quad 2^{\aleph_0} = \mathbf{c}.$$

The famous *continuum hypothesis* is that there is no cardinal number α satisfying $\aleph_0 < \alpha < \mathbf{c}$. All efforts to prove this, or to disprove it by finding a set with cardinal number strictly between those of $\mathbf{N}$ and $\mathbf{R}$, had been unsuccessful. In 1963, it was shown that the existence of such a set could neither be proved nor disproved within the usual axioms of set theory. (Those 'usual' axioms have not been discussed here).

Review exercises 1.4

(1) Define a function $f \colon \mathbf{Z} \to \mathbf{N}$ by

$$f(n) = \begin{cases} 2n + 1, & n \geqslant 0, \\ -2n, & n < 0. \end{cases}$$

Show that f determines a one-to-one correspondence between $\mathbf{Z}$ and $\mathbf{N}$.

(2) Suppose X is an uncountable set and Y is a countable set. Show that $X \backslash Y$ is uncountable.

(3) Show that the set of all polynomial functions with rational coefficients is countable.

1.5 Point sets

In this section, we are concerned only with sets of real numbers. Because real numbers can conveniently be considered as points on a line, such sets are known as *point sets* and their elements as *points*.

The simplest point sets are *intervals*, for which we have special notations. Let a and b be real numbers, with $a < b$. The point set $\{x : a \leqslant x \leqslant b\}$ is a *closed* interval, denoted by $[a, b]$, and the point set $\{x : a < x < b\}$ is an *open* interval, denoted by (a, b). There are also the *half-open* intervals $\{x : a \leqslant x < b\}$ and $\{x : a < x \leqslant b\}$, denoted by $[a, b)$ and $(a, b]$, respectively. In all cases, the numbers a and b are called *endpoints* of the intervals. Closed intervals contain their endpoints as members, but open intervals do not. The following point sets are *infinite* intervals: $\{x : a < x\}$, denoted by (a, ∞); $\{x : a \leqslant x\}$, denoted by $[a, \infty)$; $\{x : x < b\}$, denoted by $(-\infty, b)$; and $\{x : x \leqslant b\}$, denoted by $(-\infty, b]$. These have only one endpoint, which may or may not be a member of the set. The use of the signs ∞ and $-\infty$ is purely conventional and does not imply that these things are numbers. The set $\mathbf{R}$ itself is sometimes referred to as the infinite interval $(-\infty, \infty)$.

A special name is given to an open interval of the form $(a - \delta, a + \delta)$, where δ is a positive number. This is called the *δ-neighbourhood* of a.

We intend to move towards a further discussion of the assumption (g) at the beginning of Section 1.2, that there are no holes when we represent real numbers as points on a line. A few more definitions are required first.

Definition 1.5.1 Let S be a point set (that is, let S be a subset of $\mathbf{R}$).

(a) A number l is called a *lower bound* for S if $l \leqslant x$ for all $x \in S$. A number u is called an *upper bound* for S if $x \leqslant u$ for all $x \in S$.

(b) If there is a lower bound for S, then S is said to be *bounded below*. If there is an upper bound for S, then S is said to be *bounded above*. If S is bounded below and bounded above, then S is said to be *bounded*.

(c) If S is bounded below, a number L is called a *greatest lower bound* for S if L is a lower bound for S and if $l \leqslant L$ for any other lower bound l. We write

$$L = \text{glb}\, S \quad \text{or} \quad L = \inf S.$$

If S is bounded above, a number U is called a *least upper bound* for S if U is an upper bound for S and if $U \leqslant u$ for any other upper bound u. We write

$$U = \text{lub}\, S \quad \text{or} \quad U = \sup S.$$

(d) If S has a greatest lower bound m and $m \in S$, then m is called a *minimum* for S, and we write

$$m = \min S.$$

If S has a least upper bound M and $M \in S$, then M is called a *maximum* for S, and we write

$$M = \max S.$$

A number of remarks need to be made.

If S is a bounded point set, then there exists a closed interval $[l, u]$ such that $l \leqslant x \leqslant u$ for all $x \in S$; that is, $S \subseteq [l, u]$. The converse is also true. Further, any number less than l is also a lower bound for S and any number greater than u is also an upper bound for S.

A greatest lower bound for S, if one exists, is unique. To see this, suppose L and L' are both greatest lower bounds for S. Then in particular both are lower bounds for S and, by definition of greatest lower bound,

$L \leqslant L'$ and $L' \leqslant L$. These imply that $L = L'$. Similarly, a least upper bound for S, if one exists, is unique.

Notice that it is not required that the greatest lower bound for a set be an element of that set. However, when it is an element of the set it may be given a special name: the minimum for the set. A similar remark applies for the least upper bound and the maximum for a set.

The notations inf and sup are abbreviations for *infimum* and *supremum*, and these notations will be preferred in this book over glb and lub. It is sometimes convenient to write

$$L = \inf_{x \in S} x,$$

rather than $L = \inf S$, and similarly for sup, min and max. Other variations in the uses of these notations will be easily identified.

The above terms, and some others to be defined, are pictured in Figure 2 on page 27. They may be illustrated most simply using intervals. For example, let S be the open interval $(0, 1)$. The numbers -37, $-\frac{1}{2}$, 0 are lower bounds for S; the numbers 1, π, 72 are upper bounds. We have $\inf S = 0$, $\sup S = 1$. Since $\inf S \notin S$ and $\sup S \notin S$, we see that $\min S$ and $\max S$ do not exist. If T is the closed interval $[0, 1]$, then $\inf T = 0 \in T$, so $\min T = 0$; $\sup T = 1 \in T$, so $\max T = 1$. The interval $(-\infty, 0)$ is bounded above but not below; its supremum is 0.

We remark finally that if S is a finite point set, then $\min S$ and $\max S$ must both exist (unless $S = \varnothing$, for the empty set has any number as a lower bound and any number as an upper bound!).

Definition 1.5.2 Let S be a nonempty point set. A number ξ is called a *cluster point* for S if every δ-neighbourhood of ξ contains a point of S other than ξ.

This definition does not imply that a cluster point for a set must be an element of that set. Put a little differently, ξ is a cluster point for S if, no matter how small δ is, there exists a point ξ' such that $\xi' \neq \xi$, $\xi' \in (\xi - \delta, \xi + \delta)$ and $\xi' \in S$.

For example, the left-hand endpoint a is a cluster point for the closed interval $[a, b]$ because there exists a point of the interval in $(a-\delta, a+\delta)$, no matter how small δ is. Such a point is $a + \frac{1}{2}\delta$ (assuming that $\delta < 2(b-a)$). Precisely the same is true for the open interval (a, b), and this time a is not an element of the set. Similar reasoning shows that every point of $[a, b]$ is a cluster point for $[a, b]$ and for (a, b). Instead of these intervals, now consider the point sets $[a, b] \cap \mathbf{Q}$ and $(a, b) \cap \mathbf{Q}$, consisting of only the rational numbers in the intervals. Again, all points of $[a, b]$ are cluster points for these sets. This follows from the fact that between any two numbers, rational or irrational, there always exists a rational number.

It should be clear that within any δ-neighbourhood of a cluster point ξ for a set S there in fact exist infinitely many points of S. That is, $S \cap (\xi - \delta, \xi + \delta)$ is an infinite set for any number $\delta > 0$. From this it follows that a finite point set cannot have any cluster points.

An infinite point set may or may not have cluster points. For example, intervals have (infinitely many) cluster points, while the set $\mathbf{Z}$ of all integers has no cluster points. (The latter follows from the preceding paragraph, since no δ-neighbourhood could contain infinitely many points of $\mathbf{Z}$.) This leads us to the Bolzano–Weierstrass theorem, which provides a criterion for an infinite point set to have a cluster point.

Theorem 1.5.3 (Bolzano–Weierstrass Theorem) *If S is a bounded infinite point set, then there exists at least one cluster point for S.*

The criterion is that the infinite set be bounded. We stress again that the cluster point need not be a point of the set. In proving this theorem, we will see arising, in a natural way, a need to formalise our assumption (g) in Section 1.2, dealing with the completeness of the real number system. The proof follows.

Since S is a bounded set, there must be an interval $[a, b]$ such that $S \subseteq [a, b]$. Bisect this interval (by the point $\frac{1}{2}(a+b)$) and consider the intervals $[a, \frac{1}{2}(a+b)]$ and $[\frac{1}{2}(a+b), b]$. If $[a, \frac{1}{2}(a+b)]$ contains infinitely many points of S, then (renaming its endpoints) let this interval be $[a_1, b_1]$; otherwise, let $[\frac{1}{2}(a+b), b]$ be $[a_1, b_1]$. Either way, $[a_1, b_1]$ contains infinitely many points of S, since S is an infinite set. Now treat $[a_1, b_1]$ similarly: let $[a_2, b_2]$ be the interval $[a_1, \frac{1}{2}(a_1+b_1)]$ if this contains infinitely many points of S, and otherwise let $[a_2, b_2]$ be $[\frac{1}{2}(a_1+b_1), b_1]$. This process may be continued indefinitely to give a set $\{[a_1, b_1], [a_2, b_2], \dots\}$ of intervals each containing infinitely many points of S and such that, by construction,

$$[a_1, b_1] \supseteq [a_2, b_2] \supseteq \cdots .$$

Notice that $b_1 - a_1 = \frac{1}{2}(b - a)$, $b_2 - a_2 = \frac{1}{2}(b_1 - a_1) = \frac{1}{4}(b - a)$, and generally

$$b_n - a_n = \frac{b - a}{2^n}, \quad n \in \mathbf{N}.$$

We ask: are there any points belonging to all these intervals? Answering this in part, suppose ξ' and ξ'' are two points, both belonging to all the intervals, with $\xi' \neq \xi''$. Then $|\xi' - \xi''| > 0$ and we can find n so that $b_n - a_n < |\xi' - \xi''|$. (We can do this by solving the inequality $(b - a)/2^n < |\xi' - \xi''|$ for n.) This means it is impossible to have both $\xi' \in [a_n, b_n]$ and $\xi'' \in [a_n, b_n]$ for such a value of n. We must conclude that at most one point can belong to all the intervals.

Here is where we need to make a crucial assumption: precisely one point belongs to all the intervals. Let this point be ξ. We show that ξ is a cluster point for S, and this proves the theorem. Choose any number $\delta > 0$. A value of n can be found so that $b_n - a_n < \delta$ and this means that $[a_n, b_n] \subseteq (\xi - \delta, \xi + \delta)$ for this n. Since $[a_n, b_n]$ contains infinitely many points of S, this δ-neighbourhood certainly contains a point of S other than perhaps ξ itself. Thus ξ is a cluster point for S, and the proof of the theorem is finished.   □

The proof rests fundamentally on our assumption that there exists exactly one point common to all the intervals $[a_n, b_n]$ constructed above. We saw that there could not be two or more such points, so the only alternative to this assumption is that there is no point common to all the intervals. Then this would be the kind of hole in the real number system which we have explicitly stated cannot occur. That is, the need to make our assumption in the above proof is a specific instance of the need for the general, if vague, statement (g) in Section 1.2.

That statement is especially made with reference to the real number system. It is important to recognise that it does not apply to the rational number system, for example. We can indicate this in terms of intervals of the type constructed above, as follows. Remembering that we are dealing only with rational numbers now, consider the set of intervals

$$\{[1, 2], [1.4, 1.5], [1.414, 1.415], [1.4142, 1.4143], \dots\}.$$

This is like the set $\{[a_1, b_1], [a_2, b_2], \dots\}$ above in that each interval contains infinitely many (rational) numbers and each is a subset of the one before it. Is there a number belonging to all the intervals? The answer is that there is not, when we consider only rational numbers. The reason is that the only candidate for inclusion in all the intervals is $\sqrt{2}$ (the intervals were constructed with this in mind) and $\sqrt{2}$ is not a rational number.

The rational number system therefore has holes in it. What we are saying is that when the irrational numbers are added, the resulting real number system no longer has any holes. Most treatments of real numbers which do not actually construct the real number system have a statement of this type, or one equivalent to it. Such a statement is generally presented as an axiom of the real number system. We end this discussion of holes by formally presenting the axiom for completeness of the real number system which has proved convenient for our treatment.

Axiom 1.5.4 (Nested Intervals Axiom) *Let* $\{[c_1, d_1], [c_2, d_2], \dots\}$ *be a set of closed intervals for which*

$$[c_1, d_1] \supseteq [c_2, d_2] \supseteq \cdots,$$

and for which, for any number $\epsilon > 0$, *a positive integer* N *exists such that* $d_n - c_n < \epsilon$ *whenever* $n > N$. *Then there exists precisely one point common to all the intervals.*

This is called the *nested intervals axiom* because intervals $[c_1, d_1]$, $[c_2, d_2], \dots$ such that $[c_1, d_1] \supseteq [c_2, d_2] \supseteq \cdots$ are said to be *nested*.

We look again to our proof of the Bolzano–Weierstrass theorem to see what more can be gleaned. It is apparent from our construction of the intervals $[a_n, b_n]$ that for each n there are only finitely many points of S less than a_n but infinitely many points of S less than b_n. Thus if there is more than one cluster point for S, there can be none less than the one we found. We have therefore proved a little more than we set out to do. The result is presented in Theorem 1.5.6, after giving the relevant definitions.

Definition 1.5.5 A *least cluster point* for an infinite point set S is a cluster point $\underline{\xi}$ for S with the property that only finitely many points of S are less than or equal to $\underline{\xi} - \delta$ for any $\delta > 0$. A *greatest cluster point* for S is a cluster point $\overline{\xi}$ for S with the property that only finitely many points of S are greater than or equal to $\overline{\xi} + \delta$ for any $\delta > 0$.

Theorem 1.5.6 *For any bounded infinite point set there exists a least cluster point and a greatest cluster point.*

The existence of a greatest cluster point is proved by varying the construction of the intervals $[a_n, b_n]$ in an obvious manner. It is clear that there can be at most one least cluster point and at most one greatest cluster point for any point set.

The points $\underline{\xi}$ and $\overline{\xi}$ of Definition 1.5.5 are also known as the *least limit* and the *greatest limit*, respectively, for S, and the following notations are used:

$$\underline{\xi} = \underline{\lim}\, S, \qquad \overline{\xi} = \overline{\lim}\, S.$$

With reference still to our proof of the Bolzano–Weierstrass theorem, let $\overline{\xi} = \overline{\lim}\, S$. It is possible that no points of S are greater than $\overline{\xi}$. Then $\overline{\xi}$ is an upper bound for S. There is no value of $\delta > 0$ such that $\overline{\xi} - \delta$ is also an upper bound for S since, $\overline{\xi}$ being a cluster point for S, there must be (infinitely many) points of S in $(\overline{\xi} - \delta, \overline{\xi} + \delta)$. Hence in this case $\overline{\xi}$ is the least upper bound for S. That is, $\overline{\xi} = \sup S$ (which may or may not be an element of S). Alternatively, if there are points of S greater than $\overline{\xi}$, let x_0 be such a point. Then, since $\overline{\xi}$ is the greatest cluster point for S, either $x \leqslant x_0$ for all $x \in S$ or else the set $T = \{x : x > x_0,\ x \in S\}$ is finite and not empty. Either way, the existence of $\max S$ is assured ($\max S$ is x_0 or $\max T$, respectively) so that in this case also the least upper bound for S exists (and must be an element of S). We have therefore all but proved the following result.

Theorem 1.5.7 *Any nonempty point set that is bounded above has a least upper bound.*

We have just proved this in the case of a bounded infinite point set. Clearly, it would be sufficient for the set only to be bounded above for the same conclusion to follow, and clearly the result is true for any nonempty finite point set. □

In a similar manner, we could prove that any nonempty point set that is bounded below has a greatest lower bound. Theorem 1.5.7 is often stated as an axiom (the *least upper bound axiom*), alternative to our Axiom 1.5.4, to ensure the completeness of the real number system. This is quite equivalent to our approach in that, if Theorem 1.5.7 were given as an axiom, then our nested intervals axiom could be proved as a theorem.

Many of the concepts defined in this section are illustrated in Figure 2, where the dots (•) indicate the (infinite) point set. We proceed with another important definition.

Definition 1.5.8 A point which is both the least cluster point and the greatest cluster point for a point set is called the *limit point* for the set.

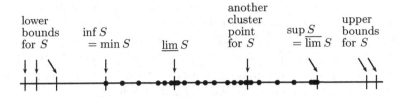

Figure 2

Such a set, in which the least cluster point and the greatest cluster point exist and are equal, has of course only one cluster point. The definition says we then call it the 'limit point' for the set.

This is not the same as saying that a set with a single cluster point must have a limit point. For example, the point set

$$S = \left\{ x : x = \frac{1}{n}, \ n \in \mathbf{N} \right\} = \left\{ 1, \frac{1}{2}, \frac{1}{3}, \frac{1}{4}, \dots \right\}$$

has 0 as its limit point, as is easily verified. The point set $S \cup \mathbf{Z}$ also has 0 as its only cluster point, but $0 \neq \underline{\lim}(S \cup \mathbf{Z})$, since infinitely many points of $S \cup \mathbf{Z}$ are less than $-\delta$ for any $\delta > 0$, and also $0 \neq \overline{\lim}(S \cup \mathbf{Z})$ for a corresponding reason.

We can look at this situation in general terms as follows. Suppose ξ is a limit point for a set S, and choose any number $\delta > 0$. Since ξ is the least cluster point for S, only finitely many points of S are less than or equal to $\xi - \delta$, and similarly only finitely many are greater than or equal to $\xi + \delta$. So all but a finite number of the points of S are within $(\xi - \delta, \xi + \delta)$. Then either $\xi - \delta$ is a lower bound for S or the set $\{ x : x \leqslant \xi - \delta, \ x \in S \}$ is not empty but is finite. Either way, S is bounded below. Similarly, S is bounded above. We have proved the following theorem.

Theorem 1.5.9 *If a point set has a limit point, then it is bounded.*

In the example above, the set $\{ 1, \frac{1}{2}, \frac{1}{3}, \frac{1}{4}, \dots \} \cup \mathbf{Z}$ is not bounded. Therefore, there is no limit point for this set.

We end this section with one more theorem.

Theorem 1.5.10 *Let S be a point set for which there exists a limit point ξ, and suppose that $l < x < u$ for all $x \in S$. Then $l \leqslant \xi \leqslant u$.*

The point to notice is that the signs $<$ become $\leqslant$ for the limit point, reflecting the fact that the limit point for a set may not be an element of that set. This happened above with the example $S = \{1, \frac{1}{2}, \frac{1}{3}, \frac{1}{4}, \dots\}$, where 0 was the limit point for S, but $0 \notin S$. Note, in the theorem, that the existence of such numbers l and u is guaranteed by Theorem 1.5.9.

To prove Theorem 1.5.10, suppose $\xi > u$ and set $\delta = \frac{1}{2}(\xi - u)$. As ξ is a cluster point for S, there must exist a point of S in $(\xi - \delta, \xi + \delta)$. Let x_0 be such a point. Then

$$x_0 > \xi - \delta = \xi - \tfrac{1}{2}\xi + \tfrac{1}{2}u = \tfrac{1}{2}\xi + \tfrac{1}{2}u > \tfrac{1}{2}u + \tfrac{1}{2}u = u;$$

that is, $x_0 > u$. This contradicts the statement that $x < u$ for all $x \in S$, so it cannot be possible to have $\xi > u$. Thus $\xi \leqslant u$. It is similarly proved that $l \leqslant \xi$. □

Review exercises 1.5

(1) Let $S = \{1 + (1/m) - (1/n) : m, n \in \mathbf{N}\}$. Find $\inf S$ and $\sup S$.

(2) Suppose a nonempty point set S is bounded below. Show that $\inf S = -\sup\{-x : x \in S\}$.

(3) Let the point sets A and B be bounded above. Show that $A \cup B$ is bounded above, and $\sup A \cup B = \max\{\sup A, \sup B\}$.

(4) (a) Show directly that $x \notin \bigcap_{n=1}^{\infty}[0, 1/n]$, for any positive real number x.

  (b) Show that $\bigcap_{n=1}^{\infty}(0, 1/n) = \varnothing$.

1.6 Open and closed sets

Topology is a branch of mathematics dealing with entities called open sets. Their properties are modelled on those defined below for real numbers. These help us with a further investigation of real numbers, including the notion of *compactness* of subsets of $\mathbf{R}$. The work in this section is sometimes called *topology of the real line*.

Definition 1.6.1 A point set S is *open* if every point $x \in S$ has a δ-neighbourhood which is a subset of S; that is, if there exists $\delta > 0$ such that $(x - \delta, x + \delta) \subseteq S$. The set S is *closed* if its complement $\sim S$ is open.

By $\sim S$ here, we mean $\mathbf{R} \backslash S$.

The set **R** itself is open because, for any $x \in$ **R**, $(x-1, x+1)$ is a δ-neighbourhood of x (with $\delta = 1$) and is a subset of **R**. The empty set $\varnothing$ is also open, 'vacuously'. Then **R** and $\varnothing$ must also both be closed sets, since $\sim$**R** $= \varnothing$ and $\sim\varnothing =$ **R**. It is easy to see that open intervals are open sets, and closed intervals are closed sets.

We can build up other examples of open sets through the next theorem. It uses yet another extension of the notion of union of sets. Let $\mathscr{T}$ be a collection of sets. (*Collection* is just another word for a set; it is useful when the elements of the set are themselves sets.) The set $\mathscr{T}$ may be finite or infinite, countable or uncountable, but we will always assume that such collections are nonempty. We define

$$\bigcup_{T \in \mathscr{T}} T = \{x : x \in T \text{ for at least one } T \in \mathscr{T}\}.$$

Theorem 1.6.2

(a) *If $\mathscr{T}$ is a collection of open sets, then $\bigcup_{T \in \mathscr{T}} T$ is open.*

(b) *If $\{T_1, T_2, \ldots, T_n\}$ is a finite collection of open sets, then $\bigcap_{k=1}^{n} T_k$ is open.*

To prove (a), put $V = \bigcup_{T \in \mathscr{T}} T$ and suppose $x \in V$. Then $x \in T$ for some $T \in \mathscr{T}$. Since T is open, there is a δ-neighbourhood of x contained in T. But $T \subseteq V$, so this δ-neighbourhood is also contained in V. So V is open.

For (b), this time put $V = \bigcap_{k=1}^{n} T_k$. If $V = \varnothing$ then it is open, so suppose $V \neq \varnothing$. Take any point $x \in V$. Then $x \in T_k$ for each k. Each T_k is open, so there are δ-neighbourhoods $(x - \delta_k, x + \delta_k)$, satisfying $(x - \delta_k, x + \delta_k) \subseteq T_k$, for each k. If $\delta = \min\{\delta_1, \delta_2, \ldots, \delta_n\}$, then $x \in (x - \delta, x + \delta) \subseteq T_k$ for all k, so $x \in (x - \delta, x + \delta) \subseteq V$. That is, V is open. $\square$

The theorem is sometimes worded as follows: arbitrary unions and finite intersections of open sets are open. Using de Morgan's laws (Theorem 1.2.5), we could write down a corresponding result for finite unions and arbitrary intersections of closed sets.

Now we introduce compactness for sets of real numbers.

Definition 1.6.3 A point set S is *compact* if any collection of open sets whose union contains S has a finite subcollection whose union contains S.

Collections of open sets like these are often called *open coverings*, and the definition of a compact set is then given as 'a set for which every open covering has a finite subcovering'. In symbols, suppose $\mathscr{T}$ is any collection of open sets such that $\bigcup_{T \in \mathscr{T}} T \supseteq S$. Then S is compact if there is a finite subcollection $\{T_1, \ldots, T_n\}$ of sets in $\mathscr{T}$ such that $\bigcup_{k=1}^{n} T_k \supseteq S$.

We wish to determine precisely which point sets are compact. We begin by establishing some properties of compact sets. It will turn out, as a consequence of the Heine–Borel theorem (Theorem 1.6.7), that the first two of these are also sufficient conditions for subsets of $\mathbf{R}$ to be compact.

Theorem 1.6.4 *If S is a compact subset of $\mathbf{R}$, then S is bounded.*

To prove this, observe that the collection $\{(-n, n) : n \in \mathbf{N}\}$ is an open covering of $\mathbf{R}$, and hence also of S. Since S is compact, a finite subset of these is a covering of S, so, for some $n \in \mathbf{N}$, $S \subseteq \bigcup_{k=1}^{n}(-k, k) = (-n, n)$. Thus S is bounded. ☐

Theorem 1.6.5 *If S is a compact subset of $\mathbf{R}$, then S is closed.*

The proof proceeds by showing that $\sim S$ is open. This is certainly the case if $\sim S = \varnothing$, so now suppose $\sim S \neq \varnothing$. Take $y \in \sim S$. Then for each $x \in S$ we set $\delta_x = \frac{1}{2}|x - y|$, and we must have $\delta_x > 0$. Clearly the collection of all δ_x-neighbourhoods is an open covering of S. As S is compact, there is a finite subcollection of these which is an open covering of S. That is, there are points x_1, x_2, $\ldots$, x_n such that $S \subseteq \bigcup_{k=1}^{n}(x_k - \delta_{x_k}, x_k + \delta_{x_k})$. Take $\delta = \min\{\delta_{x_1}, \ldots, \delta_{x_n}\}$. Our result will follow when we show that $(y - \delta, y + \delta) \subseteq \sim S$. If this is not the case, then there is a point $z \in (y - \delta, y + \delta) \cap S$. Since $z \in S$, we have $|z - x_i| < \delta_{x_i}$, for some i, and then, since $z \in (y - \delta, y + \delta)$, we have $|z - y| < \delta \leqslant \delta_{x_i}$. But then

$$|x_i - y| = |(x_i - z) + (z - y)| \leqslant |x_i - z| + |z - y| < 2\delta_{x_i} = |x_i - y|,$$

by definition of δ_{x_i}. This is a contradiction, so indeed $(y - \delta, y + \delta) \subseteq \sim S$. ☐

(Note how the inequality $|a + b| \leqslant |a| + |b|$, for $a, b \in \mathbf{R}$, was employed. This idea will be used many times in the coming pages.)

Theorem 1.6.6 *Any closed subset of a compact set is compact.*

Let T be a closed subset of a compact set S, and let $\mathscr{T}$ be an open covering of T. Since $\sim T$ is an open set, the collection $\{\sim T\} \cup \mathscr{T}$ is an open covering of S, which, since S is compact, has a finite subcovering $\{T_1, \ldots, T_n\}$, say. If this subcollection does not include $\sim T$ then it is the required subcovering of T. If it does include $\sim T$, then simply remove it from the set so that the remaining $n-1$ sets are the required subcovering of T. $\qquad\square$

Now we come to the Heine–Borel theorem. This is like the Bolzano–Weierstrass theorem in that it describes a fundamental property of the real number system—fundamental because it is very closely related to the axiomatic concept of completeness.

Theorem 1.6.7 (Heine–Borel Theorem) *A point set is compact if and only if it is closed and bounded.*

We have already proved, in Theorems 1.6.5 and 1.6.4, that compact subsets of $\mathbf{R}$ are closed and bounded, so here we must prove the converse, that subsets of $\mathbf{R}$ which are closed and bounded are compact. This will give us the required characterisation of the compact sets in $\mathbf{R}$.

Let K be a closed, bounded point set. Then, since K is bounded, $K \subseteq [a, b]$ for some interval $[a, b]$. If we can prove that $[a, b]$ is compact, then the result will follow from Theorem 1.6.6.

Let $\mathscr{T}$ be an open covering of $[a, b]$, and let S be the set

$$S = \{a\} \cup \{x : a \leqslant x \leqslant b,$$

there is an open covering of $[a, x]$ by finitely many sets in $\mathscr{T}\}.$

We have $a \in S$ and $S \subseteq [a, b]$, so S is a nonempty bounded point set. It thus has a least upper bound, c say, by Theorem 1.5.7, and $c \leqslant b$. (We have just made use of the completeness of the real number system.) The result will follow immediately if we can show that $b \in S$, and this will follow once we show that $c = b$. We suppose that $c < b$ and will obtain a contradiction.

Since $c \in [a, b]$, we have $c \in T$ for some $T \in \mathscr{T}$. Since T is open, $(c - \delta, c + \delta) \subseteq T$ for some $\delta > 0$. Let $\delta_0 = \min\{\frac{1}{2}\delta, b - c\}$. Then $\delta_0 > 0$ and $[c - \delta_0, c + \delta_0] \subseteq T$. For some $x \in S$ we must have $x > c - \delta_0$ and for this x we know there is a finite collection $\{T_1, \ldots, T_n\}$ of sets in $\mathscr{T}$ which is a covering of $[a, x]$. Then $\{T_1, \ldots, T_n, T\}$ is a finite collection of sets in $\mathscr{T}$ which is a covering of $[a, c + \delta_0]$. But, by choice of δ_0, $c + \delta_0 \leqslant b$, so $c + \delta_0 \in S$, contradicting the definition of c. Hence we have proved that $c = b$. A slight adjustment of this argument with b replacing c then shows that $b \in S$, and the theorem is proved.   □

In particular, of course, all closed intervals are compact subsets of **R**.

Review exercises 1.6

(1) Show that open intervals are open sets and closed intervals are closed sets.

(2) Let S be a point set. (a) A point $x \in \mathbf{R}$ is an *interior* point of S if, for some $\delta > 0$, $(x - \delta, x + \delta) \subseteq S$. Show that all points of **R** are interior points. (b) A point $x \in \mathbf{R}$ is an *isolated* point of S if, for some $\delta > 0$, $(x - \delta, x + \delta) \cap S = \{x\}$. Show that all points of **Z** are isolated points.

(3) Let S be a point set. A point $x \in \mathbf{R}$ is a *boundary* point of S if, for every $\delta > 0$, $(x - \delta, x + \delta) \cap S \neq \varnothing$ and $(x - \delta, x + \delta) \cap \sim S \neq \varnothing$. Show that S is closed if and only if it contains all its boundary points.

1.7 Sequences

In this section, we introduce the idea of a sequence, which, as we mentioned right at the beginning, is essential for our treatment of modern analysis: a great many of our major definitions are framed in terms of convergent sequences. This approach is adopted because it is felt that sequences are intuitively acceptable objects that are also relatively easy to manipulate.

It is time for another brief essay on what modern analysis is all about. We stated in Section 1.1 that it generalises, simplifies and unifies many classical branches of mathematics. This is accomplished essentially in two steps. First of all, everything that suggests specialisation into certain streams of mathematics needs to be discarded. For example, functions that are solutions of differential equations need to be differentiable, and matrices that help in solving certain systems of linear equations need to have inverses; but these properties are not essential to the notion of a function or a matrix. Discarding them leaves us only with some very basic entities, namely sets whose elements have no prescribed properties. The second step is then gradually to add back what was discarded. At each phase of this second step, we look around to see what bits of known mathematics can successfully be accommodated. Here is where different strands of mathematics are seen to be strikingly similar whereas originally they were thought to be distinct.

The thing that determines the order of retrieval of the various discarded bits during the second step is the real number system, for this seems to be the ideal to work towards. We said this step begins with sets alone. Retrieving the pieces is technically described as adding more and more structure to the elements of the sets: allowing the notion of a distance between pairs of elements, allowing the elements to be able to be added together, and so on. Each phase determines what is known as a space. It is not required that the elements of any of these spaces (except perhaps the ultimate one) actually be real numbers, but just that they have properties suggested by certain properties of real numbers.

This explains why up to now, and particularly in the preceding two sections, we have concentrated on properties of real numbers. We will continue to do this throughout this and the next few sections of this chapter, but largely now in the context of sequences of real numbers. Nearly everything we say about sequences here will be found generalised, either as a definition or by a theorem, somewhere in the coming chapters.

We choose to use sequences to generate much of our theory, for the reasons mentioned above, but there is a common alternative based on the more primitive notion of open sets. This approach usually begins with the concept of a topological space, which is quite an early notion in the hierarchy of spaces indicated above. We in effect will be simplifying things a little by starting some way along the hierarchy, though later, in Chapter 5, we will pull the various approaches together.

Now back to work.

Definition 1.7.1 A *sequence* is a mapping whose domain is the set **N** of positive integers.

This might more strictly be called an *infinite* sequence, but we always use the term 'sequence' alone to have this meaning. (A mapping with domain $\{1, 2, \ldots, n\}$, for some $n \in \mathbf{N}$, is a *finite* sequence, but these are not required in our work.)

Thus, a mapping $A \colon \mathbf{N} \to X$ is a sequence, whatever the set X. Being a mapping (or function), the sequence A is the set of ordered pairs $\{(n, A(n)) : n \in \mathbf{N}\}$ and is fully specified by listing the elements $A(1), A(2), A(3), \ldots$ in X. We will follow convention by writing a_n in place of $A(n)$ and denoting the sequence by $a_1, a_2, a_3, \ldots$ or by $\{a_n\}_{n=1}^{\infty}$. The latter is generally abbreviated to $\{a_n\}$, provided that this does not cause confusion with the notation for a set. The element a_n (in X) is called the *nth term* of the sequence A. A notation such as $\{a_n\}_{n=-\infty}^{\infty}$ would indicate in a similar way a mapping whose domain is $\mathbf{Z}$.

We next introduce subsequences. Generally speaking, a subsequence of a sequence $\{a_n\}$ is a subset of its terms $a_1, a_2, a_3, \ldots$ in which their original order is maintained. That is, for any positive integers $n_1, n_2, n_3, \ldots$ where $n_1 < n_2 < n_3 < \cdots$, the terms $a_{n_1}, a_{n_2}, a_{n_3}, \ldots$ form a subsequence of $\{a_n\}$. This is made precise as follows.

Definition 1.7.2 Let A be any sequence. A *subsequence* of A is a composite mapping $A \circ N$, where $N \colon \mathbf{N} \to \mathbf{N}$ is any mapping with the property that if $i, j \in \mathbf{N}$ and $i < j$ then $N(i) < N(j)$.

Notice that N is a sequence whose terms are positive integers in increasing order. Consistent with the conventional notation just described, we may write n_k for $N(k)$ ($k \in \mathbf{N}$), and then N is given alternatively as $\{n_k\}_{k=1}^{\infty}$. Since A is a mapping from $\mathbf{N}$ into some set X, the composition of N with A also maps $\mathbf{N}$ into X and so a subsequence of a sequence is itself a sequence whose terms belong to the same set as those of the original sequence. Note finally that if $A = \{a_n\}$ then

$$(A \circ N)(k) = A(N(k)) = A(n_k) = a_{n_k}, \quad k \in \mathbf{N}.$$

Thus, the kth term of $A \circ N$ is a_{n_k}, and so the subsequence $A \circ N$ of the sequence $A = \{a_n\}$ may be given alternatively as $\{a_{n_k}\}_{k=1}^{\infty}$ or, briefly, $\{a_{n_k}\}$. Examples of subsequences will be given shortly.

In this section we are interested only in sequences whose terms are all real numbers or all complex numbers. These are called *real-valued sequences* and *complex-valued sequences*, respectively. Unless we specify otherwise, we will for the time being be referring only to real-valued sequences.

An example of such a sequence is $\{1/n\}$, or $1, \frac{1}{2}, \frac{1}{3}, \frac{1}{4}, \ldots$. (Enough terms are given to indicate a natural pattern. The key word is 'natural', as you will see if you write out the first four terms of the sequence $\{(n-1)(n-2)(n-3)+1/n\}$.) One subsequence of $\{1/n\}$ is $\frac{1}{2}, \frac{1}{4}, \frac{1}{6}, \ldots$, or $\{1/2n\}$, taking every second term of the original sequence. To see how this conforms with Definition 1.7.2, write $A = \{a_n\} = \{1/n\}$ and let $N = \{n_k\}$ be the sequence $\{2k\}$. Then

$$A \circ N = \{a_{n_k}\} = \{a_{2k}\} = \left\{\frac{1}{2k}\right\}_{k=1}^{\infty} = \left\{\frac{1}{2n}\right\}_{n=1}^{\infty}.$$

Other subsequences of $\{1/n\}$ are $\{1/n^2\}$ and $\{1/n!\}$.

There may initially be confusion between the notation $\{a_n\}$ for a sequence and the notation $\{a_1, a_2, a_3, \dots\}$ for the point set which is the range of the sequence (recalling that a sequence is a function), so care is needed. Notice that by definition a sequence always has infinitely many terms, whereas its range may be a finite set. For example, the sequence $\{(-1)^n\}$ has range $\{-1, 1\}$. This sequence may also be written $-1, 1, -1, 1, \dots$. The confusion is at its worst for a *constant sequence*: a sequence whose range consists of a single element. An example is $\{5\}$, where we use (or misuse) the abbreviation for the sequence better denoted by $\{5\}_{n=1}^{\infty}$ or $5, 5, 5, \dots$. The range of this sequence is the point set $\{5\}$.

At the same time, this similarity of notations suggests how we might define a number of ideas related to sequences: we make use of the range of a sequence and employ our earlier work on point sets.

Definition 1.7.3 A point is called a *cluster point* for the (real-valued) sequence $\{a_n\}$ if it is either

(a) the element of the range of a constant subsequence of $\{a_n\}$, or

(b) a cluster point for the range of $\{a_n\}$.

The range of a sequence is a point set, so the reference to a cluster point in (b) is in the sense of Definition 1.5.2. If the sequence $\{a_n\}$ has a finite range (that is, if the range is a finite set), then there must be at least one value which is taken on by infinitely many terms of $\{a_n\}$. More precisely, there must be a subset $\{n_1, n_2, n_3, \dots\}$ of $\mathbf{N}$, with $n_1 < n_2 < n_3 < \cdots$, such that $a_{n_1} = a_{n_2} = a_{n_3} = \cdots$. Then $\{a_{n_k}\}$ is a constant subsequence of $\{a_n\}$ and, according to (a), a_{n_1} is a cluster point for $\{a_n\}$.

The range of the sequence $\{1/n\}$ is the point set $\{1, \frac{1}{2}, \frac{1}{3}, \frac{1}{4}, \dots\}$, for which 0 is a cluster point. It follows that 0 is a cluster point for the sequence. The sequence $\{(-1)^n\}$ has a finite range: it has two constant subsequences, namely $-1, -1, -1, \dots$ and $1, 1, 1, \dots$, so -1 and 1 are cluster points for this sequence. The sequence $1, \frac{1}{2}, 1, \frac{1}{3}, 1, \frac{1}{4}, \dots$ has cluster points at 1 and 0 since $1, 1, 1, \dots$ is a constant subsequence and 0 is a cluster point for the range $\{1, \frac{1}{2}, \frac{1}{3}, \frac{1}{4}, \dots\}$. Obviously, a cluster point for a sequence need not be an element of its range.

The quantities defined in Definition 1.5.1 (on lower and upper bounds, greatest lower bound and least upper bound, minimum and maximum for a point set) are carried over in a natural way for a sequence by referring to the range of the sequence. For example, a number l is a *lower bound* for a sequence $\{a_n\}$ if $l \leqslant a_n$ for all $n \in \mathbf{N}$. A sequence $\{a_n\}$

is *bounded* if there are numbers l and u such that $l \leqslant a_n \leqslant u$ for all $n \in \mathbf{N}$. The *greatest lower bound* of $\{a_n\}$ is denoted by $\mathrm{glb}\, a_n$ or $\inf a_n$, and similarly for lub, sup, max and min. The remarks immediately following Definition 1.5.1 apply also for sequences. It follows from the Bolzano–Weierstrass theorem that there exists at least one cluster point for a bounded sequence. It is the need for this statement and others like it to be true that motivates the inclusion of infinitely recurring sequence-values in the definition of a cluster point for a sequence.

Definition 1.5.5 (least cluster point, greatest cluster point) also carries over for sequences, and in this context these quantities are called the least limit or *limit inferior* and the greatest limit or *limit superior*. For the sequence $\{a_n\}$, they are denoted by

$$\underline{\lim}\, a_n \ \text{ or } \lim\inf a_n \quad \text{and} \quad \overline{\lim}\, a_n \ \text{ or } \lim\sup a_n,$$

respectively. We prefer the latter names and the latter notations, for sequences.

By Theorem 1.5.6, the limit inferior and limit superior of a sequence exist when the sequence is bounded. It follows also that if $\{a_n\}$ is a bounded sequence and ϵ is any positive number, then

$$a_n \leqslant \lim\inf a_n - \epsilon \quad \text{for finitely many } n \in \mathbf{N},$$
$$a_n \leqslant \lim\inf a_n + \epsilon \quad \text{for infinitely many } n \in \mathbf{N},$$
$$a_n \geqslant \lim\sup a_n - \epsilon \quad \text{for infinitely many } n \in \mathbf{N},$$
$$a_n \geqslant \lim\sup a_n + \epsilon \quad \text{for finitely many } n \in \mathbf{N}.$$

The following definition is suggested by Definition 1.5.8.

Definition 1.7.4 Given a sequence $\{a_n\}$, if $\lim\inf a_n$ and $\lim\sup a_n$ both exist and are equal, then their common value is called the *limit* of $\{a_n\}$, denoted by $\lim a_n$, and we say that $\{a_n\}$ is *convergent*. If $\lim a_n = \xi$, we say that $\{a_n\}$ *converges* to ξ and we write $a_n \to \xi$. If $\lim a_n$ does not exist, we say that $\{a_n\}$ is *divergent*, or that $\{a_n\}$ *diverges*.

A convergent sequence is often defined differently. The alternative is indicated in the following theorem.

Theorem 1.7.5 *A sequence $\{a_n\}$ converges to ξ if and only if for any number $\epsilon > 0$ there exists a positive integer N such that*

$$|a_n - \xi| < \epsilon \quad \text{whenever } n > N.$$

To prove this, suppose first that $\{a_n\}$ is convergent and $\lim a_n = \xi$. Then $\xi = \liminf a_n$ and $\xi = \limsup a_n$, and so

$$a_n \leqslant \xi - \epsilon \quad \text{for finitely many } n \in \mathbf{N},$$
$$a_n \geqslant \xi + \epsilon \quad \text{for finitely many } n \in \mathbf{N}.$$

Because these inequalities hold only for finitely many $n \in \mathbf{N}$, there must be some number in $\mathbf{N}$, say N, such that $a_n > \xi - \epsilon$ and $a_n < \xi + \epsilon$ whenever $n > N$. That is, $|a_n - \xi| < \epsilon$ whenever $n > N$, as required.

Now suppose the condition of the theorem is satisfied. We have to prove that $\{a_n\}$ is convergent, and that $\lim a_n = \xi$. We are given that the numbers N and ξ exist. It is possible that $a_{N+1} = a_{N+2} = \cdots = \xi$, in which case $\{a_n\}$ has a constant subsequence, so that ξ is a cluster point for $\{a_n\}$. Moreover, then $\xi = \liminf a_n = \limsup a_n$, since there are only finitely many other terms of the sequence, namely $a_1, a_2, \ldots, a_N$. If this is not the case, then the condition ensures that there is a point of the set $\{a_1, a_2, a_3, \ldots\}$, besides possibly ξ itself, lying in any ϵ-neighbourhood of ξ. Thus again ξ is a cluster point for $\{a_n\}$ and again $\xi = \liminf a_n = \limsup a_n$ since only finitely many terms of $\{a_n\}$ are less than or equal to $\xi - \epsilon$ or greater than or equal to $\xi + \epsilon$. In either case, by Definition 1.7.4, $\{a_n\}$ converges and $\xi = \lim a_n$. $\square$

The number N in the theorem generally depends on the choice made for ϵ and as a rule the smaller ϵ is chosen to be, the larger N turns out to be. This is the basis for the common rider '$n \to \infty$' when speaking of the convergence of a sequence $\{a_n\}$. The notion is superfluous with our development, but will be used whenever it helps to clarify a statement.

The following three examples serve to illustrate Definition 1.7.4 and Theorem 1.7.5.

(a) The sequence $\{1/n\}$ converges to 0 because

$$\left| \frac{1}{n} - 0 \right| = \frac{1}{n} < \epsilon$$

whenever $n > 1/\epsilon$. That is, we may choose N to be an integer greater than or equal to $1/\epsilon$.

(b) The sequence $5, 5, 5, \ldots$ converges to 5 because $|5 - 5| = 0 < \epsilon$ whenever $n > 1$. Here, and for any constant sequence, any positive integer may be chosen for N, regardless of the value of ϵ.

(c) The sequence $\{(-1)^n\}$ diverges because the requirement, for convergence, that $|(-1)^n - \xi| < \epsilon$ whenever $n > N$ cannot be satisfied: whatever value is chosen for ξ, if $\epsilon < \max\{|-1-\xi|, |1-\xi|\}$ then there is no value for N that will satisfy the condition.

Before continuing, we give the important analogues for sequences of Theorems 1.5.9 and 1.5.10. They require nothing further in the way of proof.

Theorem 1.7.6 *If a sequence converges, then it is bounded.*

Theorem 1.7.7 *Let $\{a_n\}$ be a convergent sequence, with $\lim a_n = \xi$, and suppose $l < a_n < u$ for all $n \in \mathbf{N}$. Then $l \leqslant \xi \leqslant u$.*

The following is another useful theorem, worth giving at this stage.

Theorem 1.7.8 *Let $\{a_n\}$ and $\{b_n\}$ be two convergent sequences, with $\lim a_n = \xi$ and $\lim b_n = \eta$. If $a_n \leqslant b_n$ for all $n \in \mathbf{N}$, then $\xi \leqslant \eta$.*

To prove this, suppose $\xi > \eta$ and set $\epsilon = \frac{1}{2}(\xi - \eta)$. There must exist an integer n such that $a_n > \xi - \epsilon = \frac{1}{2}(\xi + \eta)$ and $b_n < \eta + \epsilon = \frac{1}{2}(\xi + \eta)$. But then $b_n < a_n$, which is a contradiction. Hence $\xi \leqslant \eta$. $\qquad\square$

A simple but useful consequence of Theorem 1.7.6 is that a sequence which is not bounded must be divergent. In this way, the sequences $\{3n - 75\}$ and $\{2^{n-8}\}$, for example, may be shown to diverge. Thus we have a method by which some sequences may be shown to diverge without reference to the definition or Theorem 1.7.5. Simple criteria that allow conclusions like this are always worth seeking. The next theorem gives such a criterion, in this case for certain sequences to be convergent. We first define the type of sequence to which it will apply.

Definition 1.7.9 A sequence $\{a_n\}$ is said to be

  (a) *nondecreasing* if $a_n \leqslant a_{n+1}$ for all $n \in \mathbf{N}$,
  (b) *nonincreasing* if $a_n \geqslant a_{n+1}$ for all $n \in \mathbf{N}$,
  (c) *increasing* if $a_n < a_{n+1}$ for all $n \in \mathbf{N}$,
  (d) *decreasing* if $a_n > a_{n+1}$ for all $n \in \mathbf{N}$.

Any such sequence is said to be *monotone*.

The terms in (a) to (d) are very descriptive. For example, the sequence $1, 2, 2, 3, 4, 4, \ldots$ is nondecreasing, and the sequence $1, \frac{1}{2}, \frac{1}{3}, \frac{1}{4}, \ldots$ is decreasing.

Now the theorem.

Theorem 1.7.10 *If a sequence is monotone and bounded, then it is convergent.*

We will prove the theorem for a sequence $\{a_n\}$ which is nondecreasing and bounded. The proofs in the other cases are handled similarly. Note that if $\{a_n\}$ is nondecreasing, then $a_1 \leqslant a_2$, $a_2 \leqslant a_3$, $a_3 \leqslant a_4$, and so on, so that $a_1 \leqslant a_n$ for all $n \in \mathbf{N}$. Thus a nondecreasing sequence is automatically bounded below. We are assuming further that $\{a_n\}$ is bounded above. If $\{a_n\}$ has only a finite range, then the desired conclusion is easily obtained, and we omit the details. Otherwise, the point set $\{a_1, a_2, a_3, \dots\}$ is bounded and infinite and Theorem 1.5.7 may be applied: the least upper bound must exist. Write $\xi = \sup a_n$. For any $\epsilon > 0$, we must have $a_N > \xi - \epsilon$ for some $N \in \mathbf{N}$ because $\xi - \epsilon$ cannot also be an upper bound for $\{a_n\}$. But $\{a_n\}$ is nondecreasing, so that $a_N \leqslant a_{N+1} \leqslant a_{N+2} \leqslant \cdots$, implying that $a_n > \xi - \epsilon$ for all $n > N$. Furthermore, $a_n \leqslant \xi < \xi + \epsilon$ for all n, and in particular for all $n > N$. Thus $|a_n - \xi| < \epsilon$ whenever $n > N$, and hence, according to Theorem 1.7.5, $\{a_n\}$ converges (and $\lim a_n = \sup a_n$). $\qquad\square$

It is important to note the byproduct here, that $\lim a_n = \sup a_n$. Thus, to find the limit of a bounded nondecreasing (or increasing) sequence, we need only find its least upper bound. Similarly, the limit of a bounded nonincreasing or decreasing sequence is its greatest lower bound.

As an application of the theorem, suppose $\{a_n\}$ is a bounded sequence, and define a sequence $\{b_n\}$ by

$$b_n = \sup\{a_n, a_{n+1}, a_{n+2}, \dots\}, \quad n \in \mathbf{N}.$$

The point set $\{a_n, a_{n+1}, a_{n+2}, \dots\}$ is bounded for each n, so the existence of b_n for each n is guaranteed by Theorem 1.5.7. Furthermore, it is clear that $\{b_n\}$ is bounded. We will show that $\{b_n\}$ is nonincreasing. To do this, note that for any $n \in \mathbf{N}$ either $a_n \geqslant a_k$ for all $k > n$, or $a_n < a_k$ for at least one $k > n$. Thus, either $b_n = a_n$ or $b_n = \sup\{a_{n+1}, a_{n+2}, \dots\} = b_{n+1}$, so that certainly $b_n \geqslant b_{n+1}$ for all $n \in \mathbf{N}$. That is, $\{b_n\}$ is nonincreasing. Hence we may apply the preceding theorem to the bounded sequence $\{b_n\}$: we are assured that $\{b_n\}$ converges (whether or not $\{a_n\}$ does).

Write $\overline{\xi} = \lim b_n$, which we now know exists. We will show that in fact $\overline{\xi} = \limsup a_n$. To this end, choose any number $\epsilon > 0$. Suppose that $a_n > \overline{\xi} - \epsilon$ for only finitely many $n \in \mathbf{N}$. Then there exists $N \in \mathbf{N}$ such that $a_n \leqslant \overline{\xi} - \epsilon$ for all $n > N$, and in that case, by definition of $\{b_n\}$, we have $b_n \leqslant \bar{\xi} - \epsilon$ for all $n > N$. Suppose also that $a_n \geqslant \bar{\xi} + \epsilon$ for infinitely many $n \in \mathbf{N}$. Then this would imply that $b_n \geqslant \bar{\xi} + \epsilon$ for infinitely many $n \in \mathbf{N}$. Both of these possibilities are contradicted by the fact that $\bar{\xi} = \lim b_n$. Hence we must have $a_n > \bar{\xi} - \epsilon$ for infinitely many $n \in \mathbf{N}$, and $a_n \geqslant \bar{\xi} + \epsilon$ for only finitely many $n \in \mathbf{N}$. These mean that $\bar{\xi} = \limsup a_n$, as we set out to show.

In this way, we see the justification for the notation $\limsup a_n$ for the greatest limit of $\{a_n\}$. That is, we have $\limsup a_n = \lim b_n$, where $b_n = \sup\{a_n, a_{n+1}, a_{n+2}, \dots\}$ $(n \in \mathbf{N})$: the greatest limit is indeed a limit of suprema. Some authors bring this out explicitly with the notation $\lim_{n\to\infty} \sup_{k\geqslant n} a_k$. A similar justification can be given for the notation $\liminf a_n$ for the least limit of $\{a_n\}$.

We move on now to prove two theorems which share with the preceding theorem a fundamental property: the three theorems are all dependent on the completeness of the real number system. Corresponding results stated in the context of rational numbers only would not be true.

The first is often referred to as the *Bolzano–Weierstrass theorem for sequences*.

Theorem 1.7.11 *Every bounded sequence has a convergent subsequence.*

Consider the sequence $1, 1.4, 1.41, 1.414, 1.4142, \dots$ of partial decimals of $\sqrt{2}$. Within the rational number system, this sequence is not convergent because $\sqrt{2}$ is not rational. It is a bounded monotone sequence, demonstrating that Theorem 1.7.10 is not true for rational numbers alone. It demonstrates the same thing for Theorem 1.7.11, because, as is easy to see, if there were a convergent subsequence its limit would also have to be $\sqrt{2}$.

For the proof of Theorem 1.7.11, let $\{a_n\}$ be a bounded sequence, and, as above, set $b_n = \sup\{a_n, a_{n+1}, a_{n+2}, \dots\}$ for $n \in \mathbf{N}$. We consider two possibilities. First, it may be that $\max\{a_n, a_{n+1}, a_{n+2}, \dots\}$ exists for all $n \in \mathbf{N}$. In that case, the sequence $\{b_n\}$ is a subsequence of $\{a_n\}$, and, as we have seen above, it is convergent. The second possibility is that for some $N \in \mathbf{N}$, $\max\{a_N, a_{N+1}, a_{N+2}, \dots\}$ does not exist. In this case, set $n_1 = N$; then let n_2 be the smallest integer such that $n_2 > n_1$ and $a_{n_2} > a_{n_1}$ (n_2 must exist, for otherwise $a_n \leqslant a_{n_1}$ for all $n \geqslant N$ so that $b_N = a_{n_1}$); then let n_3 be the smallest integer such that $n_3 > n_2$ and $a_{n_3} > a_{n_2}$ (n_3 must exist, for otherwise $a_n \leqslant a_{n_2}$ for all $n \geqslant N$ so that $b_N = a_{n_2}$). Proceeding in this way, we obtain a subsequence $\{a_{n_k}\}$ which, being increasing and bounded, is convergent by Theorem 1.7.10.

This completes the proof of the theorem, since both possibilities lead to the existence of a convergent subsequence of $\{a_n\}$. ☐

Note the bonus in this proof: we have in fact shown that for any bounded sequence there exists a convergent subsequence which is monotone.

The next theorem is known as the *Cauchy convergence criterion*. As the name implies, and like Theorem 1.7.10, it provides a criterion for a sequence to converge. Unlike Theorem 1.7.10, it does not require the sequence to be monotone. It is a test based on the terms themselves of the sequence and does not require any foreknowledge, or any guess, of what the limit of the sequence may be, as is required in Theorem 1.7.5. Essentially the criterion is that the further we progress along the sequence, the smaller the distances between terms must become. The example above of a sequence of rational numbers converging to an irrational number shows that this too is not a property of the system of rational numbers alone.

Theorem 1.7.12 (Cauchy Convergence Criterion) *A sequence $\{a_n\}$ is convergent if and only if for any number $\epsilon > 0$ there exists a positive integer N such that*

$$|a_n - a_m| < \epsilon \quad \text{whenever } m, n > N.$$

It is easy to see that the condition is necessary. To do so, suppose $\{a_n\}$ is convergent and $\lim a_n = \xi$. Then, given $\epsilon > 0$, we know by Theorem 1.7.5 that a positive integer N exists such that $|a_n - \xi| < \frac{1}{2}\epsilon$ whenever $n > N$. If n and m are both integers greater than N, then both $|a_n - \xi| < \frac{1}{2}\epsilon$ and $|a_m - \xi| < \frac{1}{2}\epsilon$. Hence

$$|a_n - a_m| = |(a_n - \xi) + (\xi - a_m)|$$
$$\leqslant |a_n - \xi| + |a_m - \xi| < \tfrac{1}{2}\epsilon + \tfrac{1}{2}\epsilon = \epsilon,$$

proving the necessity of the condition. (The use here of $\frac{1}{2}\epsilon$ instead of ϵ is a common practice designed to make the analysis look a little tidier.)

Proving the theorem in the opposite direction is more difficult. We are now assuming the condition: for any $\epsilon > 0$ there exists N such that $|a_n - a_m| < \epsilon$ whenever $m, n > N$, and we must prove that $\{a_n\}$ converges. We will show first that this condition implies that the sequence is bounded, so that, by the preceding theorem, it has a convergent subsequence. Using the condition again, this will then be shown to imply that the sequence $\{a_n\}$ itself is convergent.

Taking $\epsilon = 1$ (for convenience) in the condition, let a corresponding integer N be determined so that the condition is satisfied. Then, for any $n > N$,

$$|a_n| = |(a_n - a_{N+1}) + a_{N+1}|$$
$$\leqslant |a_n - a_{N+1}| + |a_{N+1}| < 1 + |a_{N+1}|.$$

(We have taken $m = N + 1$ in the condition, again for convenience.) This provides upper and lower bounds for those terms a_n with $n > N$. Hence certainly

$$|a_n| \leqslant \max\{|a_1|, |a_2|, \ldots, |a_N|, 1 + |a_{N+1}|\}$$

for all $n \in \mathbf{N}$, so the sequence $\{a_n\}$ is bounded. Therefore it has a convergent subsequence, $\{a_{n_k}\}$ say. Let ξ be the limit of this subsequence. We will show that $a_n \to \xi$. Let $\epsilon > 0$ be given and let N and K be integers such that

$$|a_n - a_m| < \tfrac{1}{2}\epsilon \quad \text{whenever } m, n > N,$$
$$|a_{n_k} - \xi| < \tfrac{1}{2}\epsilon \quad \text{whenever } k > K.$$

Then, provided k is such that $k > K$ and $n_k > N$, we have, whenever $n > N$,

$$|a_n - \xi| = |(a_n - a_{n_k}) + (a_{n_k} - \xi)|$$
$$\leqslant |a_n - a_{n_k}| + |a_{n_k} - \xi| < \tfrac{1}{2}\epsilon + \tfrac{1}{2}\epsilon = \epsilon.$$

By Theorem 1.7.5, this means that the sequence $\{a_n\}$ converges, as required. $\qquad\qquad\qquad\qquad\qquad\qquad\qquad\qquad\qquad\qquad\qquad$ □

We turn briefly now to complex-valued sequences.

It is quite possible to develop a point set theory for sets of complex numbers, each number being thought of as a point in the plane. In this way, a cluster point can be defined leading to a form of the Bolzano–Weierstrass theorem after adapting an axiom of completeness like the Nested Intervals Axiom. However, it is important to realise that we could not go much further with a development parallel to that for real numbers, because there is no notion of upper and lower bounds or of least and greatest limits for sets of complex numbers. For real numbers, these notions depended on the fact that the real number system is ordered (by the ordering symbol $<$ and its properties). But no such ordering idea exists for complex numbers.

It is not possible therefore to arrive at a definition for convergence of complex-valued sequences like that of Definition 1.7.4 for real-valued sequences, and we must look elsewhere for a satisfactory way to proceed. It is highly significant that we look no further than Theorem 1.7.5, adapting this almost verbatim as a definition, not a theorem, for convergence of complex-valued sequences. This is strongly indicative of the method to be adopted later in much more general contexts.

Definition 1.7.13 A complex-valued sequence $\{z_n\}$ is said to be *convergent* to ζ if for any number $\epsilon > 0$ there exists a positive integer N such that

$$|z_n - \zeta| < \epsilon \quad \text{whenever } n > N.$$

We then write $\lim z_n = \zeta$ or $z_n \to \zeta$ and call ζ the *limit* of $\{z_n\}$.

Of course, ζ may be a complex number. The rider '$n \to \infty$' is often added for clarification.

There is no need to say more at this stage specifically about complex-valued sequences. The point has been made that we are not able to set up a definition of convergence which exactly parallels that for real-valued sequences, but nonetheless it is the real-valued theory which subsequently suggests an adequate definition. The adequacy can be seen by showing that analogues of Theorem 1.7.11 and Theorem 1.7.12 can be deduced using Definition 1.7.13. This can be done by first showing that the convergence of a complex-valued sequence $\{z_n\}$ is equivalent to the convergence of both real-valued sequences $\{a_n\}$ and $\{b_n\}$, where we set $z_n = a_n + ib_n$ for each $n \in \mathbf{N}$. But all this will be seen as byproducts of our more general theory in the coming chapters.

Only one thing remains to complete this section. The following theorem allows us to reduce considerably the work involved in determining or estimating the limit of a convergent sequence.

Theorem 1.7.14 *Let $\{s_n\}$ and $\{t_n\}$ be convergent sequences (real-valued or complex-valued) with $\lim s_n = s$ and $\lim t_n = t$. Then*

(a) *if $u_n = s_n + t_n$ for all $n \in \mathbf{N}$, then the sequence $\{u_n\}$ is convergent and $\lim u_n = s + t$;*

(b) *if $v_n = s_n t_n$ for all $n \in \mathbf{N}$, then the sequence $\{v_n\}$ is convergent and $\lim v_n = st$; and*

(c) *when $t_n \neq 0$ for any $n \in \mathbf{N}$ and $t \neq 0$, if $w_n = s_n/t_n$ for all $n \in \mathbf{N}$, then the sequence $\{w_n\}$ is convergent and $\lim w_n = s/t$.*

We will not prove this theorem. The proof is standard and is available in most textbooks on the subject. The relevance of the theorem to our development is that it provides the first occurrence of the need to add, multiply or divide terms of sequences. Up to this point the only arithmetic operation we have used on sequence-values has been the taking of absolute differences, in expressions such as $|a_n - \xi|$ and $|a_n - a_m|$. This operation has an important alternative interpretation: we have only been concerned with the distance between numbers. It is the recognition of this fact that prompts the whole theory of metric spaces that we begin in the next chapter: a *metric space* is a set where the only additional notion we are given is that of a distance between pairs of its elements.

This will be the first space treated here in the hierarchy of spaces that we have spoken of. It will be some time later (Chapter 6) when we first introduce the notion of adding elements of a set together.

Review exercises 1.7

(1) Find a positive integer N such that $|(2n - 3)/(n + 1) - 2| < \epsilon$ for all $n > N$, when (a) $\epsilon = 10^{-1}$, (b) $\epsilon = 10^{-6}$.

(2) Suppose $\{a_n\}$ and $\{b_n\}$ are sequences for which there exist numbers ξ, B and N ($B \in \mathbf{R}_+$, $N \in \mathbf{N}$) so that $|a_n - \xi| \leqslant B|b_n|$ for all $n > N$. Suppose also that $\lim b_n = 0$. Prove that $\lim a_n = \xi$.

(3) Suppose $\{a_n\}$ is a sequence of nonnegative numbers for which $\{(-1)^n a_n\}$ converges. Show that $\{a_n\}$ converges.

(4) Define a sequence $\{a_n\}$ by $a_1 = \sqrt{2}$ and $a_{n+1} = \sqrt{2 + a_n}$, for $n \geqslant 1$. Show that $\{a_n\}$ is bounded and increasing. Hence show that $a_n \to 2$.

(5) Let the sequence $\{a_n\}$ be such that $|a_{n+1} - a_n| < r^n$ for all $n \in \mathbf{N}$, where $0 < r < 1$. Take any $\epsilon > 0$. Show that there exists a positive integer N such that $|a_n - a_m| < \epsilon$ whenever $m, n > N$.

1.8 Series

The definition of a series involves the adding together of terms of a sequence so that, as suggested at the end of the preceding section, we will not see any generalisation of the notion of a series until Chapter 6. However, series of real numbers will occur quite early in the next chapter, and series of complex numbers will arise in Chapter 8, so this section serves to review the latter concepts and to suggest relevant definitions when we come to the more general context.

Until we point out a change, the following definitions and results apply equally to real or complex numbers.

Definition 1.8.1 By a *series* (or *infinite series*), we mean an ordered pair $(\{a_n\}, \{s_n\})$ of sequences, where $\{a_n\}$ is any sequence of numbers and

$$s_n = a_1 + a_2 + \cdots + a_n = \sum_{k=1}^{n} a_k$$

for $n \in \mathbf{N}$. The series is more commonly denoted by

$$\sum_{k=1}^{\infty} a_k \quad \text{or} \quad a_1 + a_2 + a_3 + \cdots,$$

or simply $\sum a_k$, when there is no likelihood of confusion. The number a_n is called the *nth term* of the series $\sum a_k$. The number s_n is called the *nth partial sum* of the series $\sum a_k$.

The series $\sum a_k$ is said to *converge*, or to be *convergent*, if the sequence $\{s_n\}$ converges, and then the number $\lim s_n$ is called the *sum* of the series, also denoted by $\sum_{k=1}^{\infty} a_k$ or $\sum a_k$. If the sequence $\{s_n\}$ diverges, then the series $\sum a_k$ is said to *diverge*, or to be *divergent*.

The 'more common' notation is in fact used universally, because of its suggestion that a series is a limiting sum of a sequence. Given a sequence $\{a_n\}$, we form the sequence $\{s_n\}$ of partial sums (where $s_1 = a_1$, $s_2 = a_1 + a_2$, $s_3 = a_1 + a_2 + a_3$, etc.). Then the convergence or divergence of the series $\sum a_k$ is determined by the convergence or divergence, respectively, of the sequence $\{s_n\}$, and the limit of the sequence $\{s_n\}$, if it exists, is the sum of the series $\sum a_k$.

Since the convergence or divergence of a series is defined in terms of the convergence or divergence of a sequence, many of our results on sequences carry over without further proof to give results on series. For example, with only a slight adjustment, applying Theorem 1.7.12 to the sequence of partial sums, leads to the following theorem. It is also often referred to as a *Cauchy convergence criterion*.

Theorem 1.8.2 *A series* $\sum a_k$ *is convergent if and only if for any number* $\epsilon > 0$ *there exists a positive integer* N *such that*

$$\left| \sum_{k=m}^{n} a_k \right| < \epsilon \quad \text{whenever } n \geqslant m > N.$$

Using the earlier result, it is only necessary to observe that

$$\sum_{k=m}^{n} a_k = a_m + a_{m+1} + \cdots + a_n = s_n - s_{m-1},$$

where $\{s_n\}$ is the associated sequence of partial sums. $\square$

This theorem quickly allows us to conclude as follows that the *harmonic series* $\sum_{k=1}^{\infty} 1/k$ is divergent. We notice that

$$\sum_{k=m}^{2m} \frac{1}{k} = \frac{1}{m} + \frac{1}{m+1} + \cdots + \frac{1}{2m}$$

$$> \frac{1}{2m} + \frac{1}{2m} + \cdots + \frac{1}{2m} = \frac{m+1}{2m} > \frac{1}{2}.$$

Then, choosing ϵ to satisfy $0 < \epsilon \leqslant \frac{1}{2}$, we see that no matter what value we try for N we cannot have $\sum_{k=m}^{n} 1/k < \epsilon$ whenever $n \geqslant m > N$. That is, we cannot satisfy the convergence criterion, so the series is divergent.

In Theorem 1.8.2, if the series is convergent then the criterion must hold in particular for $m = n$. That observation immediately gives us the following.

Theorem 1.8.3 *If a series $\sum a_k$ is convergent, then for any number $\epsilon > 0$ there exists a positive integer N such that*

$$|a_n| < \epsilon \quad \text{whenever } n > N.$$

To paraphrase this: if $\sum a_k$ is convergent, then $a_n \to 0$. Importantly, we can put this still another way (the *contrapositive* way) and say that if $\{a_n\}$ is a sequence not converging to zero, then the series $\sum a_k$ cannot be convergent.

The converse is not true: nothing can be said about the convergence or divergence of the series $\sum a_k$ if we know only that $a_n \to 0$. The harmonic series is an example of this: we have $1/n \to 0$, but $\sum 1/k$ diverges.

We also use Theorem 1.8.2 to provide a simple proof that a series converges if it is absolutely convergent. The latter term must be defined:

Definition 1.8.4   A series $\sum a_k$ is called *absolutely convergent* if the series $\sum |a_k|$ is convergent. A series which is convergent but not absolutely convergent is called *conditionally convergent*.

Suppose $\sum a_k$ is absolutely convergent. This means that $\sum |a_k|$ is convergent, so that, by Theorem 1.8.2, for any $\epsilon > 0$ there is a positive integer N such that

$$\left| \sum_{k=m}^{n} |a_k| \right| = \sum_{k=m}^{n} |a_k| < \epsilon \quad \text{whenever } n \geqslant m > N.$$

Using the extension of the inequality $|a + b| \leqslant |a| + |b|$ to a sum of more than two numbers, we then have

$$\left| \sum_{k=m}^{n} a_k \right| \leqslant \sum_{k=m}^{n} |a_k| < \epsilon \quad \text{whenever } n \geqslant m > N,$$

and, again applying Theorem 1.8.2, this implies that $\sum a_k$ is convergent. As required, we have proved the following theorem.

Theorem 1.8.5 *A series is convergent if it is absolutely convergent.*

It is interesting to trace the chain of theorems that led to this result. Look at Figure 3. All the numbers refer to theorems, except the one at the beginning of the chain, which is our Nested Intervals Axiom, and the one in the centre, which is our definition of the limit of a sequence. Rather dramatically, this shows the supreme role played by the notion of completeness of the real number system and the central role played by the notion of convergence of a sequence. The main point to be made at this time is the ultimate dependence of Theorem 1.8.5 on our assumption that there are no holes in the real number system (at least, according to our treatment of this topic), and this is an assumption that would appear to be totally unrelated to the content of Theorem 1.8.5.

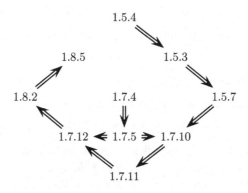

Figure 3

Returning to that theorem, we point out that the converse is not true (a convergent series certainly need not be absolutely convergent), and indeed provision was made in Definition 1.8.4 for convergent series which are not absolutely convergent: they are termed 'conditionally' convergent. A simple example is the series $\sum(-1)^{k+1}/k$. This is shown to be convergent in most standard texts. For absolute convergence we would require the series $\sum|(-1)^{k+1}/k|$, that is $\sum 1/k$, to be convergent, and this is not the case.

The remainder of this section (except the last paragraph) applies only to series $\sum a_k$ where $\{a_n\}$ is a real-valued sequence.

By a *positive series*, we mean a series $\sum a_k$ in which $a_n > 0$ for all $n \in \mathbf{N}$. The advantage in working with positive series is that there are numerous tests which allow us to determine whether they converge or diverge, without recourse to the definition. All these tests use the basic *comparison test*, to be given below, and this in turn relies on the fact that, for a positive series $\sum a_k$, the associated sequence $\{s_n\}$ of partial sums is increasing (since $s_{n+1} = s_n + a_{n+1} > s_n$ for all $n \in \mathbf{N}$). Hence Theorem 1.7.10 may be employed to assure us that a positive series is convergent if its sequence of partial sums is bounded.

Theorem 1.8.6 (Comparison Test) *Let $\sum a_k$ and $\sum b_k$ be two positive series, with $a_n \leqslant b_n$ for all n greater than some integer N. Then*

(a) *if $\sum b_k$ converges, so does $\sum a_k$;*
(b) *if $\sum a_k$ diverges, so does $\sum b_k$.*

To prove (a), set

$$s_n = \sum_{k=1}^{n} a_k, \quad t_n = \sum_{k=1}^{n} b_k,$$

for all $n \in \mathbf{N}$, so that, with $n > N$,

$$s_n - s_N = a_{N+1} + a_{N+2} + \cdots + a_n,$$
$$t_n - t_N = b_{N+1} + b_{N+2} + \cdots + b_n.$$

Then $0 < s_n - s_N \leqslant t_n - t_N$ for all $n > N$ since $0 < a_m \leqslant b_m$ for all $m > N$. Since we are given in (a) that $\sum b_k$ converges, we have by definition that the sequence $\{t_n\}$ is convergent, and hence bounded (Theorem 1.7.6). That is, there is a number K such that $0 < t_n \leqslant K$ for all $n \in \mathbf{N}$, and so $0 < s_n \leqslant K - t_N + s_N$ for $n > N$. Hence

$$0 < s_n \leqslant \max\{s_1, s_2, \ldots, s_N, K - t_N + s_N\}$$

for all $n \in \mathbf{N}$, and so the sequence $\{s_n\}$ is also bounded. As we just stated, this implies that $\{s_n\}$, and thus the series $\sum a_k$, is convergent. This proves (a), and then (b) follows immediately since if the series $\sum b_k$ were convergent then so would be the series $\sum a_k$, giving a contradiction. □

As an example, we will show that the series $\sum 1/k^2$ is convergent. It is not easy to apply the definition directly to show this, but we can show directly that the series $\sum 2/k(k+1)$ converges, and can then use this in a comparison test. We have only to note that

$$\sum_{k=1}^{n} \frac{2}{k(k+1)} = 2\sum_{k=1}^{n} \left(\frac{1}{k} - \frac{1}{k+1}\right)$$

$$= 2\left(\left(1 - \frac{1}{2}\right) + \left(\frac{1}{2} - \frac{1}{3}\right) + \cdots + \left(\frac{1}{n} - \frac{1}{n+1}\right)\right)$$

$$= 2\left(1 - \frac{1}{n+1}\right) \to 2.$$

That is, the sequence of partial sums of $\sum 2/k(k+1)$ converges, so the series converges (and its sum is 2). Then we note that, for $k \geqslant 1$,

$$\frac{1}{k^2} \leqslant \frac{2}{k(k+1)},$$

since this is equivalent to $k + 1 \leqslant 2k$. Hence, by the comparison test, $\sum 1/k^2$ converges.

The series $\sum 1/k$, which we know diverges, and the series $\sum 1/k^2$, which we have just shown converges, are very commonly used in comparison tests to show that other series are divergent or convergent. Another series which is used very often in this regard, and with which we assume familiarity, is the *geometric series*

$$\sum_{k=0}^{\infty} ax^k = a + ax + ax^2 + \cdots,$$

where a and x are real. This converges for any x in the open interval $(-1, 1)$, and diverges otherwise. Its sum, when it converges, is $a/(1-x)$. (Note that '$k = 0$' in the summation, instead of the usual '$k = 1$', has the natural meaning indicated.)

We will consider here only one of the tests of convergence and divergence deducible from the comparison test. (Others are given in the exercises at the end of this section.)

Let $\sum a_k$ be a positive series and set

$$r_n = \frac{a_{n+1}}{a_n}$$

for all $n \in \mathbf{N}$. If there exists a number r, with $0 < r < 1$, such that $r_n \leqslant r$ for all n greater than some positive integer N, then the series $\sum a_k$ is convergent; if $r_n \geqslant 1$ for all n greater than some integer N, then the series $\sum a_k$ is divergent.

This is known as the *ratio test*. It is proved, in the test for convergence, by noting that

$$a_n = \frac{a_n}{a_{n-1}} \frac{a_{n-1}}{a_{n-2}} \frac{a_{n-2}}{a_{n-3}} \cdots \frac{a_{N+2}}{a_{N+1}} \frac{a_{N+1}}{a_N} a_N \leqslant r^{n-N} a_N,$$

when $n > N$. Then a favourable comparison may be made between the series $\sum a_k$ and the geometric series $\sum a_N r^{-N} r^k$, which converges since $0 < r < 1$. To prove the test for divergence, we note that $r_n \geqslant 1$ when $n > N$ so that $a_{n+1} \geqslant a_n \geqslant \cdots \geqslant a_{N+1} > 0$. Hence we cannot possibly have $a_n \to 0$ (which, by Theorem 1.8.3, is necessary for the convergence of $\sum a_k$).

As an application of the ratio test, we prove that the series

$$\sum_{k=0}^{\infty} \frac{x^k}{k!}$$

converges for any value of x. Since the test applies only to positive series, we consider instead the series

$$\sum_{k=0}^{\infty} \frac{|x|^k}{k!},$$

for $x \neq 0$, and may set $r = \frac{1}{2}$ (for convenience). We have

$$\frac{|x|^{n+1}}{(n+1)!} \bigg/ \frac{|x|^n}{n!} = \frac{|x|}{n+1} \leqslant \frac{1}{2}$$

whenever $n \geqslant 2|x| - 1$, and so (choosing N to be an integer greater than $2|x| - 1$), we have proved the convergence of the latter series. Convergence for $x = 0$ is clear, so that in effect we have proved the original series to be absolutely convergent. Hence, by Theorem 1.8.5, that series converges for any value of x.

This means that we may define a real-valued function, traditionally denoted by exp, by the equation

$$\exp(x) = \sum_{k=0}^{\infty} \frac{x^k}{k!}, \quad x \in \mathbf{R}.$$

We will assume here the fact, found in calculus texts, that

$$\exp(x) = e^x$$

for all $x \in \mathbf{R}$.

In a similar way, functions sin and cos are defined by the equations

$$\sin(x) = \sum_{k=0}^{\infty} \frac{(-1)^k x^{2k+1}}{(2k+1)!}, \quad x \in \mathbf{R},$$

$$\cos(x) = \sum_{k=0}^{\infty} \frac{(-1)^k x^{2k}}{(2k)!}, \quad x \in \mathbf{R}.$$

These are just the familiar sine and cosine functions.

Though not relevant to our development, we end this section by recalling the *binomial theorem*, which is to be used in Chapter 6. It states that, for any numbers a and b and any positive integer n,

$$(a+b)^n = a^n + \binom{n}{1} a^{n-1}b + \binom{n}{2} a^{n-2}b^2 + \cdots + \binom{n}{n-1} ab^{n-1} + b^n,$$

where we have used the *binomial coefficient*

$$\binom{n}{r} = \frac{n!}{r!(n-r)!}, \quad r = 0, 1, 2, \ldots, n.$$

(Recall that $0! = 1$.)

Review exercises 1.8

(1) Show that

$$\sum_{k=1}^{\infty} \frac{1}{k(k+2)} = \frac{3}{4}, \qquad \sum_{k=1}^{\infty} \frac{1}{k(k+1)(k+2)} = \frac{1}{4}.$$

(2)   (a) Let $\sum a_k$ and $\sum b_k$ be two positive series, with $a_n \leqslant C b_n$ for all $n \in \mathbf{N}$ and some positive number C. Show that if $\sum b_k$ converges then $\sum a_k$ converges, and if $\sum a_k$ diverges then $\sum b_k$ diverges.

       (b) Let $\sum a_k$ and $\sum b_k$ be two positive series, with the property that $\lim(a_n/b_n) = 1$. Use (a) to show that $\sum a_k$ converges if and only if $\sum b_k$ converges. (This is the *limit comparison test*.)

(3)   (a) Prove the *root test* for convergence: Let $\sum a_k$ be a positive series. If $a_n^{1/n} < r$ for all $n \in \mathbf{N}$ and some number r, $0 < r < 1$, then $\sum a_k$ is convergent.

(b) Hence show that the series $a+b+a^2+b^2+a^3+b^3+a^4+\cdots$, where $0 < a < b < 1$, is convergent.

(4) Determine whether the following series are absolutely convergent, conditionally convergent or divergent:

$$\sum_{k=1}^{\infty} \frac{(-1)^{k+1}}{\sqrt{k^2+1}}, \qquad \sum_{k=1}^{\infty} \frac{(-1)^{k+1}}{k^2+1}, \qquad \sum_{k=1}^{\infty} \frac{(-1)^{k+1}k}{\sqrt{k^2+1}}.$$

1.9 Functions of a real variable

We are mainly concerned in this section with certain properties of a function $f\colon D \to \mathbf{R}$, where usually $D \subseteq \mathbf{R}$. These are the classical real-valued functions of a real variable. The more important results for our purpose require D to be a compact set, but most comments will be valid for any point set D. We recall that the *graph* of f is the subset of $\mathbf{R}^2$ consisting of points $(x, f(x))$ for x in the domain D of f. This has a common pictorial representation, the details of which will be assumed.

There will be a brief reference, at the end of the section, to real-valued functions of two or more real variables and to complex-valued functions of a real variable. However, unless we specify otherwise, the domain and range of any function are to be taken as sets of real numbers.

We begin by giving the definition of a continuous function. Continuity of a function (and of a mapping generally) is one of the most important notions of analysis, so the following discussion paves the way for our use of continuous functions in applications and also indicates apt definitions to come in the following chapters.

Definition 1.9.1 A function f is said to be *continuous* at a point x_0 in its domain if for any number $\epsilon > 0$ there exists a number $\delta > 0$ such that, whenever x is in the domain of f and $|x - x_0| < \delta$, we have $|f(x) - f(x_0)| < \epsilon$.

This is the usual definition, to be thought of in rough terms as saying that $f(x)$ will be close in value to $f(x_0)$ whenever x is close to x_0. For our purposes, with our emphasis on the convergence of sequences, the alternative provided by the following theorem is often more useful.

Theorem 1.9.2 *A function f is continuous at a point x_0 in its domain if and only if, whenever $\{x_n\}$ is a convergent sequence in the domain of f with $\lim x_n = x_0$, then $\{f(x_n)\}$ is also a convergent sequence and $\lim f(x_n) = f(x_0)$.*

Briefly: f is continuous at x_0 if and only if $f(x_n) \to f(x_0)$ whenever $x_n \to x_0$.

To prove this, suppose first that f is continuous at x_0 and let $\{x_n\}$ be any sequence in the domain of f, convergent to x_0. For each $n \in \mathbf{N}$, $f(x_n)$ is a point in the range of f, so $\{f(x_n)\}$ is a well-defined sequence, which we need to show converges to $f(x_0)$. Let $\epsilon > 0$ be given. Since f is continuous at x_0, there exists $\delta > 0$ such that $|f(x) - f(x_0)| < \epsilon$ whenever $|x - x_0| < \delta$ (and when x is in the domain of f). Also, since $\{x_n\}$ is a convergent sequence in the domain of f, and $\lim x_n = x_0$, there exists a positive integer N such that $|x_n - x_0| < \delta$ whenever $n > N$. Therefore, provided $n > N$, we have $|f(x_n) - f(x_0)| < \epsilon$ and this proves that the sequence $\{f(x_n)\}$ converges, with limit $f(x_0)$.

Suppose next, in proving the converse, that f is not continuous at x_0. We will show that there exists a sequence $\{x_n\}$ in the domain of f, converging to x_0, but such that the sequence $\{f(x_n)\}$ is not convergent. This will complete the proof of the theorem. To say that f is not continuous at x_0 means that there exists a number $\epsilon_0 > 0$ such that, whatever the number $\delta > 0$, there is a number x in the domain of f with $|x - x_0| < \delta$ but for which $|f(x) - f(x_0)| \geqslant \epsilon_0$. For this ϵ_0, choose $\delta = 1/n$, and let x_n be such a number x, so that $|x_n - x_0| < \delta$ but $|f(x_n) - f(x_0)| \geqslant \epsilon_0$. In this way, we have constructed sequences $\{x_n\}$ and $\{f(x_n)\}$: the sequence $\{x_n\}$ converges to x_0 but the sequence $\{f(x_n)\}$ cannot be convergent. This is what we set out to do. $\square$

We say that a function is *continuous on a subset* of its domain if it is continuous at every point of that subset. Such a subset, which may be the whole domain, is commonly an open or closed interval. The function is said to be *discontinuous* at any point of its domain at which it is not continuous, and such a point is called a *discontinuity* of the function.

As an example, we can introduce here the *greatest-integer function*. It has domain $\mathbf{R}$ and range $\mathbf{Z}$ and is denoted by $[x]$, where $x \in \mathbf{R}$. This is defined to be the integer in the half-open interval $(x-1, x]$. Thus, $[3.24]$ is the integer in $(2.24, 3.24]$, namely 3, and similarly $[-7.8] = -8$, $[\pi] = 3$, $[28] = 28$. The greatest-integer function $[x]$ is discontinuous when $x \in \mathbf{Z}$. To see this, let c be any integer and consider the sequence $\{c - 1/n\}$, or $c - 1, c - \frac{1}{2}, c - \frac{1}{3}, \dots$. Clearly, $c - 1/n \to c$, but $[c - 1/n] = c - 1$ for all $n \in \mathbf{N}$ since c is an integer. At any other value, not an integer, the greatest-integer function is continuous.

The sum, product and quotient of two functions are defined as follows.

Definition 1.9.3 If f and g are two functions, then their *sum* $f + g$, *product* fg and *quotient* f/g are functions defined by the equations

$$(f + g)(x) = f(x) + g(x),$$
$$(fg)(x) = f(x)g(x),$$
$$\left(\frac{f}{g}\right)(x) = \frac{f(x)}{g(x)}.$$

Their domains are the intersection of the domains of f and g, excluding, in the case of f/g, points x where $g(x) = 0$.

We can use precisely the same definitions for functions $f\colon X \to \mathbf{R}$ and $g\colon X \to \mathbf{R}$, where X is any set. Then $f+g$, fg and f/g are also functions from X into $\mathbf{R}$, except that points x where $g(x) = 0$ are excluded from the domain of f/g.

A *constant function* is a function k with domain $\mathbf{R}$ such that $k(x) = c$ for all $x \in \mathbf{R}$ and some number c. The preceding definition of the product of two functions includes the case where one of the functions is the constant function k. Thus kg is the function, whose domain is the domain of g, such that $(kg)(x) = cg(x)$. This function is usually written simply as cg, such as $3g$ or $(-5)g$. When $c = -1$, we write $-g$ instead of $(-1)g$.

By simply combining Theorems 1.7.14 and 1.9.2, we obtain the following.

Theorem 1.9.4 *If f and g are functions which are continuous on their domains, then the functions $f + g$, fg and f/g are continuous on their domains.*

This result is useful in determining whether a complicated-looking function is continuous, since we may be able to decompose it into sums, products or quotients of simpler functions which are known to be continuous.

We also note the following. If f is a given function, then by $|f|$ we mean the function given by

$$|f|(x) = |f(x)|,$$

and having the same domain as f. It is easy to show that $|f|$ is continuous at any points where f is.

As we have mentioned, we are particularly interested in functions whose domains are compact sets, but for the discussion here we will take a slightly simpler approach and suppose those sets are closed intervals. When the functions are continuous, they possess properties which are made use of in a vast number of applications, as we will see. Moreover, these are properties which may readily be generalised and we carry out that generalisation in Chapter 4. The interest in closed intervals rests on the fact that if $\{x_n\}$ is a real-valued sequence such that $x_n \in [a, b]$, say, for all n, then $\lim x_n$, when it exists, is also a point of $[a, b]$. This follows from Theorem 1.7.7. In contrast, we cannot always say that the limit of a sequence of points in an open interval also belongs to the interval.

The following theorems give two of those properties. The first refers to *bounded* functions: a function $f \colon D \to \mathbf{R}$ is bounded if $|f(x)| \leqslant M$ for some positive number M and all $x \in D$.

Theorem 1.9.5 *If the domain of a function is a closed interval and the function is continuous on the interval, then it is bounded.*

Theorem 1.9.6 *If the domain of a function is a closed interval and the function is continuous on the interval, then it attains its minimum and maximum values.*

To say that f attains its minimum and maximum values means that there exist points x_m and x_M in the domain, $[a, b]$ say, such that

$$f(x_m) = \min_{x \in [a,b]} f(x) \quad \text{and} \quad f(x_M) = \max_{x \in [a,b]} f(x).$$

We will discuss the theorems before giving their proofs. Theorems like these two should be looked on as useful not only for the conclusions they state, but for the conditions they give as sufficient for those conclusions to hold. Drop either of the conditions (the domain being a closed interval and the function being continuous), and the conclusion can no longer be guaranteed.

For example, consider the functions

$$g_1(x) = \frac{1}{x}, \quad 0 < x \leqslant 1;$$

$$g_2(x) = x, \quad 0 < x < 1;$$

$$g_3(x) = \begin{cases} \dfrac{1}{x}, & |x| \leqslant 1,\ x \neq 0, \\ 0, & x = 0. \end{cases}$$

The domain of g_1 is the half-open interval $(0, 1]$. This function is continuous on its domain, since if x_0 is any point in it (so $0 < x_0 \leqslant 1$)

and $\{x_n\}$ is a sequence in the domain converging to x_0 (so $0 < x_n \leqslant 1$ for all $n \in \mathbf{N}$ and $x_n \to x_0$), then

$$g_1(x_n) = \frac{1}{x_n} \to \frac{1}{x_0} = g_1(x_0),$$

that is, the sequence $\{g_1(x_n)\}$ converges to $g_1(x_0)$. However, the function is not bounded, since we cannot have $|g_1(x)| \leqslant M$ for all $x \in (0,1]$ no matter what the value of $M > 0$: just take $x \in (0, 1/M)$ (assuming $M \geqslant 1$) so that

$$|g_1(x)| = \left|\frac{1}{x}\right| = \frac{1}{x} > M \quad \text{if } 0 < x < \frac{1}{M}.$$

Also, the function does not attain its maximum value, since in fact not even $\sup_{x \in (0,1]} g_1(x)$ exists. However, g_1 does attain its minimum; it is the value $g_1(1) = 1$.

The function g_2 is continuous, but its domain is not a closed interval. It is easy to see that $\inf_{x \in (0,1)} g_2(x) = 0$ and $\sup_{x \in (0,1)} g_2(x) = 1$, but we do not have $g_2(x) = 0$ or $g_2(x) = 1$ for any $x \in (0,1)$. So g_2 is bounded, but does not attain its maximum or minimum values.

For g_3, the domain is the closed interval $[-1,1]$, but the function is discontinuous at 0. To see that it is discontinuous at 0, consider the sequence $\{1/n\}$, all of whose terms are in the domain of g_3, and which converges to 0. However, $g_3(1/n) = n$ for all $n \in \mathbf{N}$, so certainly $\{g_3(1/n)\}$ does not converge to $g_3(0) = 0$. Like g_1, this function is not bounded.

For the proof of Theorem 1.9.5, to be specific consider the function $f \colon [a,b] \to \mathbf{R}$ and suppose that f is continuous on $[a,b]$ but not bounded. We will obtain a contradiction. Since f is not bounded, for any $n \in \mathbf{N}$ there exists a point $x_n \in [a,b]$ such that $|f(x_n)| > n$. This gives rise to a bounded sequence $\{x_n\}$. (Do not confuse the different uses of the word 'bounded'.) It follows from the Bolzano–Weierstrass theorem for sequences (Theorem 1.7.11) that there is a convergent subsequence $\{x_{n_k}\}$ of this sequence, and its limit, x_0 say, must belong to $[a,b]$. Since f is continuous at x_0, we have $x_{n_k} \to x_0$ (as $k \to \infty$), and hence $f(x_{n_k}) \to f(x_0)$. The convergent sequence $\{f(x_{n_k})\}$ must be bounded (Theorem 1.7.6), so we cannot have $|f(x_{n_k})| > n_k$ for all $k \in \mathbf{N}$, since n_k may be as large as we please. This is the desired contradiction arising from the assumption that f is not bounded. $\qquad \square$

We now use this result in the proof of Theorem 1.9.6. The proof will be given only in the case of the maximum value, the proof for the minimum value being analogous.

Let the continuous function be $f : [a, b] \to \mathbf{R}$. By the preceding result, we know that f is bounded on $[a, b]$. That is, the set $\{f(x) : a \leqslant x \leqslant b\}$ is bounded, so its least upper bound, M say, exists (Theorem 1.5.7). Thus $f(x) \leqslant M$ for all $x \in [a, b]$. The theorem will follow if we can show that $f(x_M) = M$ for some $x_M \in [a, b]$. If this is not true, so that $M - f(x) > 0$ for all $x \in [a, b]$, then the function g, where $g(x) = 1/(M - f(x))$, $a \leqslant x \leqslant b$, is continuous by Theorem 1.9.4. Then it too is bounded, by the preceding result, so $1/(M - f(x)) \leqslant C$, say, with $C > 0$. It follows that $f(x) \leqslant M - 1/C$ for all $x \in [a, b]$, and this contradicts the fact that M is the least upper bound of the set $\{f(x) : a \leqslant x \leqslant b\}$. $\qquad \square$

Again it is worth noting the ultimate dependence of these results on our axiom of completeness (Axiom 1.5.4), via Theorems 1.7.11 and 1.5.7.

There are corresponding definitions and results for real-valued functions of two or more real variables. Without going into much detail, we will give some of the theory for functions of two variables. Such a function is $f : D \times E \to \mathbf{R}$, where D and E are sets of real numbers so that the domain is a set of ordered pairs of real numbers. The image of (x, y) under f is written as $f(x, y)$, rather than the more strictly correct $f((x, y))$.

The function f is continuous at a point (x_0, y_0) in its domain if for any number $\epsilon > 0$ there exists a number $\delta > 0$ such that, whenever (x, y) is in the domain of f and both $|x - x_0| < \delta$ and $|y - y_0| < \delta$, then $|f(x, y) - f(x_0, y_0)| < \epsilon$. An equivalent formulation may be given in terms of sequences (but will be omitted here). When D and E are closed intervals, it may be shown that if f is continuous on its domain then it is bounded. Here, that means there exists a number $M > 0$ such that $|f(x, y)| \leqslant M$ for all points (x, y) in the domain of f. We will make considerable use of this result in our examples and applications. It is of course the obvious analogue of Theorem 1.9.5, and is also a special instance of a theorem to be proved in Chapter 4.

It is necessary to mention also in this section that in many of our examples and applications we will make use of elementary properties of the following.

(a) The functions exp, log, sin and cos.

(b) The *derivative* of a function, and its left and right derivatives. (A function is said to be *differentiable* on a closed interval $[a, b]$ if its right derivative exists at a, its derivative exists at each point of (a,b) and its left derivative exists at b.)

(c) The *definite integral* of a function over an interval. (A function is said to be *integrable* over an interval if its integral exists on that interval. All integrals in this book are *Riemann* integrals, which may be thought of as the usual integrals of a first calculus course.)

(d) *Partial derivatives.*

(e) *Double integrals.*

(f) *Ordinary differential equations.*

These topics are too large to be able to review them adequately here. In any case, such a review would not be pertinent to our mainstream since our general theory will not specifically use any of these concepts and no generalisations of them will be given (though many have been developed). Other than for the simplest properties, we will however carefully describe whatever result is being used at its first occurrence. Our notation for derivatives and integrals will be quite standard.

A topic that we will be generalising in Chapter 9 is that of Fourier series, and some acquaintance with the classical treatment of trigonometric Fourier series will be helpful there.

There is little that needs to be said about complex-valued functions of a real variable. The imaginary unit i may be treated as an ordinary constant and any property common to the real-valued functions given by the real and imaginary parts of the original function may be taken as true for that function also. For example, for the function $f\colon [a,b] \to \mathbf{C}$, if the functions $f_r\colon [a,b] \to \mathbf{R}$ given by $f_r(x) = \operatorname{Re} f(x)$ ($x \in [a,b]$) and $f_j\colon [a,b] \to \mathbf{R}$ given by $f_j(x) = \operatorname{Im} f(x)$ ($x \in [a,b]$) are both integrable over $[a,b]$, then f will be integrable over $[a,b]$, and

$$\int_a^b f(x)\,dx = \int_a^b f_r(x)\,dx + i \int_a^b f_j(x)\,dx.$$

Although we will generally be precise in our handling of functions, speaking of 'the function f', for example, there will be occasions where a looser (and common) approach is more immediately suggestive and elegant, and we will drop our formalities on those occasions. For example, it is easier to speak of the set of functions $\{1, x, x^2, \dots\}$ than the set $\{f_k : f_k(x) = x^k,\ k = 0,\ 1,\ 2, \dots\}$. And we have already used the notation $[x]$ for the greatest-integer function.

Review exercises 1.9

(1) Let the function $f\colon D \to \mathbf{R}$ be continuous at $x_0 \in D \subseteq \mathbf{R}$, and suppose $f(x_0) > 0$. Show that there exists a number $\delta > 0$ such that $f(x) > 0$ for $x \in (x_0 - \delta, x_0 + \delta) \cap D$.

(2) Let the function $f\colon [a,b] \to \mathbf{R}$ be continuous on $[a,b]$. Suppose there is a number c, $0 < c < 1$, with the property that for every $x \in [a,b]$ there exists $y \in [a,b]$ such that $|f(y)| \leqslant c|f(x)|$. Show that $f(x_0) = 0$ for some $x_0 \in [a,b]$.

(3) Take $D \subseteq \mathbf{R}$. For a function $f\colon D \to \mathbf{R}$ such that $f(x) \geqslant 0$ for all $x \in D$, the function $\sqrt{f}\colon D \to \mathbf{R}$ is defined by $\sqrt{f}(x) = \sqrt{f(x)}$, for $x \in D$. If, further, f is continuous at $x_0 \in D$, show that $\sqrt{f}$ is continuous at x_0.

(4) Use trigonometric identities and the fact that $|\sin x| \leqslant |x|$ for all $x \in \mathbf{R}$ to show that the functions sin and cos are continuous on $\mathbf{R}$.

1.10 Uniform convergence

A sequence was defined as a mapping whose domain is the set $\mathbf{N}$. We said that the range of a sequence may be any set. Until now, we have only considered sequences where the range was a set of real or complex numbers, but we intend in this section to look at sequences which have as their range a set of real-valued functions of a real variable. All functions in this section will be of that type.

Let $C[a,b]$ denote the set of all real-valued functions whose domain is the closed interval $[a,b]$ and which are continuous on that domain. We could, for example, consider properties of a sequence $F\colon \mathbf{N} \to C[a,b]$. Writing $F(1) = f_1$, $F(2) = f_2$, and so on, in the usual way, this is a sequence $\{f_n\}$ in which every term is a function continuous on $[a,b]$.

We have as yet no notion of convergence for such a sequence. It may seem strange at first that we are soon to define two different ways in which a sequence of functions may be said to converge. It may be possible for a sequence to be deemed convergent under one definition but not under the other, but if it is convergent under both definitions then it will turn out that the limit is a function which is the same in both cases. In our example where the range is a subset of $C[a,b]$, it will be of interest to know whether the limit, if it exists, is again a member of $C[a,b]$. We will see that this is assured under one of the definitions but not the other. This question has some similarity with our earlier concern as to whether the limit of every convergent sequence of real numbers chosen from some interval was also a member of that interval: the answer was 'yes' only when the interval was closed. Such questions are typical of those that will be asked, and answered, in more general contexts later on.

Let $\{f_n\}$ be a sequence of functions, all having the same domain D. For any $x_0 \in D$, the numbers $f_1(x_0)$, $f_2(x_0)$, $f_3(x_0)$, ..., that is, the images of x_0 under the terms of the sequence, themselves form a sequence of real numbers. This sequence $\{f_n(x_0)\}$ is precisely the same type of sequence as those we have considered earlier, and of course we have a definition of convergence for such sequences. Possibly, whatever the point $x \in D$, the real-valued sequence $\{f_n(x)\}$ will converge. In that case, there exists a function f, with domain D, such that

$$f(x) = \lim f_n(x).$$

This suggests the first of the definitions: in this situation, the sequence of functions is termed convergent and the function f is called the limit of the sequence. Because we have another definition coming up, this one is distinguished by referring to *pointwise convergence* and the *pointwise limit*. We notice that for pointwise convergence we need nothing more than our earlier idea of convergence of real-valued sequences.

This definition may be written formally as follows. A sequence $\{f_n\}$ of functions with domain D is said to converge pointwise to a function f with domain D if, given a number $\epsilon > 0$, for each $x \in D$ there exists a positive integer $N(x)$ such that

$$|f_n(x) - f(x)| < \epsilon \quad \text{whenever } n > N(x).$$

We write $\lim f_n = f$ or $f_n \to f$ and call f the pointwise limit of $\{f_n\}$. Otherwise the sequence is termed divergent.

Compare this with the second definition.

Definition 1.10.1 A sequence $\{f_n\}$ of functions with domain D is said to converge *uniformly* to a function f with domain D if, given a number $\epsilon > 0$, there exists a positive integer N such that, for all $x \in D$,

$$|f_n(x) - f(x)| < \epsilon \quad \text{whenever } n > N.$$

We write $f_n \rightrightarrows f$ and call f the *uniform limit* of $\{f_n\}$.

There is a crucial difference in the wording of this second form of convergence, which we note is to be called 'uniform convergence' to distinguish

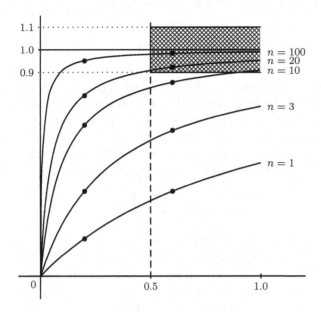

Figure 4

it from pointwise convergence. This is in the phrases 'for each $x \in D$ ... $N(x)$' and 'N ... for all $x \in D$'. By $N(x)$ in the definition of pointwise convergence we mean, in the usual way, the value of a function $N: D \to \mathbf{N}$ at x. That is, the number N required in showing that the sequence $\{f_n(x)\}$ converges depends on the number $x \in D$ (as well as on the choice of ϵ). For uniform convergence, however, the N that needs to be determined may depend on the choice of ϵ but must not depend on the number $x \in D$.

As an example, consider the sequence $\{f_n\}$, where

$$f_n(x) = \frac{nx}{nx+1}, \quad 0 \leqslant x \leqslant 1.$$

Graphs of f_n, for $n = 1, 3, 10, 20, 100$, are given in Figure 4. The sequence is pointwise convergent, with pointwise limit f where

$$f(x) = \begin{cases} 0, & x = 0, \\ 1, & 0 < x \leqslant 1. \end{cases}$$

To see this, we observe that, given $\epsilon > 0$, $|f_n(0) - f(0)| = 0 < \epsilon$ for n

larger than any positive integer, while, when $0 < x \leqslant 1$,

$$|f_n(x) - f(x)| = \left| \frac{nx}{nx+1} - 1 \right| = \frac{1}{nx+1} < \frac{1}{nx} < \epsilon$$

provided $n > 1/\epsilon x$. Then, in the definition of pointwise convergence, we can define the function N by $N(0) = 1$ and $N(x) = 1 + [1/\epsilon x]$ for $0 < x \leqslant 1$. (We have made use of the greatest-integer function $[x]$.) Although here N depends explicitly on x, this does not deny that a different approach might come up with an expression for N which does not depend on x. That is, we have not shown that $\{f_n\}$ is not uniformly convergent. We can do this by first noting that

$$\left| f_n \left(\frac{1}{n} \right) - f \left(\frac{1}{n} \right) \right| = \frac{1}{2}.$$

Then, setting $\epsilon = \frac{1}{2}$, we cannot possibly find N so that $|f_n(x) - f(x)| < \frac{1}{2}$ for all x in the domain $[0, 1]$ and all $n > N$.

The dots in Figure 4 indicate the pointwise convergence of $\{f_n\}$ to f, showing terms from the real-valued sequences $\{f_n(0.2)\}$ and $\{f_n(0.6)\}$.

Suppose now we take the sequence $\{g_n\}$, where

$$g_n(x) = \frac{nx}{nx+1}, \quad \tfrac{1}{2} \leqslant x \leqslant 1.$$

Like the former sequence, this one is pointwise convergent, with pointwise limit g, where

$$g(x) = 1, \quad \tfrac{1}{2} \leqslant x \leqslant 1,$$

but moreover $g_n \rightrightarrows g$; the sequence is uniformly convergent. This time we make the following calculation: given $\epsilon > 0$, then, whenever $n > 2/\epsilon$, we have

$$|g_n(x) - g(x)| = \frac{1}{nx+1} < \frac{1}{nx} < \frac{\epsilon}{2} \cdot 2 = \epsilon,$$

for all x in $[\frac{1}{2}, 1]$, making explicit use of the fact that $x \geqslant \frac{1}{2}$. That is, we may take N to be any positive integer greater than $2/\epsilon$, and this number is independent of x. The right-hand half of Figure 4 shows the graphs of five terms of $\{g_n\}$. The figure illustrates the basic idea of uniform convergence: once a value of ϵ is given there must be a positive integer N so that all the terms $g_{N+1}, g_{N+2}, \dots$ have graphs lying in the strip of width 2ϵ about the graph of the limit function g. We have set $\epsilon = 0.1$ in the figure and it is apparent that we may take $N = 19$, since $g_{20}, g_{21}, \dots$ all have their graphs in the shaded portion. A corresponding picture was not possible for the former sequence $\{f_n\}$, with any value of ϵ less than 1.

It should be observed that we could similarly prove the uniform convergence of any sequence $\{h_n\}$, where

$$h_n(x) = \frac{nx}{nx+1}, \quad 0 < a \leqslant x \leqslant 1,$$

the point being that we must avoid allowing x to be too close to 0, which is a discontinuity of the limit f of the first example.

A simple comparison of the two definitions shows that any sequence of functions that is uniformly convergent must also be pointwise convergent, but we have just seen that a pointwise convergent sequence need not be uniformly convergent.

The following theorem gives a useful test for determining whether a sequence of functions is uniformly convergent.

Theorem 1.10.2 *A sequence $\{f_n\}$ of functions with domain D converges uniformly to the function f if there is a real-valued sequence $\{a_n\}$ such that $a_n \to 0$ and*

$$|f_n(x) - f(x)| \leqslant |a_n|$$

for all $x \in D$ and all $n \in \mathbf{N}$.

The proof of this theorem is easy. Given $\epsilon > 0$, we know (since $a_n \to 0$) that there exists a positive integer N such that $|a_n| < \epsilon$ whenever $n > N$. This N is independent of $x \in D$, and $|f_n(x) - f(x)| \leqslant |a_n| < \epsilon$ for all $x \in D$ whenever $n > N$, so the sequence $\{f_n\}$ converges uniformly to f, as required. $\square$

In the example above of the sequence $\{h_n\}$, we have $h_n \to h$, where $h(x) = 1$ $(0 < a \leqslant x \leqslant 1)$, and

$$|h_n(x) - h(x)| = \frac{1}{nx+1} \leqslant \frac{1}{na+1}$$

for all $n \in \mathbf{N}$, since $a \leqslant x$. But $1/(na+1) \to 0$, so the sequence is uniformly convergent.

We now return to the question posed at the beginning of this section: whether the limit of a sequence of continuous functions, when it exists, is again a continuous function. Our example of the sequence $\{f_n\}$, where $f_n(x) = nx/(nx+1)$ $(0 \leqslant x \leqslant 1)$, shows that pointwise convergence of the sequence is not sufficient to ensure continuity of the limit, because each term f_n here is continuous (as is easily shown) whereas the limit function has a discontinuity at $x = 0$. However, the next result shows that whenever the convergence is uniform then the limit function is continuous.

Theorem 1.10.3 *Let the sequence* $\{f_n\}$ *of functions with domain* D *be uniformly convergent, with limit* f. *If* f_n *is continuous on* D *for each* $n \in \mathbf{N}$, *then also* f *is continuous on* D.

To prove this, let x_0 be any point in D and let $\{x_m\}_{m=1}^\infty$ be a sequence in D such that $x_m \to x_0$. Choose any number $\epsilon > 0$. Because $f_n \rightrightarrows f$, there exists a positive integer N_1 such that

$$|f(x_m) - f_n(x_m)| < \tfrac{1}{3}\epsilon \quad \text{for all } m \in \mathbf{N} \text{ and all } n > N_1$$

and

$$|f(x_0) - f_n(x_0)| < \tfrac{1}{3}\epsilon \quad \text{for all } n > N_1.$$

(The fact that the convergence is uniform means that the single integer N_1 may be used for all the points x_0 and x_m, $m \in \mathbf{N}$.) Choose any $n > N_1$. Then, since f_n is continuous at x_0, there exists a positive integer N such that

$$|f_n(x_m) - f_n(x_0)| < \tfrac{1}{3}\epsilon \quad \text{whenever } m > N,$$

and so

$$
\begin{aligned}
|f(x_m) - f(x_0)| &= |(f(x_m) - f_n(x_m)) \\
&\quad + ((f_n(x_m) - f_n(x_0)) + (f_n(x_0) - f(x_0))| \\
&\leqslant |f(x_m) - f_n(x_m)| \\
&\quad + |f_n(x_m) - f_n(x_0)| + |f_n(x_0) - f(x_0)| \\
&< \tfrac{1}{3}\epsilon + \tfrac{1}{3}\epsilon + \tfrac{1}{3}\epsilon = \epsilon,
\end{aligned}
$$

whenever $m > N$. This proves that $f(x_m) \to f(x_0)$, so f is continuous at x_0. Thus f is continuous at all points of D, as required. $\square$

We have proved that, under the conditions of the theorem, when $\{x_m\}$ is any convergent sequence in D whose limit is in D,

$$\lim_{n \to \infty} \left(\lim_{m \to \infty} f_n(x_m) \right) = \lim_{m \to \infty} \left(\lim_{n \to \infty} f_n(x_m) \right).$$

That is, the interchange of limit operations is valid. When $\{f_n\}$ is the non-uniformly convergent sequence of our earlier example, in Figure 4, and $x_m \to 0$, the left-hand side here is 0 and the right-hand side is 1.

The next two theorems show that uniform convergence enters also into other questions involving an interchange of limit operations.

Theorem 1.10.4 *Let* $\{f_n\}$ *be a sequence of functions, integrable on their domain* $[a, b]$ *and uniformly convergent with limit* f. *For each* $n \in \mathbf{N}$, *define a function* g_n *by*

$$g_n(x) = \int_a^x f_n(t)\, dt, \quad a \leqslant x \leqslant b.$$

Then the sequence $\{g_n\}$ *also converges uniformly on* $[a, b]$. *Furthermore,* $\lim g_n = g$, *where*

$$g(x) = \int_a^x f(t)\, dt.$$

That is, under the conditions of the theorem, for each $x \in [a, b]$,

$$\lim \int_a^x f_n(t)\, dt = \int_a^x \lim f_n(t)\, dt.$$

The proof follows. Choose $\epsilon > 0$. Since $f_n \rightrightarrows f$, there exists a positive integer N such that

$$|f_n(x) - f(x)| < \frac{\epsilon}{b-a}$$

for all $x \in [a, b]$ and all $n > N$. Then

$$
\begin{aligned}
|g_n(x) - g(x)| &= \left| \int_a^x f_n(t)\, dt - \int_a^x f(t)\, dt \right| \\
&= \left| \int_a^x (f_n(t) - f(t))\, dt \right| \\
&\leqslant \int_a^x |f_n(t) - f(t)|\, dt \\
&< \frac{\epsilon}{b-a} \int_a^x dt = \frac{\epsilon}{b-a}(x - a) \\
&\leqslant \frac{\epsilon}{b-a}(b - a) = \epsilon,
\end{aligned}
$$

for all $x \in [a, b]$, provided $n > N$. Since N is independent of $x \in [a, b]$, this proves that $g_n \rightrightarrows g$ on $[a, b]$. $\square$

Notice our use in this proof of the following two results. They will occur many times in the rest of this book without special mention.

(a) If f and g are integrable functions on $[a,b]$ and $f(x) \leqslant g(x)$ for all $x \in [a,b]$, then

$$\int_a^b f(x)\,dx \leqslant \int_a^b g(x)\,dx.$$

(b) If a function f is integrable on $[a,b]$, then so is the function $|f|$, and

$$\left| \int_a^b f(x)\,dx \right| \leqslant \int_a^b |f(x)|\,dx.$$

The inequality in (b) is an integral version of $|a+b| \leqslant |a| + |b|$, where a and b are any numbers. Assuming the integrability of $|f|$, it is proved using (a) and the fact that $-|f(x)| \leqslant f(x) \leqslant |f(x)|$ for all $x \in [a,b]$. In the proof of Theorem 1.10.4, we also made the assumption that the limit function f is integrable on $[a,b]$ when each f_n is.

Theorem 1.10.5 *Let $\{f_n\}$ be a sequence of functions, differentiable on their domain $[a,b]$ and pointwise convergent with limit f. If the derivatives f_n' are continuous on $[a,b]$ for all $n \in \mathbf{N}$ and if the sequence $\{f_n'\}$ is uniformly convergent, then $\lim f_n' = f'$.*

That is, under the conditions of the theorem (which should be carefully noted, particularly that it is the sequence of derivatives that is required to be uniformly convergent), the limit of the derivatives is the derivative of the limit.

The proof follows. There must exist a function h such that $f_n' \rightrightarrows h$. We will show that $h = f'$. By Theorem 1.10.3, h is continuous on $[a,b]$, and, by Theorem 1.10.4, $g_n \to g$, where

$$g_n(x) = \int_a^x f_n'(t)\,dt = f_n(x) - f_n(a), \quad a \leqslant x \leqslant b,$$

$$g(x) = \int_a^x h(t)\,dt, \quad a \leqslant x \leqslant b.$$

However, for each $x \in [a,b]$,

$$g_n(x) = f_n(x) - f_n(a) \to f(x) - f(a),$$

so $g(x) = f(x) - f(a)$, for each x, since (as we will prove later in a more general context) the limit of a convergent sequence is unique. By the Fundamental Theorem of Calculus, since h is continuous the function g is differentiable on $[a,b]$ and $g' = h$. Then, in turn, we have that f is differentiable on $[a,b]$, and $f' = g' = h$, as required. $\square$

We move on to consider next corresponding results for series of functions. Given a sequence $\{f_n\}$ of functions with a common domain, the series $\sum f_k$ is said to be pointwise or uniformly convergent on that domain if the sequence of partial sums $\{s_n\}$ is pointwise or uniformly convergent, respectively. (As usual, $s_n = f_1 + f_2 + \cdots + f_n$ for each $n \in \mathbf{N}$, but now, of course, s_n is a function for each n.) Since convergence of a series of functions is defined in terms of convergence of a certain sequence of functions, there is little required to extend the three preceding theorems to series.

Theorem 1.10.6 *Let the series $\sum f_k$ of functions with domain D be uniformly convergent with sum s. If f_n is continuous on D for each $n \in \mathbf{N}$, then also s is a continuous function on D.*

We only need to note that since f_n is continuous on D for each n, then also $s_n = \sum_{k=1}^{n} f_k$ is continuous on D for each n (using an extension of Theorem 1.9.4), and then Theorem 1.10.3 may be applied. $\qquad\square$

Theorem 1.10.7 *Let $\sum f_k$ be a series of functions, each integrable on the domain $[a, b]$, and let the series be uniformly convergent with sum s. Then s is integrable on $[a, b]$, and*

$$\int_a^x s(t)\, dt = \sum_{k=1}^{\infty} \int_a^x f_k(t)\, dt$$

for each $x \in [a, b]$.

This result is expressed roughly by saying that a uniformly convergent series of functions may be integrated term by term, or that summing a series and then integrating the sum is the same as integrating each term and then summing the integrals. It is proved by defining functions g_n by

$$g_n(x) = \int_a^x s_n(t)\, dt = \int_a^x \sum_{k=1}^{n} f_k(t)\, dt = \sum_{k=1}^{n} \int_a^x f_k(t)\, dt$$

$(a \leqslant x \leqslant b, \, n \in \mathbf{N})$ and then using Theorem 1.10.4. $\qquad\square$

There is also an analogue of Theorem 1.10.5, which we need not reproduce.

Finally, we give a useful test by which a series of functions may sometimes be shown to be uniformly convergent.

Theorem 1.10.8 (Weierstrass M-test) *Let $\{f_n\}$ be a sequence of functions with domain D, and let $\sum M_k$ be a convergent positive series for which*

$$|f_n(x)| \leqslant M_n$$

for all $x \in D$ and each $n \in \mathbf{N}$. Then the series $\sum f_k$ is uniformly convergent on D.

To prove this, we note first that, by the comparison test (Theorem 1.8.6), the series $\sum f_k(x)$ is absolutely convergent, and hence convergent (Theorem 1.8.5), for each $x \in D$. Therefore, the series $\sum f_k$ is pointwise convergent on D, with sum s, say. Set

$$s_n = f_1 + f_2 + \cdots + f_n, \qquad t_n = M_1 + M_2 + \cdots + M_n,$$

for each $n \in \mathbf{N}$. (Note that each s_n is a function, each t_n a real number.) Then $s_n \to s$ and $t_n \to t$, say. For each $x \in D$, if $n > m$,

$$|s_n(x) - s_m(x)| = \left| \sum_{k=m+1}^{n} f_k(x) \right|$$

$$\leqslant \sum_{k=m+1}^{n} |f_k(x)| \leqslant \sum_{k=m+1}^{n} M_k = t_n - t_m.$$

The sequences $\{|s_n(x) - s_m(x)|\}_{n=1}^{\infty}$ and $\{t_n - t_m\}_{n=1}^{\infty}$ are both convergent, so, by Theorem 1.7.8,

$$|s(x) - s_m(x)| \leqslant t - t_m,$$

for each $m \in \mathbf{N}$ and all $x \in D$. Since $t - t_m \to 0$ (as $m \to \infty$), the sequence $\{s_m\}_{m=1}^{\infty}$ is uniformly convergent on D, by Theorem 1.10.2. This completes the proof. $\square$

Review exercises 1.10

(1)   (a) Find $\lim f_n$, where $f_n(x) = x^n/(1+x^n)$, for $0 \leqslant x \leqslant 1$ and $n \in \mathbf{N}$, and show that the convergence of $\{f_n\}$ to its limit is not uniform.

   (b) Find $\lim g_n$, where $g_n(x) = x^n/(1+x^n)$, for $0 \leqslant x \leqslant a$, where $0 < a < 1$, and $n \in \mathbf{N}$, and show that the convergence of $\{g_n\}$ to its limit is uniform.

(2) Define a sequence $\{f_n\}$ of functions by $f_n(x) = xe^{-nx}$, for $x \geqslant 0$ and $n \in \mathbf{N}$. Show that $0 \leqslant f_n(x) \leqslant (en)^{-1}$ for all x and n, and hence that $f_n \rightrightarrows 0$.

(3) Let a sequence $\{f_n\}$ of functions be defined by $f_n(x) = x^n/n$, for $0 \leqslant x \leqslant 1$ and $n \in \mathbf{N}$. Show that $f_n \rightrightarrows f$, say, and $f_n' \to g$, say, but $g(1) \neq f'(1)$.

(4) Let $f_n(x) = x(1-x)^{n-1}$, for $0 \leqslant x \leqslant 1$ and $n \in \mathbf{N}$. Show that $\sum_{k=1}^{\infty} f_k$ converges, but not uniformly.

(5) Let $\{f_n\}$ be a sequence of functions with domain D. Show that if $\sum f_k$ is uniformly convergent on D then $f_n \rightrightarrows 0$ on D.

1.11 Some linear algebra

We have indicated a few times our intention to 'add' elements of a set together. This step enriches the basic structure of a set and so allows more statements to be made about sets. These statements can then be applied in areas where a corresponding notion is already present. The groundwork will be given here briefly. In Chapter 6 and subsequent chapters, the strength of this idea will become apparent.

A simple way to proceed, and the one we will adopt, is to suppose our sets to be vector spaces. A vector space is a set on the elements of which two operations have been defined. The operations must satisfy a list of properties designed to make them accord with our experience of addition and multiplication by scalars in a number of areas. Reversing the line of thought, those areas then become examples of the abstract notion of a vector space, and any further properties found in the abstract setting may be given concrete forms in the examples.

A prime example, and the reason behind the name, is the set of ordinary three-dimensional vectors. If $\mathbf{u}$ and $\mathbf{v}$ are such vectors, then Figure 5(a) shows other related vectors, namely $-\mathbf{u}$, $2\mathbf{v}$ and $\mathbf{u} + \mathbf{v}$. If f and g are two functions defined and continuous on the interval $[a, b]$, then we can speak of the related functions $-f$, $2g$ and $f + g$, which are also continuous on $[a, b]$. (See Definition 1.9.3 and Theorem 1.9.4, and the graphs of these functions in Figure 5(b).) These two different subject areas have one aspect in common: their elements are combined together in exactly corresponding ways.

More examples will follow the precise definition of a vector space. By *scalars* in this definition, we mean complex numbers. This will be commented on below.

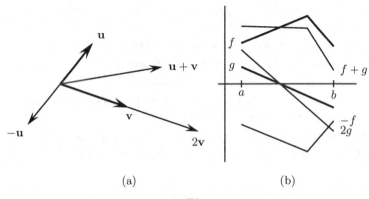

(a)   (b)

Figure 5

Definition 1.11.1 A *vector space* (or *linear space*) is a nonempty set V of objects, called *vectors*, *elements* or *points*, such that

(a) for any $x, y \in V$, there exists a unique vector in V called the *sum* of x and y, and denoted by $x + y$,

(b) for any $x \in V$ and any scalar α, there exists a unique vector in V called the *scalar multiple* of x by α, and denoted by αx.

It is required that (for any $x, y, z \in V$ and scalars α, β),

(i) there exists a vector in V, called a *zero vector* and denoted by θ, for which $x + \theta = x$,

(ii) there exists a vector in V, called a *negative* of x and denoted by $-x$, for which $x + (-x) = \theta$,

(iii) $x + y = y + x$,

(iv) $x + (y + z) = (x + y) + z$,

(v) $\alpha(x + y) = \alpha x + \alpha y$,

(vi) $(\alpha + \beta)x = \alpha x + \beta x$,

(vii) $(\alpha\beta)x = \alpha(\beta x)$,

(viii) $1x = x$.

The requirements (iii) to (viii) are simply properties that together ensure the ability to carry out in this general setting any of the manipulations commonly done with, say, three-dimensional vectors or continuous functions with a common domain. Many similar-looking results can be obtained very quickly: for example,

(ix) $0x = \theta$,

(x) $\alpha\theta = \theta$,

(xi) $(-1)x = -x$.

We will prove (ix) shortly, as an example of the method. Commenting on (xi), notice that this implies that a negative of a vector is unique, since here we have $-x$ expressed as $(-1)x$ and such a scalar multiple is unique, by (b). In the same vein,

(xii) if θ' is a vector in V such that $x + \theta' = x$ for all $x \in V$, then $\theta' = \theta$. That is, the zero vector is unique.

The proof of (ix) follows from that of (xii). We note that if θ' has the stated property, then

$$\theta = \theta + \theta' \quad \text{by hypothesis}$$
$$= \theta' + \theta \quad \text{by (iii)}$$
$$= \theta' \quad \text{by (i)},$$

proving (xii). Now (ix) is proved as follows. For any $x \in V$,

$$x + 0x = 1x + 0x \quad \text{by (viii)}$$
$$= (1 + 0)x \quad \text{by (vi)}$$
$$= 1x$$
$$= x \quad \text{by (viii)},$$

so $0x = \theta$, by (xii).

The properties (iii) to (xii) and other simple manipulative results will be used from here on without special reference to this list.

Two comments on notation: we will denote the vector space itself by V, since this is unlikely to cause confusion with the set V on which the operations are defined, and we will write $x - y$ for $x + (-y)$ $(x, y \in V)$.

Notice that the terms 'sum' and 'scalar multiple' and words like 'addition', and the notations for these, are merely based on habit. They could have been avoided by talking of the existence of two mappings, $f: V \times V \to V$ and $g: \mathbf{C} \times V \to V$, and then agreeing to write $x + y$ for $f(x, y)$ and αx for $g(\alpha, x)$ $(x, y \in V, \alpha \in \mathbf{C})$. We could write (v) for example as $g(\alpha, f(x, y)) = f(g(\alpha, x), g(\alpha, y))$, but this is hardly very suggestive.

More strictly, what we have defined is known as a *complex* vector space, since all the scalars in the definition are complex numbers. If the scalars are restricted to be real numbers, then the resulting system is called a *real vector space*. We will therefore use the latter term when we are specifically concerned with a vector space in which the scalars are to be real numbers, but otherwise it will be assumed that the scalars are complex numbers.

Whenever an example of a vector space is proposed, it needs to be tested whether sums and scalar multiples of its elements again belong to the space, to accord with (a) and (b) of the definition, and whether the axioms (i) to (viii) are satisfied.

For instance, on the set $\mathbf{C}^2$ of ordered pairs of complex numbers, we may define addition and scalar multiplication by

$$(x_1, x_2) + (y_1, y_2) = (x_1 + y_1, x_2 + y_2),$$
$$\alpha(x_1, x_2) = (\alpha x_1, \alpha x_2)$$

$(\alpha, x_1, x_2, y_1, y_2 \in \mathbf{C})$. The right-hand sides are again ordered pairs of complex numbers, as required by (a) and (b). We have

$$(x_1, x_2) + (0, 0) = (x_1 + 0, x_2 + 0) = (x_1, x_2)$$

for any $(x_1, x_2) \in \mathbf{C}^2$, so a zero vector, namely $(0, 0)$, exists; we have

$$(x_1, x_2) + (-x_1, -x_2) = (x_1 - x_1, x_2 - x_2) = (0, 0),$$

so a negative exists for each $(x_1, x_2) \in \mathbf{C}^2$, namely $(-x_1, -x_2)$; and so on down the list verifying (iii) to (viii). Hence we are entitled to call $\mathbf{C}^2$ a vector space when addition and scalar multiplication are defined as above. Such verifications are generally tedious and, as here, we often omit checking (iii) to (viii) and trust our instincts instead.

For vectors in the set $\mathbf{C}^n$, we define

$$(x_1, x_2, \ldots, x_n) + (y_1, y_2, \ldots, y_n) = (x_1 + y_1, x_2 + y_2, \ldots, x_n + y_n),$$
$$\alpha(x_1, x_2, \ldots, x_n) = (\alpha x_1, \alpha x_2, \ldots, \alpha x_n), \ \alpha \in \mathbf{C},$$

and, as for $\mathbf{C}^2$, we may verify that we have a vector space. On the set $\mathbf{R}^n$, whose elements are n-tuples of real numbers, we may define addition and scalar multiplication in precisely the same way. But this does not give us a (complex) vector space, since $i(x_1, x_2, \ldots, x_n) = (ix_1, ix_2, \ldots, ix_n)$ does not belong to $\mathbf{R}^n$ when $(x_1, x_2, \ldots, x_n)$ does. However, $\mathbf{R}^n$ is a real vector space with the above definitions, the scalars now being real, too.

By the vector space $\mathbf{C}^n$ and the real vector space $\mathbf{R}^n$ we will always mean the spaces just defined, namely, with addition and scalar multiplication precisely as given here. It is important to realise that these operations could be defined differently on the same sets $\mathbf{C}^n$ and $\mathbf{R}^n$ but these would result either in different vector spaces, or things that are not vector spaces at all.

The set $C[a, b]$ of continuous functions on the closed interval $[a, b]$ is a real vector space when we define $f + g$ and αf by

$$(f + g)(x) = f(x) + g(x), \qquad (\alpha f)(x) = \alpha f(x)$$

($f, g \in C[a, b]$, $\alpha \in \mathbf{R}$). This of course conforms with the earlier use of these operations. The zero vector in the space is the function θ where $\theta(x) = 0$ for $a \leqslant x \leqslant b$, and the negative of any $f \in C[a, b]$ is the function $-f$ where $(-f)(x) = -f(x)$ for $a \leqslant x \leqslant b$. It is crucial to notice how (a) and (b) of the definition are satisfied: whenever $f, g \in C[a, b]$ we also have $f + g \in C[a, b]$ and $\alpha f \in C[a, b]$, by Theorem 1.9.4. (We could well have chosen some other criterion, such as that the functions be differentiable on $[a, b]$. Sums and scalar multiples of such functions are again differentiable on $[a, b]$. But this leads to different vector spaces.)

As a final example at this stage, consider the set c of all convergent complex-valued sequences. If, for any two such sequences $\{x_n\}$ and $\{y_n\}$, we define

$$\{x_n\} + \{y_n\} = \{z_n\}, \quad \text{where } z_n = x_n + y_n \text{ for each } n \in \mathbf{N},$$
$$\alpha\{x_n\} = \{w_n\}, \quad \text{where } w_n = \alpha x_n \text{ for each } n \in \mathbf{N}, \ \alpha \in \mathbf{C},$$

then c may readily be shown to be a vector space, by virtue of Theorem 1.7.14.

We specify now that whenever in this book we use vector spaces whose elements are n-tuples, functions or sequences, then the operations of addition and multiplication by scalars will always be defined as we have defined them for the spaces $\mathbf{C}^n$, $C[a, b]$ and c.

We next define the concept of a vector subspace.

Definition 1.11.2 Let V be a vector space and let W be a nonempty subset of V. Then W is called a *subspace* of V if $x + y \in W$ whenever $x, y \in W$, and $\alpha x \in W$ whenever $x \in W$, $\alpha \in \mathbf{C}$.

Thus, a subspace of a vector space is a subset which contains as members all sums and scalar multiples of any of its elements. Under this definition, the vector space V is certainly a subspace of itself. Also, the subset $\{\theta\}$, consisting of only the zero vector of V, is a subspace of V because $\theta + \theta = \theta \in \{\theta\}$ and $\alpha\theta = \theta \in \{\theta\}$ for any $\alpha \in \mathbf{C}$.

Since $0x = \theta$, it follows that θ is in fact an element of any subspace W of V. And, since $(-1)x = -x$, it follows that any subspace contains the negatives of all its elements. The axioms (iii) to (viii) of Definition 1.11.1 hold for elements in the subspace W, since those elements belong also to V. Putting these statements together, we see that the subspace W is a vector space in its own right. The converse is also true: if W and V are both vector spaces with addition and scalar multiplication defined in the same way, and if W (as a set) is a subset of V (as a set), then W is a subspace of V.

We can give many examples of subspaces of the vector spaces given above.

The vector space W of ordered triples of complex numbers of the form $(x_1, x_2, 0)$ $(x_1, x_2 \in \mathbf{C}$, the third element of every triple being zero) is a subspace of $\mathbf{C}^3$, since

$$(x_1, x_2, 0) + (y_1, y_2, 0) = (x_1 + y_1, x_2 + y_2, 0) \in W,$$
$$\alpha(x_1, x_2, 0) = (\alpha x_1, \alpha x_2, 0) \in W, \ \alpha \in \mathbf{C}.$$

There is obviously a natural connection between this subspace W of $\mathbf{C}^3$ and the vector space $\mathbf{C}^2$, though the spaces cannot be called the same: the elements of $\mathbf{C}^2$ are ordered pairs, not triples. The connection is properly described by noting that there is a one-to-one correspondence $f \colon W \to \mathbf{C}^2$ (see Section 1.4) such that

$$f(x + y) = f(x) + f(y),$$
$$f(\alpha x) = \alpha f(x),$$

for $x, y \in W$, $\alpha \in \mathbf{C}$. This is the mapping defined by

$$f(x_1, x_2, 0) = (x_1, x_2), \quad x_1, x_2 \in \mathbf{C}.$$

Through the mapping f, or its inverse f^{-1}, all manipulations carried out in one of the spaces may be precisely reflected in the other. Such a mapping is termed a *vector space isomorphism* and the spaces W and $\mathbf{C}^2$ are called *isomorphic*. These terms will be discussed more fully in Chapter 9.

The set of all differentiable functions defined on the interval $[a, b]$ is easily shown to be a real vector space: we will denote it by $C^{(1)}[a, b]$. This is a subspace of $C[a, b]$, since every differentiable function is continuous. Another useful real vector space is the set of all polynomial functions restricted to the interval $[a, b]$. This space is denoted by $P[a, b]$ and is a subspace of $C^{(1)}[a, b]$ since every polynomial function is differentiable. It is easily checked that if U, V, W are vector spaces and U is a subspace of W and W is a subspace of V, then U is also a subspace of V. Hence, here, $P[a, b]$ is also a subspace of $C[a, b]$; this can readily be seen directly.

The vector space c of all convergent sequences has a number of subspaces which will be referred to throughout the book. One which we may mention now is the space of all sequences which converge with limit 0. This vector space is denoted by c_0.

All the remaining definitions relevant to our purposes are gathered in the following.

Definition 1.11.3 Let $\{v_1, v_2, \ldots, v_n\}$ be a set of vectors in a vector space V.

(a) Any vector of the form

$$\alpha_1 v_1 + \alpha_2 v_2 + \cdots + \alpha_n v_n,$$

where $\alpha_1, \alpha_2, \ldots, \alpha_n \in \mathbf{C}$, is called a *linear combination* of v_1, $v_2, \ldots, v_n$ and the scalar α_k is called the *coefficient* of v_k, for $k = 1, 2, \ldots, n$.

(b) If a linear combination of the vectors $v_1, v_2, \ldots, v_n$ equals the zero vector in V and at least one coefficient is not 0, then the set $\{v_1, v_2, \ldots, v_n\}$ is called *linearly dependent*. Otherwise the set is called *linearly independent*.

(c) The set of all linear combinations of the vectors $v_1, v_2, \ldots, v_n$ is a subspace of V called the *span* of $\{v_1, v_2, \ldots, v_n\}$, denoted by $\mathrm{Sp}\{v_1, v_2, \ldots, v_n\}$. This subspace is said to be *spanned* or *generated* by $v_1, v_2, \ldots, v_n$.

(d) If the set $\{v_1, v_2, \ldots, v_n\}$ is linearly independent and

$$\mathrm{Sp}\{v_1, v_2, \ldots, v_n\} = V,$$

then it is called a *basis* for V. In that case, V is said to have *dimension* n, and to be *finite-dimensional*. If there does not exist any finite set of vectors that is a basis for V, then V is called *infinite-dimensional*. The dimension of the vector space $\{\theta\}$ is 0.

There are a number of comments that need to be made. In particular, we must justify some statements occurring in (c) and (d).

Rephrasing the second part of (b), the set $\{v_1, v_2, \ldots, v_n\}$ is linearly independent if the equation

$$\alpha_1 v_1 + \alpha_2 v_2 + \cdots + \alpha_n v_n = \theta$$

can only be true when $\alpha_1 = \alpha_2 = \cdots = \alpha_n = 0$. For example, in $\mathbf{C}^3$ the vectors $\{(1,0,0), (0,1,0), (0,0,1)\}$ are linearly independent, because if

$$\alpha_1(1,0,0) + \alpha_2(0,1,0) + \alpha_3(0,0,1) = \theta,$$

then, equivalently, $(\alpha_1, \alpha_2, \alpha_3) = \theta = (0, 0, 0)$, so that $\alpha_1 = \alpha_2 = \alpha_3 = 0$. Notice that if $\{v_1, v_2, \ldots, v_n\}$ is a linearly independent set of vectors, then it cannot include the zero vector, for if $v_1 = \theta$, say, then

$$1v_1 + 0v_2 + 0v_3 + \cdots + 0v_n = \theta,$$

and the first coefficient is not 0.

It needs to be verified that the span of a set of vectors in a vector space is indeed a subspace of the space, as asserted in (c). Consider the set $S = \{v_1, v_2, \ldots, v_n\}$ of vectors in a vector space V. If $x, y \in \mathrm{Sp}\, S$, then for some scalars α_k, β_k $(k = 1, 2, \ldots, n)$,

$$x = \alpha_1 v_1 + \alpha_2 v_2 + \cdots + \alpha_n v_n, \qquad y = \beta_1 v_1 + \beta_2 v_2 + \cdots + \beta_n v_n,$$

and so

$$x + y = (\alpha_1 + \beta_1)v_1 + (\alpha_2 + \beta_2)v_2 + \cdots + (\alpha_n + \beta_n)v_n \in \mathrm{Sp}\, S,$$
$$\alpha x = (\alpha\alpha_1)v_1 + (\alpha\alpha_2)v_2 + \cdots + (\alpha\alpha_n)v_n \in \mathrm{Sp}\, S,\ \alpha \in \mathbf{C},$$

That is, $x + y$ and αx are again linear combinations of v_1, v_2, $\ldots$, v_n and so belong to $\mathrm{Sp}\, S$. Thus $\mathrm{Sp}\, S$ is a subspace of V.

Turning to (d), we need to show that if $\{v_1, v_2, \ldots, v_n\}$ is a basis for a vector space V, then any other basis for V has the same number of elements. Otherwise, the definition of dimension for a vector space does not make much sense. Suppose the set $\{u_1, u_2, \ldots, u_m\}$ of vectors in V is also a basis for V, and suppose $m > n$. Each vector u_j is a linear combination of v_1, v_2, $\ldots$, v_n since the set $\{v_1, v_2, \ldots, v_n\}$ is a basis for V, so we may write in particular

$$u_1 = \alpha_1 v_1 + \alpha_2 v_2 + \cdots + \alpha_n v_n$$

for some scalars $\alpha_1, \alpha_2, \ldots, \alpha_n$, which cannot all be 0. We may suppose $\alpha_1 \neq 0$, so that

$$v_1 = \frac{1}{\alpha_1} u_1 - \frac{\alpha_2}{\alpha_1} v_2 - \cdots - \frac{\alpha_n}{\alpha_1} v_n.$$

This gives v_1 as a linear combination of the vectors u_1, v_2, v_3, $\ldots$, v_n. Every vector in V is a linear combination of v_1, v_2, $\ldots$, v_n, so now every vector in V may be expressed as a linear combination of u_1, v_2, v_3, $\ldots$, v_n. In particular, this applies to the vector u_2:

$$u_2 = \gamma_1 u_1 + \beta_2 v_2 + \beta_3 v_3 + \cdots + \beta_n v_n$$

for some scalars γ_1, β_2, β_3, $\ldots$, β_n. In this, it cannot be the case that

$\beta_2 = \beta_3 = \cdots = \beta_n = 0$, for then $u_2 = \gamma_1 u_1$ and the equation

$$\gamma_1 u_1 + (-1)u_2 + 0u_3 + \cdots + 0u_m = \theta$$

denies the linear independence of $\{u_1, u_2, \ldots, u_m\}$, since at least one coefficient here is nonzero. We may suppose $\beta_2 \neq 0$, so that

$$v_2 = -\frac{\gamma_1}{\beta_2}u_1 + \frac{1}{\beta_2}u_2 - \frac{\beta_3}{\beta_2}v_3 - \cdots - \frac{\beta_n}{\beta_2}v_n.$$

As before, this allows us to express any vector in V, in particular u_3, as a linear combination of u_1, u_2, v_3, v_4, $\ldots$, v_n. This process may be continued until we conclude that every vector in V may be expressed as a linear combination of u_1, u_2, $\ldots$, u_n. But this is not possible, since expressing u_{n+1} as such a linear combination contradicts the linear independence of u_1, u_2, $\ldots$, u_n, $\ldots$, u_m. Our assumption that $m > n$ must therefore be wrong, so we cannot have two bases for V with one having more elements than the other. This implies that all bases of a finite-dimensional space contain the same number of elements. (That number is the number we call the dimension of the space.)

We have shown that the vectors $\{(1,0,0), (0,1,0), (0,0,1)\}$ are linearly independent in $\mathbf{C}^3$. They also span the space, for if (x_1, x_2, x_3) is any vector in $\mathbf{C}^3$, then

$$(x_1, x_2, x_3) = x_1(1,0,0) + x_2(0,1,0) + x_3(0,0,1),$$

expressing the vector as a linear combination of $(1,0,0)$, $(0,1,0)$ and $(0,0,1)$. Hence these vectors are a basis for $\mathbf{C}^3$, which is therefore of dimension 3. Likewise, the real vector space $\mathbf{R}^3$ also has dimension 3 (the above three vectors again being a basis) and this agrees with the common usage of the term 'three-dimensional space'.

In the same way, we may show that $\mathbf{C}^n$ is a finite-dimensional space: it has dimension n and the set

$$\{(1,0,\ldots,0), (0,1,0,\ldots,0), \ldots, (0,0,\ldots,0,1)\}$$

of n-tuples (in which the kth has kth component equal to 1 and the others equal to 0, for $k = 1, 2, \ldots, n$) is a basis.

A convenient way to show that a particular vector space is infinite-dimensional is to show that it has an infinite-dimensional subspace. We need a little preparation before verifying this, and will then apply it to the spaces $C[a,b]$ and c.

It is necessary to prove that, for a vector space of dimension n, any set of n linearly independent vectors in the space is a basis. Suppose V

is a vector space with dimension n, and let $S = \{u_1, u_2, \ldots, u_n\}$ be a set of n linearly independent vectors in V. These vectors will be shown to be a basis for V if we can show that $\operatorname{Sp} S = V$. Suppose there is a vector $u \in V$ such that $u \notin \operatorname{Sp} S$, and consider the equation

$$\alpha_1 u_1 + \alpha_2 u_2 + \cdots + \alpha_n u_n + \alpha u = \theta$$

for scalars $\alpha_1, \alpha_2, \ldots, \alpha_n, \alpha$. We must have $\alpha = 0$ (otherwise $u \in \operatorname{Sp} S$). That leaves us with $\alpha_1 u_1 + \alpha_2 u_2 + \cdots + \alpha_n u_n = \theta$, so we must also have $\alpha_1 = \alpha_2 = \cdots = \alpha_n = 0$, since S is a linearly independent set. This implies that the set $\{u, u_1, u_2, \ldots, u_n\}$ is linearly independent, and then, precisely as in the discussion concerning Definition 1.11.3(d), this leads to a contradiction. (Take $m = n + 1$ and $u = u_{n+1}$ in that discussion.) Hence $u \in \operatorname{Sp} S$ whenever $u \in V$, so the set S is a basis for V.

Next we prove that in an infinite-dimensional vector space there exists a set of n vectors which is linearly independent, regardless of the value of the positive integer n. If this is not the case, then there is an integer N such that $\{v_1, v_2, \ldots, v_N\}$ is linearly independent, while $\{v_1, v_2, \ldots, v_N, v\}$ is linearly dependent, for all other vectors v in the space. This means that all other vectors in the space are linear combinations of $v_1, v_2, \ldots, v_N$, or that these vectors span the space. Hence they are a basis for the space, contradicting the fact that it is infinite-dimensional.

Now we can prove the result indicated.

Theorem 1.11.4 *A vector space is infinite-dimensional if it has an infinite-dimensional subspace.*

There is little more to do. Let W be an infinite-dimensional subspace of a vector space V, and suppose that V is finite-dimensional, with dimension n, say. By what was just said, there exists a set of n linearly independent vectors in W, which, since they belong also to V, must be a basis for V. Every vector in V, which includes all those in W, is expressible as a linear combination of these basis vectors, so they span W. Hence that set of n vectors is also a basis for W, contradicting the fact that W is infinite-dimensional. $\qquad\Box$

Now to our examples.

It is easy to see that the real vector space $P[a, b]$ of polynomial functions defined on $[a, b]$ is infinite-dimensional. We simply note that any proposed basis must be a set containing a finite number of polynomial functions, but no polynomial function with degree higher than any of those in the basis could possibly be expressed as a linear combination of them. Since $P[a, b]$ is a subspace of $C[a, b]$, Theorem 1.11.4 immediately implies that $C[a, b]$ is also infinite-dimensional.

To see that the space c of convergent sequences is infinite-dimensional, we consider the subspace of sequences all of whose terms are zero after some finite number of terms. It is evident that these indeed constitute a subspace of c. It is an infinite-dimensional subspace since, no matter how many sequences a proposed basis may contain, we may always find another sequence with more nonzero terms than any of those in the proposed basis. Such a sequence could never be a linear combination of the others. Then Theorem 1.11.4 may be applied.

A little thought shows that the space $P[a, b]$ and the subspace of c constructed above are not as unlike as they might at first appear. Let c_R be the latter subspace, where we restrict the sequences to be real-valued and use real scalars only, so that c_R is a real vector space. A typical member of c_R is the sequence

$$a_0, a_1, a_2, a_3, \ldots, a_{n-1}, a_n, 0, 0, 0, \ldots$$

where $a_0, a_1, \ldots, a_n$ are any real numbers. Compare this with a typical element of $P[a, b]$: the polynomial p, where

$$p(t) = a_0 + a_1 t + a_2 t^2 + a_3 t^3 + \cdots + a_{n-1} t^{n-1} + a_n t^n, \quad a \leqslant t \leqslant b.$$

Adding elements of c_R and multiplying them by scalars can in fact be accomplished in the space $P[a, b]$ by suppressing all but the coefficients in the polynomials. The reverse is similarly true. We have here an example of isomorphic vector spaces, as introduced following Definition 1.11.2. This explains in part the applicability of Theorem 1.11.4 to the two examples.

In this section, we wish also to mention a few simple properties of matrices. A *matrix* is a set of mn numbers, called *elements*, arranged in m rows and n columns, and indicated in general fashion as

$$\begin{pmatrix} a_{11} & a_{12} & \cdots & a_{1n} \\ a_{21} & a_{22} & \cdots & a_{2n} \\ \vdots & \vdots & & \vdots \\ a_{m1} & a_{m2} & \cdots & a_{mn} \end{pmatrix}$$

Here, a_{jk} is the element in the jth row and kth column ($j = 1, 2, \ldots, m$; $k = 1, 2, \ldots, n$). This matrix may be given more simply as (a_{jk}). The *size* of the matrix is written as $m \times n$, indicating the numbers of its rows and columns. An $m \times 1$ matrix (having only one column) is also referred to as a *column vector*; a $1 \times n$ matrix (having only one row) as a *row vector*. The elements of a matrix may be real or complex numbers, and it is convenient to think of an $m \times 1$ column vector as an element of $\mathbf{R}^m$ or $\mathbf{C}^m$.

The *transpose* of an $m \times n$ matrix (a_{jk}) is the $n \times m$ matrix (a_{kj}) obtained by writing all the rows as columns. In particular, the transpose of a row vector is a column vector. The operation of taking the transpose is indicated by a superscript T. Thus $(a_{jk})^T = (a_{kj})$ and

$$
\begin{pmatrix} b_1 & b_2 & \dots & b_m \end{pmatrix}^T = \begin{pmatrix} b_1 \\ b_2 \\ \vdots \\ b_m \end{pmatrix}.
$$

In text, like right here, we will write row and column vectors more conveniently as $(a_1, a_2, \ldots, a_n)$ and $(b_1, b_2, \ldots, b_m)^T$, for example.

The *conjugate* of the matrix $A = (a_{jk})$ is the matrix $\overline{A}$ defined by $\overline{A} = (\overline{a}_{jk})$. That is, to obtain the conjugate of a matrix, take the conjugate of each of its elements.

The set of all matrices of a given size is a vector space under the definitions

$$(a_{jk}) + (b_{jk}) = (c_{jk}), \quad \text{where } c_{jk} = a_{jk} + b_{jk},$$

$$\alpha(a_{jk}) = (d_{jk}), \quad \text{where } d_{jk} = \alpha a_{jk}, \ \alpha \in \mathbf{C}.$$

That is, matrices (of the same size) are added by adding corresponding elements, and a matrix is multiplied by a scalar by multiplying all elements by that scalar. The zero of this vector space is the matrix, of the same size as all matrices in the space, having all elements 0. (If the matrices are restricted to having real elements only, then a real vector space is obtained by restricting the scalars to be real numbers only.)

Two matrices may be multiplied together according to the following rule. If (a_{jk}) is an $m \times n$ matrix and (b_{jk}) is an $n \times p$ matrix, then (and only for such sizes) the product exists, and

$$(a_{jk})(b_{jk}) = (c_{jk})$$

where (c_{jk}) is an $m \times p$ matrix, and

$$c_{jk} = \sum_{i=1}^{n} a_{ji} b_{ik}, \quad j = 1, \ 2, \ \ldots, \ m; \ k = 1, \ 2, \ \ldots, \ p.$$

Notice that the product $(b_{jk})(a_{jk})$ is not defined unless $p = m$. When A is a *square* matrix (having the same number of rows and columns), we may form the product AA, denoted by A^2, and extend this to obtain any power A^n, $n \in \mathbf{N}$.

It is not difficult to prove that $(BC)^T = C^T B^T$, for any matrices B and C for which the product BC exists.

Illustrating these definitions, we have

$$\begin{pmatrix} 1 & 2 \\ 3 & 4 \\ 5 & 6 \end{pmatrix}^T = \begin{pmatrix} 1 & 3 & 5 \\ 2 & 4 & 6 \end{pmatrix},$$

$$\overline{\begin{pmatrix} 1 & i \\ 2-i & 3+4i \end{pmatrix}} = \begin{pmatrix} 1 & -i \\ 2+i & 3-4i \end{pmatrix},$$

$$\begin{pmatrix} 1 & 2 \\ 3 & 4 \\ 5 & 6 \end{pmatrix} + \begin{pmatrix} 7 & 8 \\ 9 & 10 \\ 11 & 12 \end{pmatrix} = \begin{pmatrix} 8 & 10 \\ 12 & 14 \\ 16 & 18 \end{pmatrix},$$

$$-2 \begin{pmatrix} 1 & 2 \\ 3 & 4 \\ 5 & 6 \end{pmatrix} = \begin{pmatrix} -2 & -4 \\ -6 & -8 \\ -10 & -12 \end{pmatrix},$$

$$\begin{pmatrix} 1 & 2 \\ 3 & 4 \\ 5 & 6 \end{pmatrix} \begin{pmatrix} 7 & 8 \\ 9 & 10 \end{pmatrix} = \begin{pmatrix} 25 & 28 \\ 57 & 64 \\ 89 & 100 \end{pmatrix}.$$

Systems of linear equations may very conveniently be expressed in terms of matrices. The system

$$a_{11}x_1 + a_{12}x_2 + \cdots + a_{1n}x_n = b_1,$$
$$a_{21}x_1 + a_{22}x_2 + \cdots + a_{2n}x_n = b_2,$$
$$\vdots$$
$$a_{m1}x_1 + a_{m2}x_2 + \cdots + a_{mn}x_n = b_m,$$

of m equations in n unknowns $x_1, x_2, \ldots, x_n$ may be written

$$Ax = b$$

where

$$A = \begin{pmatrix} a_{11} & a_{12} & \cdots & a_{1n} \\ a_{21} & a_{22} & \cdots & a_{2n} \\ \vdots & \vdots & & \vdots \\ a_{m1} & a_{m2} & \cdots & a_{mn} \end{pmatrix}, \quad x = \begin{pmatrix} x_1 \\ x_2 \\ \vdots \\ x_n \end{pmatrix}, \quad b = \begin{pmatrix} b_1 \\ b_2 \\ \vdots \\ b_m \end{pmatrix}.$$

A square matrix is called a *unit matrix*, or an *identity matrix*, if all its elements are 0 except those on the main diagonal (top left to bottom right) which are all 1. An identity matrix is commonly denoted by I, or I_n when it is necessary to show explicitly that its size is $n \times n$. This matrix has the property

$$IA = AI = A,$$

when A is a square matrix of the same size as I.

If A is a given square matrix, then a matrix B (of the same size as A) is called an *inverse* of A if

$$AB = BA = I.$$

We commonly write A^{-1} for B and it is easy to show that the inverse of a square matrix, if one exists, is unique. It is shown in books on linear algebra that a condition for an $n \times n$ matrix to have an inverse is that its columns, considered as elements of $\mathbf{R}^n$ or $\mathbf{C}^n$, be linearly independent. An equivalent condition is that the determinant of the matrix be nonzero. (There is one instance, in Chapter 9, where we need some knowledge of determinants, but we will not review that theory here.)

If $Ax = b$ is the system of linear equations mentioned above, where now A is a square matrix possessing an inverse, then the system has a solution, given by $x = A^{-1}b$. This is easily checked: $A(A^{-1}b) = (AA^{-1})b = Ib = b$. We have used here the associative property of matrix multiplication: $A(BC) = (AB)C$, where A, B, C are matrices and all the indicated products exist. Furthermore, the solution $A^{-1}b$ is unique. Putting $b = \theta$ here (θ is a zero matrix, of size $n \times 1$ if A is $n \times n$), we see that the only solution of $Ax = \theta$ is $x = \theta$ when the inverse A^{-1} exists. On the other hand, if the inverse does not exist then it can be shown that there are infinitely many other solutions (called *nontrivial* solutions) of the system of equations.

Determining the inverse of a matrix is rarely easy, and often methods of approximation are used to solve systems of equations such as that above, to a given degree of accuracy. This will be one of our major examples in Chapter 3.

Review exercises 1.11

(1)   (a) Show that the set V of all 2×2 matrices (a_{jk}) is a vector space of dimension 4, by finding a basis for V with four elements.

(b) Show that the subset of V consisting of those 2×2 matrices (a_{jk}) with $a_{11} + a_{22} = 0$ is a subspace of V, and find a basis for this subspace.

(2) Let P_2 be the real vector space of all polynomial functions on $\mathbf{R}$, of degree at most 2. Let $t \in \mathbf{R}$ be fixed. Show that $\{p_1, p_2, p_3\}$, where $p_1(x) = 1$, $p_2(x) = x + t$, $p_3(x) = (x + t)^2$, is a basis for P_2. Express $a + bx + cx^2$ $(a, b, c \in \mathbf{R})$ as a linear combination of these basis vectors.

(3) Let S and T be subspaces of a vector space V. Their sum is defined as $S + T = \{s + t : s \in S, \ t \in T\}$. Show that $S + T$ is also a subspace of V. If S and T are finite-dimensional, show that $S + T$ is finite-dimensional and that the union of bases of S and T is a basis of $S + T$.

(4) Deduce a condition for the matrix $A = \begin{pmatrix} a & b \\ c & d \end{pmatrix}$ to have an inverse and then obtain a formula for A^{-1} as an explicit 2×2 matrix.

1.12 Setting off

We are about to begin our journey through space. We will visit many spaces: topological spaces, metric spaces, normed vector spaces, Banach spaces, inner product spaces, Hilbert spaces. These are not the foreign worlds of the fictional space traveller. Instead, they offer real, down-to-earth means by which our own world may be explored a little further. We have discussed previously how these are some of a vast hierarchy of spaces, each containing a little more structure than the one before, and how each item of structure may be related to some property of the real numbers. This is an important principle to be kept in mind as we proceed 'through space'. The other principle of importance that we have talked of is the similarity that becomes apparent in various fields of mathematics when they are bared to their essentials. This underlies the applicability of abstract methods and should also be kept continually in view.

2

Metric Spaces

2.1 Definition of a metric space

In Chapter 1 we went into some detail regarding properties of convergent sequences of real or complex numbers. The essential idea of convergence is that distances between points of the sequence and some other point become smaller and smaller as we proceed along the sequence. We need not restrict this notion to sequences of numbers and indeed, in discussing uniform convergence of sequences of real-valued functions with a common domain, we have already extended it. All that is required to speak of convergence of a sequence of elements of any particular set is that a meaning be given to the concept of the distance between points of that set. If we can come up with an adequate definition of 'distance between points' that is applicable in a totally general setting, then any consequences of that definition will be reflected in particular examples.

Thus we arrive at our first instance of an abstract space (apart from our introduction to vector spaces). A metric space is an arbitrary set X together with a real-valued mapping d defined on pairs of elements x and y in X such that the number $d(x, y)$ suitably represents the idea of the distance between the points x and y. Defining d so that it does this job is not easy: we wish to ensure that the single definition can, in its various applications, fully account for what we already understand by distances between numbers on a line or between points in the plane, and that it can distinguish functions whose graphs are close together or far apart. An example of a desirable property is: $d(x, y) = d(y, x)$ for all $x, y \in X$. That is, the distance between points x and y in X must be the same as the distance between y and x. This may seem pedantic, but this approach is vital when we move to abstract settings so that full

84

applicability and generality are available. A formal definition of metric space follows.

Definition 2.1.1 A *metric space* is a nonempty set X together with a mapping $d\colon X \times X \to \mathbf{R}_+$ with the properties

(M1) $d(x,y) = 0$ if and only if $x = y$ $(x,y \in X)$,

(M2) $d(x,y) = d(y,x)$ for all $x,y \in X$,

(M3) $d(x,z) \leqslant d(x,y) + d(y,z)$ for all $x,y,z \in X$.

This metric space is denoted by (X,d) and the mapping d is called the *metric* (or *distance function*) for the space.

The properties (M1), (M2) and (M3) must be viewed with regard to the above discussion. We have already predicted the appearance of (M2). The property (M1) says that points are zero distance apart if and only if they coincide. Notice that $d(x,y) > 0$ when $x \neq y$ since the range of the mapping d is a subset of $\mathbf{R}_+$, the set of all nonnegative real numbers. The property (M3) says that the distance between any two points is never greater than the sum of their distances to a third point. Thinking of this in terms of points in the plane, it is simply the statement that the length of any side of a triangle is never greater than the sum of the lengths of the other two sides, and for this reason the inequality of (M3) is known as the *triangle inequality*.

A metric space is not fully described unless both the set and the metric are given. This accounts for the notation (X,d). It is quite possible, as examples below will show, to define different metrics for the same set X. If d_1 and d_2 are different metrics defined on X, then (X,d_1) and (X,d_2) are different metric spaces. However, when there is no possibility of confusion about which metric is being used for a given set X at a given time, then X alone is often used to denote the metric space as well.

2.2 Examples of metric spaces

In each of the following examples, it needs to be checked that the proposed metric indeed satisfies Definition 2.1.1. The checking is omitted in some examples, since it is a consequence of that for more general examples that come later, and in others it is left as an exercise.

(1) Let X be any nonempty set of real numbers and define d by

$$d(x,y) = |x - y|, \quad x,y \in X.$$

This is the usual definition of distance between two points on a line, and is called the *natural metric* for such a set X.

(2) Let X be any nonempty set of points in the plane (so X may be considered as a subset of $\mathbf{R}^2$) and define d by

$$d(x,y) = \sqrt{(x_1 - y_1)^2 + (x_2 - y_2)^2},$$

where $x = (x_1, x_2)$ and $y = (y_1, y_2)$ are any two points of X. This is the usual definition of distance between two points in a plane. The triangle inequality of (M3) says very explicitly here that the length of any side of a plane triangle is not greater than the sum of the lengths of the other two sides,

(3) For the same set X as in (2), we may define a different metric d' by

$$d'(x,y) = \max\{|x_1 - y_1|, |x_2 - y_2|\}.$$

Under this metric, we are saying that the length of any line segment is to be understood as the larger of its projections on the coordinate axes. This gives an indication of the possible distortions that can occur in our intuitive notions of 'length': the 'square' in Figure 6 is in fact a circle in the metric space $(\mathbf{R}^2, d')$, the circle with centre (x_1, y_1) and radius a, since the distance between the centre and any point on it is a.

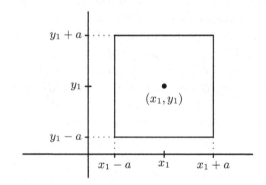

Figure 6

Since this metric d' and the metric d of Example (2) may have different values for the same points $x, y \in X$, the metric spaces (X, d) and (X, d') are different, though they use the same set X of points in the plane.

We will carry out the verification that d' is a metric. The definition of absolute value implies immediately that $d'(x,y) \geqslant 0$ for all $x, y \in X$ so certainly the range of d' is a subset of $\mathbf{R}_+$. Also, (M1) and (M2) are easily seen to hold. The only problem, and this is commonly but not always the case, is to verify (M3). The reasoning in the following should be watched carefully. The same method is used in many other examples.

Let $z = (z_1, z_2)$ be any third point in X. Letting j be either 1 or 2, we have

$$|x_j - z_j| = |(x_j - y_j) + (y_j - z_j)|$$
$$\leqslant |x_j - y_j| + |y_j - z_j|$$
$$\leqslant \max_{k=1,2} |x_k - y_k| + \max_{k=1,2} |y_k - z_k|.$$

Since this is true for both $j = 1$ and $j = 2$, we have

$$\max_{k=1,2} |x_k - z_k| \leqslant \max_{k=1,2} |x_k - y_k| + \max_{k=1,2} |y_k - z_k|,$$

or

$$d'(x, z) \leqslant d'(x, y) + d'(y, z),$$

verifying (M3) for the mapping d'.

(4) Let X be any nonempty set in $\mathbf{R}^n$, so that X consists of ordered n-tuples of real numbers, and define d by

$$d(x, y) = \sqrt{\sum_{k=1}^{n} (x_k - y_k)^2},$$

where $x = (x_1, x_2, \ldots, x_n)$, $y = (y_1, y_2, \ldots, y_n)$ are points of X. This is a generalisation of Examples (1) and (2), which correspond to the special cases $n = 1$ and $n = 2$, respectively. The mapping d here is known as the *Euclidean metric* for such a set X, and we will now specify that whenever we refer to the metric space $\mathbf{R}^n$ (rather than just the set $\mathbf{R}^n$) then we mean the metric space (X, d) of this example with $X = \mathbf{R}^n$. Putting this another way, reference to the metric space $\mathbf{R}^n$ will always imply that the Euclidean metric is being used. The term *Euclidean space* is often used for the metric space $\mathbf{R}^n$.

Verification that this d is in fact a metric again comes down to checking (M3). That is, we must prove that

$$\sqrt{\sum_{k=1}^{n} (x_k - z_k)^2} \leqslant \sqrt{\sum_{k=1}^{n} (x_k - y_k)^2} + \sqrt{\sum_{k=1}^{n} (y_k - z_k)^2},$$

where $z = (z_1, z_2, \ldots, z_n)$ is any third point of X. This is a consequence of the *Cauchy–Schwarz inequality*, which we give now as a theorem.

We will need another form of this very useful inequality shortly, and in Chapter 8 a general form will be given that includes these earlier ones as special cases.

Theorem 2.2.1 (Cauchy–Schwarz Inequality) *Let* $(a_1, a_2, \ldots, a_n)$ *and* $(b_1, b_2, \ldots, b_n)$ *be any points in* $\mathbf{R}^n$. *Then*

$$\left(\sum_{k=1}^{n} a_k b_k\right)^2 \leqslant \left(\sum_{k=1}^{n} a_k^2\right)\left(\sum_{k=1}^{n} b_k^2\right).$$

This is proved by the following device. We introduce the function ψ defined by

$$\psi(u) = \sum_{k=1}^{n} (a_k u + b_k)^2, \quad u \in \mathbf{R}.$$

Then

$$\psi(u) = \left(\sum_{k=1}^{n} a_k^2\right) u^2 + 2\left(\sum_{k=1}^{n} a_k b_k\right) u + \sum_{k=1}^{n} b_k^2,$$

and we see that $\psi(u)$ is a quadratic form in u. That is, it has the form $Au^2 + 2Bu + C$. Being a sum of squares, $\psi(u) \geqslant 0$ for all u. Hence the discriminant $(2B)^2 - 4AC$ cannot be positive. Divide by 4: $B^2 - AC \leqslant 0$, or

$$\left(\sum_{k=1}^{n} a_k b_k\right)^2 - \left(\sum_{k=1}^{n} a_k^2\right)\left(\sum_{k=1}^{n} b_k^2\right) \leqslant 0.$$

This proves the theorem.   □

We need another inequality, based on this one.

Theorem 2.2.2 *For any points* $(a_1, a_2, \ldots, a_n)$, $(b_1, b_2, \ldots, b_n)$ *in* $\mathbf{R}^n$, *we have*

$$\sqrt{\sum_{k=1}^{n} (a_k + b_k)^2} \leqslant \sqrt{\sum_{k=1}^{n} a_k^2} + \sqrt{\sum_{k=1}^{n} b_k^2}.$$

Taking square roots of both sides of the Cauchy–Schwarz inequality gives

$$\left|\sum_{k=1}^{n} a_k b_k\right| \leqslant \sqrt{\sum_{k=1}^{n} a_k^2} \sqrt{\sum_{k=1}^{n} b_k^2},$$

so certainly

$$\sum_{k=1}^{n} a_k b_k \leqslant \sqrt{\sum_{k=1}^{n} a_k^2} \sqrt{\sum_{k=1}^{n} b_k^2}.$$

But then

$$\sum_{k=1}^{n} a_k^2 + 2 \sum_{k=1}^{n} a_k b_k + \sum_{k=1}^{n} b_k^2 \leqslant \sum_{k=1}^{n} a_k^2 + 2 \sqrt{\sum_{k=1}^{n} a_k^2} \sqrt{\sum_{k=1}^{n} b_k^2} + \sum_{k=1}^{n} b_k^2$$

or

$$\sum_{k=1}^{n} (a_k + b_k)^2 \leqslant \left(\sqrt{\sum_{k=1}^{n} a_k^2} + \sqrt{\sum_{k=1}^{n} b_k^2} \right)^2.$$

Taking square roots now gives the inequality of Theorem 2.2.2. □

To check that the triangle inequality holds for the Euclidean metric, we simply put $a_k = x_k - y_k$ and $b_k = y_k - z_k$ in the second theorem.

(5) Another metric for the set $\mathbf{R}^n$ is the mapping d_1, where

$$d_1(x, y) = \sum_{k=1}^{n} |x_k - y_k|.$$

This also reduces to the metric of Example (1) when $n = 1$.

(Both the Euclidean metric and the metric d_1 just defined are special cases of the metric d_p, where

$$d_p(x, y) = \left(\sum_{k=1}^{n} |x_k - y_k|^p \right)^{1/p},$$

with $p \geqslant 1$. The verification of (M3) for this mapping for general values of p requires a discussion of the *Hölder inequality* and the *Minkowski inequality*, which are generalisations of the inequalities in Theorems 2.2.1 and 2.2.2, respectively. We will not be making use of these metric spaces $(\mathbf{R}^n, d_p)$.)

(6) A third metric for the set $\mathbf{R}^n$ is given by the mapping d_∞, where

$$d_\infty(x, y) = \max_{1 \leqslant k \leqslant n} |x_k - y_k|.$$

When $n = 1$, we again obtain the metric of Example (1), while when $n = 2$ we obtain that of Example (3). The method of Example (3) is used in showing that d_∞ is a metric.

(The notation d_∞ is used because the sequence $\{d_p(x,y)\}_{p=1}^\infty$ has limit $d_\infty(x,y)$ for any $x, y \in \mathbf{R}^n$, where d_p is the metric just mentioned, with $p \in \mathbf{N}$. It is left as an exercise to prove this statement.)

(7) We may obtain metrics for the set $\mathbf{C}^n$ (or for nonempty subsets of $\mathbf{C}^n$) by simple adjustments to Examples (4), (5) and (6). The metric d, where

$$d(x,y) = \sqrt{\sum_{k=1}^{n} |x_k - y_k|^2},$$

where $x = (x_1, x_2, \ldots, x_n)$, $y = (y_1, y_2, \ldots, y_n)$ are now n-tuples of complex numbers, is again referred to as the Euclidean metric and again is the metric implied by reference to $\mathbf{C}^n$ as a metric space. Verification of (M3) for this metric will follow from some work below.

(8) We now introduce one of the most important spaces of modern analysis, the metric space l_2. This is a generalisation of the metric space $\mathbf{C}^n$ in which, loosely speaking, we allow n to be arbitrarily large. A little thought will suggest that 'arbitrarily large n-tuples' are no more than (infinite) sequences. The Euclidean metric then becomes an infinite series and we therefore need some constraints to ensure that the series converges for all pairs of points in the space.

Definition 2.2.3 Denote by l_2 the set of all complex-valued sequences $x_1, x_2, \ldots$ for which the series $\sum_{k=1}^\infty |x_k|^2$ converges. Define a metric d on l_2 by

$$d(x,y) = \sqrt{\sum_{k=1}^{\infty} |x_k - y_k|^2},$$

where x and y are the sequences $x_1, x_2, \ldots$ and $y_1, y_2, \ldots$, respectively. This metric space is itself denoted by l_2.

We must justify this definition by showing that $d(x,y)$ is always finite for any x and y in the set l_2, and that the requirements of a metric are satisfied by d.

To show that $d(x,y)$ is finite, we recall the inequality of Theorem 2.2.2 and set $a_k = |x_k|$, $b_k = |y_k|$. Since $|x_k - y_k|^2 \leqslant (|x_k| + |y_k|)^2 = (a_k + b_k)^2$, we obtain

$$\sqrt{\sum_{k=1}^{n} |x_k - y_k|^2} \leqslant \sqrt{\sum_{k=1}^{n} |x_k|^2} + \sqrt{\sum_{k=1}^{n} |y_k|^2}.$$

On the right-hand side, we have partial sums for the series $\sum_{k=1}^{\infty} |x_k|^2$ and $\sum_{k=1}^{\infty} |y_k|^2$ and these series converge, since this is the condition that $x, y \in l_2$. As the terms of these series are nonnegative, we have

$$\sqrt{\sum_{k=1}^{n} |x_k - y_k|^2} \leqslant \sqrt{\sum_{k=1}^{\infty} |x_k|^2} + \sqrt{\sum_{k=1}^{\infty} |y_k|^2},$$

showing that the partial sums of the series $\sum_{k=1}^{\infty} |x_k - y_k|^2$ form a bounded sequence. This ensures the convergence of the latter series and furthermore we see that

$$d(x, y) = \sqrt{\sum_{k=1}^{\infty} |x_k - y_k|^2} \leqslant \sqrt{\sum_{k=1}^{\infty} |x_k|^2} + \sqrt{\sum_{k=1}^{\infty} |y_k|^2},$$

so that $d(x, y)$ is finite. This is a common form of argument which we will considerably abbreviate in future.

It remains to verify that d is indeed a metric. The definition of the modulus of a complex number answers all questions except, once again, the truth of the triangle inequality. We use the same basic inequality as above (in Theorem 2.2.2), this time setting $a_k = |x_k - y_k|$ and $b_k = |y_k - z_k|$, where $z_1, z_2, \ldots$ is any third element of l_2. Noting that

$$|x_k - z_k| = |(x_k - y_k) + (y_k - z_k)| \leqslant |x_k - y_k| + |y_k - z_k| = a_k + b_k,$$

we have

$$\sqrt{\sum_{k=1}^{n} |x_k - z_k|^2} \leqslant \sqrt{\sum_{k=1}^{n} |x_k - y_k|^2} + \sqrt{\sum_{k=1}^{n} |y_k - z_k|^2}$$

(which is all that is required to prove (M3) for the metric space $\mathbf{C}^n$) and then, by a similar argument to that above,

$$\sqrt{\sum_{k=1}^{n} |x_k - z_k|^2} \leqslant \sqrt{\sum_{k=1}^{\infty} |x_k - y_k|^2} + \sqrt{\sum_{k=1}^{\infty} |y_k - z_k|^2}$$

so that

$$\sqrt{\sum_{k=1}^{\infty} |x_k - z_k|^2} \leqslant \sqrt{\sum_{k=1}^{\infty} |x_k - y_k|^2} + \sqrt{\sum_{k=1}^{\infty} |y_k - z_k|^2}.$$

This verifies (M3) for the metric of l_2.

(9) Let l_1 be the set of all complex-valued sequences $x_1, x_2, \ldots$ for which the series $\sum_{k=1}^{\infty} |x_k|$ converges. Define d by

$$d(x, y) = \sum_{k=1}^{\infty} |x_k - y_k|,$$

where x and y are the sequences $x_1, x_2, \ldots$ and $y_1, y_2, \ldots$, respectively, in l_1. It is easy to show that $d(x, y)$ is finite for all $x, y \in l_1$ and that d defines a metric for l_1. This metric space is itself also denoted by l_1, and may be thought of as a generalisation of the metric space of Example (5).

(It should be evident that there is a further generalisation of the metric spaces l_2 and l_1 along the lines of that in the remark following Example (5). This leads to metric spaces known as the l_p spaces. We will only require the special cases $p = 1$ and $p = 2$.)

(10) In the theory of functions of a complex variable, use is sometimes made of the so-called *chordal metric*. This is the metric d, where

$$d(x, y) = \frac{|x - y|}{\sqrt{(1 + |x|^2)(1 + |y|^2)}}, \quad x, y \in \mathbf{C}.$$

Thinking of x and y as points in the complex plane, the significance of the name may be seen as follows. Place a sphere of unit diameter above the plane, just touching it at the origin, and join the north pole of the sphere to the points x, y in the plane. It may be shown (for example, using ordinary vector methods) that the chord joining the points where these lines intersect the sphere has ordinary (or Euclidean) length $d(x, y)$. With this interpretation, the triangle inequality is intuitively clear.

(11) The next three examples concern the set of all continuous functions defined on the closed interval $[a, b]$. This set was denoted by $C[a, b]$ at the beginning of Section 1.10. Define d by

$$d(x, y) = \max_{a \leqslant t \leqslant b} |x(t) - y(t)|,$$

where x and y are any two functions in $C[a, b]$, t being used for the independent variable. It is by virtue of Theorem 1.9.6 that we know $d(x, y)$ is a finite number for all $x, y \in C[a, b]$: the function $|x - y|$ is continuous when x and y are and, since its domain is a closed interval, it attains its maximum value at at least one point of the domain. Some writers replace 'max' in the definition of d by 'sup' though this is not necessary here, but this has led to the name *sup metric* for d. We will refer to d by its alternative common name: the *uniform metric*. There

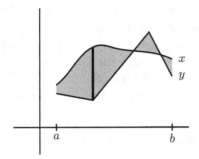

Figure 7

will be many subsequent references to this metric space $(C[a, b], d)$, which from now on we will denote by $C[a, b]$ alone.

We are now introducing what may be a novel notion of 'distance' between functions. For the functions x and y whose graphs are shown in Figure 7, this 'distance', under the uniform metric, is the length of the heavy vertical line segment. There is perhaps little advantage to be gained by still thinking in terms of distance. It may be preferable to consider the metric as measuring the 'degree of difference' between the functions. The essential notion remains that the closer the functions are to each other, the smaller this difference is.

We must verify that d above is indeed a metric. By definition of absolute value, certainly $d(x, y) \in \mathbf{R}_+$ for any $x, y \in C[a, b]$. If $x = y$, then $d(x, y) = 0$; if $x \neq y$, then for some t_0 in $[a, b]$, $x(t_0) \neq y(t_0)$, so

$$d(x, y) \geqslant |x(t_0) - y(t_0)| > 0,$$

and hence $d(x, y) \neq 0$. Thus (M1) is verified. Easily, (M2) is also true. For (M3), if z is any third function in $C[a, b]$ and t is any point in $[a, b]$, then

$$|x(t) - z(t)| \leqslant |x(t) - y(t)| + |y(t) - z(t)|$$
$$\leqslant \max_{a \leqslant t \leqslant b} |x(t) - y(t)| + \max_{a \leqslant t \leqslant b} |y(t) - z(t)|.$$

In particular, this is true wherever the function $|x - z|$ attains its maximum value, so $d(x, z) \leqslant d(x, y) + d(y, z)$, as required.

(12) Another metric for the same set $C[a, b]$ is given by

$$d(x, y) = \int_a^b |x(t) - y(t)| \, dt.$$

A function that is continuous over a closed interval is integrable over that interval, so $d(x,y)$ certainly exists for any $x, y \in C[a,b]$. The verification of the axioms for this metric is easy, though care must be taken with (M1): note how essential it is that the functions be continuous. In Figure 7, the degree of difference between the functions x and y, under this metric, is the area of the shaded region. The metric space of this example will be denoted by $C_1[a,b]$.

(13) A third metric for the set $C[a,b]$ is given by

$$d(x,y) = \sqrt{\int_a^b (x(t) - y(t))^2\, dt}.$$

In verifying (M1), the same note as in Example (12) is relevant. For the triangle inequality, an integral version of the Cauchy–Schwarz inequality must first be obtained. See Exercise 2.4(6). We will denote this metric space by $C_2[a,b]$.

(14) Our final example shows that a metric may be defined for any nonempty set X, without any specification as to the nature of its elements. We define d by

$$d(x,y) = \begin{cases} 0, & x = y, \\ 1, & x \neq y, \end{cases}$$

where $x, y \in X$. It is a simple matter to check that (M1), (M2) and (M3) are satisfied. This metric is called the *discrete metric*, or the *trivial metric*, for X, and serves a useful purpose as a provider of counterexamples. What is not true in this metric space cannot be true in metric spaces generally.

2.3 Solved problems

(1) Let (X,d) be a metric space. For any points $x, y, z \in X$, prove that

$$|d(x,z) - d(y,z)| \leqslant d(x,y).$$

Solution. By property (M3) of a metric, $d(x,z) \leqslant d(x,y) + d(y,z)$, so that

$$d(x,z) - d(y,z) \leqslant d(x,y).$$

This is half the desired result. Then, interchanging x and y, we have

$d(y, z) - d(x, z) \leqslant d(y, x)$, which, using property (M2), is equivalent to

$$d(x, z) - d(y, z) \geqslant -d(x, y),$$

and this is the other half. □

(2) Let (X, d) be a metric space. Show that the mapping d', where

$$d'(x, y) = \frac{d(x, y)}{1 + d(x, y)}, \quad x, y \in X,$$

is also a metric for X.

Solution. All properties of a metric, except (M3), are immediately true for d' since they are true for d. So we only have to show that the triangle inequality holds for d'. Since d is a metric for X, we know, from (M3), that

$$d(x, z) \leqslant d(x, y) + d(y, z),$$

for $x, y, z \in X$. Then

$$d(x, z) + d(x, z)d(x, y) + d(x, z)d(y, z)$$
$$\leqslant d(x, y) + d(y, z) + d(x, z)d(x, y) + d(x, z)d(y, z).$$

Rearranging this, we have

$$\frac{d(x, z)}{1 + d(x, z)} \leqslant \frac{d(x, y) + d(y, z)}{1 + d(x, y) + d(y, z)},$$

so

$$\frac{d(x, z)}{1 + d(x, z)} \leqslant \frac{d(x, y)}{1 + d(x, y) + d(y, z)} + \frac{d(y, z)}{1 + d(x, y) + d(y, z)}$$
$$\leqslant \frac{d(x, y)}{1 + d(x, y)} + \frac{d(y, z)}{1 + d(y, z)},$$

since $d(y, z) \geqslant 0$ and $d(x, y) \geqslant 0$. Thus

$$d'(x, z) \leqslant d'(x, y) + d'(y, z),$$

and (M3) is proved for d'. □

Another approach to this solution uses the function f, defined by

$$f(u) = \frac{u}{1 + u}, \quad u \in \mathbf{R}_+.$$

This can be shown, by the methods of calculus if you like, to be a non-decreasing function on $\mathbf{R}_+$. That is, if $0 \leqslant u_1 \leqslant u_2$, then $f(u_1) \leqslant f(u_2)$. Taking $u_1 = d(x, z)$ and $u_2 = d(x, y) + d(y, z)$, we can then finish off the solution as above.

2.4 Exercises

(1) If (X, d) is a metric space, and $x, y, z, u \in X$, prove that

$$|d(x, z) - d(y, u)| \leqslant d(x, y) + d(z, u).$$

(2) If (X, d) is a metric space, and $x_1, x_2, \ldots, x_n \in X$ $(n \geqslant 2)$, prove that

$$d(x_1, x_n) \leqslant d(x_1, x_2) + d(x_2, x_3) + \cdots + d(x_{n-1}, x_n).$$

(3) Let d_1 and d_2 be two metrics for the same set X. Show that d_3 and d_4, where

$$d_3(x, y) = d_1(x, y) + d_2(x, y),$$
$$d_4(x, y) = \max\{d_1(x, y), d_2(x, y)\}$$

$(x, y \in X)$, are also metrics for X.

(4) Refer to Examples 2.2(5) and 2.2(6). Verify that d_1 and d_∞ are metrics for $\mathbf{R}^n$.

(5) Refer to Example 2.2(9). Show that $d(x, y)$ is finite for all $x, y \in l_1$ and that d defines a metric for l_1.

(6) Let f and g be continuous functions defined on $[a, b]$.

    (a) Derive the integral form of the Cauchy–Schwarz inequality:

$$\left(\int_a^b f(t)g(t)\, dt \right)^2 \leqslant \left(\int_a^b (f(t))^2\, dt \right) \left(\int_a^b (g(t))^2\, dt \right).$$

    (b) Show that there is equality if and only if $f = \beta g$ for some number β.

    (c) Use this Cauchy–Schwarz inequality to deduce the triangle inequality for the mapping d of Example 2.2(13).

(7) Let X be the set of all continuous functions defined on the whole real line which are zero outside some interval (not necessarily the same interval for different functions). Show that

$$d(x, y) = \max_{t \in \mathbf{R}} |x(t) - y(t)|, \quad x, y \in X,$$

defines a metric for X.

(8) Take any $n \in \mathbf{N}$ and let X be the set of all $n \times n$ matrices

with complex elements. If $A = (a_{jk})$ and $B = (b_{jk})$ are any two members of X, show that d_1 and d_2, where

$$d_1(A, B) = \max_{1 \leqslant j \leqslant n, 1 \leqslant k \leqslant n} |a_{jk} - b_{jk}|,$$

$$d_2(A, B) = \max_{1 \leqslant k \leqslant n} \sum_{j=1}^{n} |a_{jk} - b_{jk}|,$$

are both metrics for X.

. . . .[1]

(9) Let X be the set of all complex-valued sequences. Show that the mapping d, where

$$d(x, y) = \sum_{k=1}^{\infty} \frac{1}{2^k} \frac{|x_k - y_k|}{1 + |x_k - y_k|},$$

with $x = \{x_n\} \in X$ and $y = \{y_n\} \in X$, is a metric for X. This metric space is commonly denoted by s.

(10) Let $(x_1, x_2, \ldots, x_n)$ and $(y_1, y_2, \ldots, y_n)$ be two fixed elements of $\mathbf{R}^n$. Prove that

$$\lim_{p \to \infty} \left(\sum_{k=1}^{n} |x_k - y_k|^p \right)^{1/p} = \max_{1 \leqslant k \leqslant n} |x_k - y_k|.$$

(11) There are different and more economical ways of defining the axioms for a metric space. Let X be any nonempty set and let $\rho \colon X \times X \to \mathbf{R}$ be a mapping such that

(a) $\rho(x, y) = 0$ if and only if $x = y$ $(x, y \in X)$, and
(b) $\rho(x, y) \leqslant \rho(z, x) + \rho(z, y)$ $(x, y, z \in X)$.

Show that ρ is a metric for X.

(12) A weaker set of axioms than those for a metric is sufficient for many applications. Often the 'only if' requirement in (M1) is omitted, so that distinct elements may be zero distance apart. Then the mapping $d \colon X \times X \to \mathbf{R}_+$ satisfying $d(x, x) = 0$ $(x \in X)$ and (M2) and (M3) is called a *semimetric* for X and (X, d) is called a *semimetric space*. Show that (X, d) is a semimetric space, but not a metric space, when

[1] Exercises before the dotted line, here and later, are designed to assist an understanding of the preceding concepts and in some cases are referred to subsequently. Those after the line are either harder practice exercises or introduce theoretical ideas not later required.

(a) X is the set of all integrable functions on $[a, b]$, and

$$d(f, g) = \int_a^b |f(t) - g(t)| \, dt \quad (f, g \in X);$$

(b) X is the set of all differentiable functions on $[a, b]$, and

$$d(f, g) = \max_{a \leqslant t \leqslant b} |f'(t) - g'(t)| \quad (f, g \in X).$$

2.5 Convergence in a metric space

A sequence has been defined in Definition 1.7.1 as a mapping from **N** into some set X. If a metric d has been defined for X, we may speak then of sequences in the metric space (X, d).

Because we will often be dealing with metric spaces whose elements are themselves sequences, it is useful to adopt the following convention on notation. If an element of a metric space is itself a sequence (such as occurs in the spaces l_1 and l_2), then it will be denoted, for example, by $(x_1, x_2, \dots)$, and may be thought of as an extended n-tuple. A sequence of elements of a metric space will continue to be denoted as $\{x_n\}_{n=1}^\infty$ or $\{x_n\}$ or $x_1, x_2, \dots$, for example. Thus, a sequence denoted by $(x_1, x_2, \dots)$ is a particular element of a particular metric space and each x_k is a 'component' of this element, whereas a sequence denoted by $\{x_n\}$ is a mapping from **N** into the space and each x_k is an element of the space.

At the beginning of this chapter, it was pointed out that the idea of a metric is all that is required in order to speak generally of convergence of a sequence. Theorem 1.7.5 and Definition 1.7.13 suggest the following.

Definition 2.5.1   A sequence $\{x_n\}$ in a metric space (X, d) is said to *converge* to an element $x \in X$ if for any number $\epsilon > 0$ there exists a positive integer N such that

$$d(x_n, x) < \epsilon \quad \text{whenever } n > N.$$

Then x is called the *limit* of the sequence, and we write $x_n \to x$ or $\lim x_n = x$ (adding '$n \to \infty$' when needed for clarification).

An alternative way of putting this is to require that the real-valued sequence $\{d_n\}$, where $d_n = d(x_n, x)$, converge with limit 0. Thus $x_n \to x$ if and only if $d(x_n, x) \to 0$.

Two important points must be noticed about the definition. First, the element x to which the sequence $\{x_n\}$ in X converges must itself be an element of X. Secondly, the metric by which the convergence is defined must be the metric of the metric space (X, d): the fact that $d(x_n, x) \to 0$ does not imply that $d'(x_n, x) \to 0$, where d' is a different metric for the same set X.

To illustrate the first point, suppose (X, d) is the open interval $(0, 1)$, with the natural metric. The sequence $\frac{1}{2}, \frac{1}{3}, \frac{1}{4}, \ldots$, all of whose terms belong to this space, cannot be called convergent since the only candidate for its limit, namely 0, does not belong to the space. For the second point, consider the same sequence as a subset this time of the closed interval $[0, 1]$. Under the natural metric, the sequence converges to 0 (which now is an element of the space), but if d' is the discrete metric of Example 2.2(14), then the sequence does not converge to 0, since $d'(x_n, 0) = 1$ for every term x_n in the sequence.

The following observation, which we have alluded to previously, is elementary in nature but useful in tidying up proofs of other results.

Theorem 2.5.2 *If a sequence in a metric space is convergent, then the limit is unique.*

We anticipated this in Definition 2.5.1 when we spoke of 'the' limit of a sequence (but see Exercise 2.9(14)). To prove the theorem, we suppose that $\{x_n\}$ is a convergent sequence in a metric space (X, d) and both $x_n \to x$ and $x_n \to y$ $(x, y \in X)$. It follows from the properties of a metric that, for any $n \in \mathbf{N}$,

$$0 \leqslant d(x, y) \leqslant d(x, x_n) + d(x_n, y) = d(x_n, x) + d(x_n, y).$$

Since $d(x_n, x) \to 0$ and $d(x_n, y) \to 0$, we must have $d(x, y) = 0$, or $x = y$, proving the uniqueness of the limit. □

We investigate now how convergence in some particular metric spaces may be related to our earlier ideas of convergence.

Let (X, d) be the metric space $\mathbf{C}^m$ and let $\{x_n\}$ be a sequence in this space. Each term of the sequence is an ordered m-tuple of complex numbers: we will write $x_n = (x_{n1}, x_{n2}, \ldots, x_{nm})$ so that x_{nk} is the kth component of the nth term of the sequence $\{x_n\}$ ($k = 1, 2, \ldots, m$; $n \in \mathbf{N}$). Suppose the sequence converges to an element x (in $\mathbf{C}^m$, of course) and write $x = (x_{.1}, x_{.2}, \ldots, x_{.m})$. Then

$$d(x_n, x) = \sqrt{\sum_{k=1}^{m} |x_{nk} - x_{.k}|^2} \to 0.$$

Since

$$0 \leqslant |x_{nk} - x_{\cdot k}| \leqslant \sqrt{\sum_{k=1}^{m} |x_{nk} - x_{\cdot k}|^2}$$

for each $k = 1, 2, \ldots, m$, we must also have $x_{nk} \to x_{\cdot k}$ (as $n \to \infty$) for each k. That is, all the ordinary complex-valued sequences $\{x_{nk}\}_{n=1}^{\infty}$ are convergent. Conversely, if $x_{nk} \to x_{\cdot k}$ for each k, then $d(x_n, x) \to 0$. This may be expressed by saying that convergence of a sequence in $\mathbf{C}^m$ is equivalent to convergence by components. The same may clearly be said of the metric space $\mathbf{R}^m$.

However, now let (X, d) be the metric space l_2, introduced in Definition 2.2.3, and let $\{x_n\}$ be a convergent sequence in l_2, with $\lim x_n = x$. Each term x_n is a complex-valued sequence, as is the limit x: we write $x_n = (x_{n1}, x_{n2}, \ldots)$, for each $n \in \mathbf{N}$, and $x = (x_{\cdot 1}, x_{\cdot 2}, \ldots)$, in accordance with the note at the beginning of this section. For each n, the condition that $x_n \in l_2$ is that the series $\sum_{k=1}^{\infty} |x_{nk}|^2$ converges. Since the sequence $\{x_n\}$ converges,

$$d(x_n, x) = \sqrt{\sum_{k=1}^{\infty} |x_{nk} - x_{\cdot k}|^2} \to 0$$

and again it follows that $x_{nk} \to x_{\cdot k}$ (as $n \to \infty$) for each $k \in \mathbf{N}$. Thus, convergence of a sequence in l_2 implies convergence by components.

The following example shows that this time the converse is not true. Consider the sequence $\{e_n\}$, where

$$e_1 = (1, 0, 0, 0, \ldots),$$
$$e_2 = (0, 1, 0, 0, \ldots),$$
$$e_3 = (0, 0, 1, 0, \ldots),$$

and so on, all components of e_n being 0 except for the nth component which is 1 ($n \in \mathbf{N}$). The sequence of kth components converges to 0 for each k, but $x = (0, 0, 0, 0, \ldots)$ is certainly not $\lim e_n$, since $d(e_n, x) = 1$ for each n. That $\lim e_n$ does not exist will follow immediately from some work below.

Finally here we consider a sequence $\{x_n\}$ in the metric space $C[a, b]$. If the sequence is convergent, and $\lim x_n = x$, then, given $\epsilon > 0$, we can find a positive integer N so that

$$\max_{a \leqslant t \leqslant b} |x_n(t) - x(t)| < \epsilon$$

whenever $n > N$. Then certainly, when $n > N$, $|x_n(t) - x(t)| < \epsilon$ for all t in $[a, b]$. This N is independent of the choice of t in $[a, b]$, so, recalling Definition 1.10.1, the sequence $\{x_n\}$ is uniformly convergent on $[a, b]$. This works also in reverse, so we conclude that convergence of a sequence in $C[a, b]$ is equivalent to uniform convergence of the sequence on $[a, b]$. This is why the metric for $C[a, b]$ is called the uniform metric. We summarise these results.

Theorem 2.5.3

(a) *A sequence in $\mathbf{C}^n$ or $\mathbf{R}^n$ converges if and only if the sequence of kth components converges for each $k = 1, 2, \ldots, n$.*

(b) *If a sequence in l_2 converges, then the sequence of kth components converges for each $k \in \mathbf{N}$.*

(c) *A sequence in $C[a, b]$ converges if and only if the sequence converges uniformly on $[a, b]$.*

Look again now at Definition 2.5.1, on convergence of a sequence in a metric space. This definition has an unfortunate drawback in that to test a sequence for convergence we must beforehand make at least an educated guess as to whether or not it converges and to what its limit might be. A similar situation was noted for real-valued sequences and there a useful alternative was provided by the Cauchy convergence criterion in Theorem 1.7.12. This provides a test for convergence that depends only on the actual terms of the sequence. If the test works it provides no information on the limit of the sequence but this is often of secondary importance to the basic question of the existence of that limit. It would be easy to write down an exact analogue of that test for metric spaces in general, but unfortunately the analogue would not be true for all metric spaces. Those in which it is true are called *complete*. We now lead up to a precise definition of that term.

Definition 2.5.4 A sequence $\{x_n\}$ in a metric space (X, d) is called a *Cauchy sequence* if for any number $\epsilon > 0$ there exists a positive integer N such that

$$d(x_n, x_m) < \epsilon \quad \text{whenever } m, n > N.$$

Therefore, by the Cauchy convergence criterion, we can state that every Cauchy sequence in the metric space $\mathbf{R}$ is convergent.

Definition 2.5.5 If every Cauchy sequence in a metric space converges, then the space is said to be *complete*.

Hence we say that **R** is complete, in agreement with our earlier discussions of the completeness of the real number system. As we have indicated before, the set **Q** of rational numbers, on which we impose the natural metric, is not complete. An example of a Cauchy sequence in **Q** which does not converge is the sequence

$$0.1, 0.101, 0.101001, 0.1010010001, \ldots .$$

This is clearly a Cauchy sequence, but since the only conceivable limit is a number whose decimal expansion is neither terminating nor periodic, the sequence cannot have a limit which is a rational number. Other examples of metric spaces which are not complete will be given shortly. We can however make the following general statement.

Theorem 2.5.6 *If a sequence in a metric space is convergent, then it is a Cauchy sequence.*

To prove this, we suppose that $\{x_n\}$ is a convergent sequence in a metric space (X, d), with $\lim x_n = x$. Let $\epsilon > 0$ be given. We know that there exists an integer N such that, when $m, n > N$, both $d(x_n, x) < \frac{1}{2}\epsilon$ and $d(x_m, x) < \frac{1}{2}\epsilon$. Then

$$d(x_n, x_m) \leqslant d(x_n, x) + d(x, x_m)$$
$$= d(x_n, x) + d(x_m, x) < \tfrac{1}{2}\epsilon + \tfrac{1}{2}\epsilon = \epsilon,$$

whenever $m, n > N$. Hence $\{x_n\}$ is a Cauchy sequence. □

It is the fact that the converse of this theorem is not true that prompts the notion of complete metric spaces, and, as we have illustrated, all of this is suggested by the earlier work on real-valued sequences.

A little while back, we introduced the sequence $\{e_n\}$ in l_2, where $e_1 = (1, 0, 0, \ldots)$, $e_2 = (0, 1, 0, \ldots)$, $\ldots$. We can show now that this sequence does not converge. To do this, we need only note that when $n \neq m$ we have $d(e_n, e_m) = \sqrt{2}$. Hence $\{e_n\}$ is not a Cauchy sequence and so, by the preceding theorem, it is not convergent.

2.6 Examples on completeness

(1) We have shown that the metric space **R** is complete and that the set **Q**, with the natural metric, is not complete.

(2) Let (X, d) be the metric space $\mathbf{C}$, consisting of the set of all complex numbers with the natural metric $d(x, y) = |x - y|$ $(x, y \in \mathbf{C})$. We will show that $\mathbf{C}$ is a complete metric space. Let $\{x_n\}$ be a Cauchy sequence in $\mathbf{C}$. For each $n \in \mathbf{N}$, write $x_n = u_n + iv_n$, where u_n and v_n are real numbers and $i = \sqrt{-1}$. Because $\{x_n\}$ is a Cauchy sequence, for any $\epsilon > 0$ there is a positive integer N such that $|x_n - x_m| < \epsilon$ when $m, n > N$. But

$$|u_n - u_m| = |\operatorname{Re}(x_n - x_m)| \leqslant |x_n - x_m|,$$

and also $|v_n - v_m| \leqslant |x_n - x_m|$, so $\{u_n\}$ and $\{v_n\}$ are Cauchy sequences in $\mathbf{R}$. Since $\mathbf{R}$ is complete, these sequences are convergent, and we can write $\lim u_n = u$ and $\lim v_n = v$, say, for some real numbers u, v. Put $x = u + iv$. Then $x \in \mathbf{C}$. Furthermore, $x = \lim x_n$, because

$$0 \leqslant d(x_n, x) = |x_n - x| = |(u_n + iv_n) - (u + iv)|$$
$$= |(u_n - u) + i(v_n - v)| \leqslant |u_n - u| + |v_n - v| < \epsilon$$

for any $\epsilon > 0$, provided n is large enough. Hence we have proved that the Cauchy sequence $\{x_n\}$ is convergent, so $\mathbf{C}$ is a complete metric space.

This proof has been written out in full detail. A similar process is followed in Examples (3) and (5) below. The general technique is to take a Cauchy sequence in the space, postulate a natural limit for the sequence, show that it is an element of the space, and then verify that it is indeed the limit.

(3) *The metric space l_2 is complete.* Let $\{x_n\}$ be a Cauchy sequence in l_2. We must show that the sequence converges. For each $n \in \mathbf{N}$, write $x_n = (x_{n1}, x_{n2}, \dots)$. By definition of the space l_2, the series $\sum_{k=1}^{\infty} |x_{nk}|^2$ converges for each n. Since $\{x_n\}$ is a Cauchy sequence, for any $\epsilon > 0$ there is a positive integer N such that

$$\sqrt{\sum_{k=1}^{\infty} |x_{nk} - x_{mk}|^2} < \epsilon$$

when $m, n > N$, using the definition of the metric for l_2. That is,

$$\sum_{k=1}^{\infty} |x_{nk} - x_{mk}|^2 < \epsilon^2, \quad m, n > N,$$

so we must have

$$|x_{nk} - x_{mk}| < \epsilon, \quad m, n > N,$$

for each $k \in \mathbf{N}$. Then, for each k, $\{x_{nk}\}$ is a Cauchy sequence in $\mathbf{C}$ so $\lim_{n \to \infty} x_{nk}$ exists since $\mathbf{C}$ is complete. Write $\lim_{n \to \infty} x_{nk} = x_{\cdot k}$ and set $x = (x_{\cdot 1}, x_{\cdot 2}, \dots)$. We will show that $x \in l_2$ and that $\{x_n\}$ converges to x. This will then mean that l_2 is complete. We note first that for any $r = 1, 2, \dots,$

$$\sum_{k=1}^{r} |x_{nk} - x_{mk}|^2 < \epsilon^2, \quad m, n > N,$$

so that, keeping n fixed and using the fact that $\lim_{m \to \infty} x_{mk} = x_{\cdot k}$,

$$\sum_{k=1}^{r} |x_{nk} - x_{\cdot k}|^2 \leqslant \epsilon^2, \quad n > N,$$

by Theorem 1.7.7. For points

$$(a_1, a_2, \dots, a_r), (b_1, b_2, \dots, b_r), (c_1, c_2, \dots, c_r) \in \mathbf{C}^r,$$

the triangle inequality in $\mathbf{C}^r$ gives us

$$\sqrt{\sum_{k=1}^{r} |a_k - c_k|^2} \leqslant \sqrt{\sum_{k=1}^{r} |a_k - b_k|^2} + \sqrt{\sum_{k=1}^{r} |b_k - c_k|^2}.$$

Replacing a_k by $x_{\cdot k}$, b_k by x_{nk} and c_k by 0, we have

$$\sqrt{\sum_{k=1}^{r} |x_{\cdot k}|^2} \leqslant \sqrt{\sum_{k=1}^{r} |x_{\cdot k} - x_{nk}|^2} + \sqrt{\sum_{k=1}^{r} |x_{nk}|^2}$$

$$\leqslant \epsilon + \sqrt{\sum_{k=1}^{r} |x_{nk}|^2} \leqslant \epsilon + \sqrt{\sum_{k=1}^{\infty} |x_{nk}|^2}$$

if $n > N$. The convergence of the final series here thus implies the convergence of $\sum_{k=1}^{\infty} |x_{\cdot k}|^2$, so that indeed $x \in l_2$. Moreover, an inequality a few lines back shows further that

$$\sqrt{\sum_{k=1}^{\infty} |x_{nk} - x_{\cdot k}|^2} \leqslant \epsilon, \quad n > N,$$

and this implies that the sequence $\{x_n\}$ converges to x. This completes the proof that l_2 is complete.

(4) *The metric spaces* $\mathbf{R}^n$ *and* $\mathbf{C}^n$ *are complete.* This is easily shown by adapting the method of Example (3).

(5) *The metric space* $C[a, b]$ *is complete.* Let $\{x_n\}$ be a Cauchy sequence

in $C[a, b]$. Then, for any $\epsilon > 0$, we can find N so that, using the definition of the metric for this space,

$$\max_{a \leqslant t \leqslant b} |x_n(t) - x_m(t)| < \epsilon$$

when $m, n > N$. Certainly then for each particular t in $[a, b]$ we have

$$|x_n(t) - x_m(t)| < \epsilon, \quad m, n > N,$$

so $\{x_n(t)\}$ is a Cauchy sequence in $\mathbf{R}$. But $\mathbf{R}$ is complete, so the sequence $\{x_n(t)\}$ converges to a real number, which we will write as $x(t)$, for each t in $[a, b]$. This determines a function x, defined on $[a, b]$. In the preceding inequality, fix n (and let $m \to \infty$) to give

$$|x_n(t) - x(t)| \leqslant \epsilon, \quad n > N.$$

The N here is independent of t in $[a, b]$, so we have shown that the sequence $\{x_n\}$ converges uniformly on $[a, b]$ to x. Using the theorem that the uniform limit of a sequence of continuous functions is itself continuous (Theorem 1.10.3), our limit function x must be continuous on $[a, b]$. That is, $x \in C[a, b]$. Furthermore, uniform convergence on $[a, b]$ is equivalent to convergence in $C[a, b]$ (Theorem 2.4.3(c)). Thus the Cauchy sequence $\{x_n\}$ converges to x, completing the proof that $C[a, b]$ is complete.

(6) The metric space $C_1[a, b]$ (defined in Example 2.2(12)) is not complete. That a metric space is not complete can always be shown by a single example of a Cauchy sequence in the space that does not converge.

We will give an example of such a sequence when $a < 0$ and $1 < b$. Similar examples could be devised for other values of a and b, but see Exercise 2.9(12) to show that some care is necessary. Let $\{x_n\}$ be the sequence of functions for which

$$x_n(t) = \begin{cases} 0, & a \leqslant t \leqslant 0, \\ nt, & 0 < t < \dfrac{1}{n}, \\ 1, & \dfrac{1}{n} \leqslant t \leqslant b, \end{cases}$$

Figure 8 shows the graphs of typical functions x_m and x_n (where we

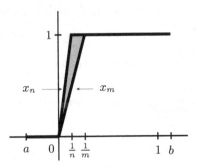

Figure 8

have taken $m < n$). Using the definition of the metric for $C_1[a, b]$,

$$d(x_n, x_m) = \int_a^b |x_n(t) - x_m(t)|\, dt \quad (\ = \text{area of shaded region})$$

$$= \frac{1}{2}\left|\frac{1}{m} - \frac{1}{n}\right| < \epsilon$$

when m and n are sufficiently large, no matter how small ϵ is. Hence $\{x_n\}$ is a Cauchy sequence in $C_1[a, b]$. However, the sequence does not converge. To see this, let g be the function defined on $[a, b]$ by

$$g(t) = \begin{cases} 0, & a \leqslant t \leqslant 0, \\ 1, & 0 < t \leqslant b, \end{cases}$$

and let f be any continuous function on $[a, b]$. Then, for any t in $[a, b]$ and any $n \in \mathbf{N}$,

$$|g(t) - f(t)| = |(g(t) - x_n(t)) + (x_n(t) - f(t))|$$
$$\leqslant |g(t) - x_n(t)| + |x_n(t) - f(t)|,$$

so

$$\int_a^b |g(t) - f(t)|\, dt \leqslant \int_a^b |g(t) - x_n(t)|\, dt + \int_a^b |x_n(t) - f(t)|\, dt.$$

The integral on the left is $\int_a^0 |f(t)|\, dt + \int_0^b |1 - f(t)|\, dt$. Since f is continuous, this sum must be positive. The first integral on the right is arbitrarily small for large enough n, as we see in the same way that $\{x_n\}$ was shown to be a Cauchy sequence. It follows from this that we cannot have $d(x_n, f) = \int_a^b |x_n(t) - f(t)|\, dt \to 0$, no matter what the function f is (remembering that f must be continuous). Hence the sequence $\{x_n\}$ does not converge.

(7) The same choice of $\{x_n\}$ as in Example (6) shows that the metric space $C_2[a, b]$, of Example 2.2(13), is not complete.

2.7 Subspace of a metric space

Let (X, d) be a metric space and let S be any nonempty subset of X. By definition, d is a mapping from $X \times X$ into $\mathbf{R}_+$. By *the restriction of d to S*, we mean the mapping $d_S \colon S \times S \to \mathbf{R}_+$ such that $d_S(x, y) = d(x, y)$, $x, y \in S$. It is immediately clear that d_S is a metric for S, so that (S, d_S) is a metric space. As d_S is nothing more than the mapping d when considered as a mapping of the points in S alone, we normally drop the subscript on d_S. This leads us to the notion of a subspace of a metric space.

Definition 2.7.1 Let (X, d) be a metric space. The metric space obtained by restricting d to a nonempty subset of X is called a *subspace* of (X, d).

We have in fact already met many subspaces. The wording of Example 2.2(2) means that the metric space (X, d) in that example is a subspace of $\mathbf{R}^2$. Similar wording was used in Examples 2.2(1), 2.2(3) and 2.2(4).

Let $\{x_n\}$ be a sequence in a subspace (S, d) of a metric space (X, d) and suppose that $x_n \to x$, when we consider $\{x_n\}$ as a sequence in X. By definition of convergence of a sequence in a metric space, we must have $x \in X$. If we also have $x \in S$, then (S, d) is called a *closed* subspace of (X, d). Putting this another way, we have the following definition.

Definition 2.7.2 A subspace S of a metric space X is said to be (*sequentially*) *closed* if it contains the limits of all the sequences in S which converge in X.

The more correct, but clumsier, term is 'sequentially closed'. We will stay with the simpler term for now, but in Chapter 5 we will see the word 'closed' used in a different way, and we will then need to be more careful with our terminology.

It will not be unexpected, because of the nomenclature used, that a closed interval with the natural metric is a closed subspace of $\mathbf{R}$. This is little more than a restatement of Theorem 1.7.7. It is left as an exercise to verify that the subset $\{z : z \in \mathbf{C}, \ |z| \leqslant c\}$ of $\mathbf{C}$ is a closed subspace of **C** for any positive number c. Such a set is referred to as a *closed disc* in **C**.

It is apparent that any metric space can be considered as a subspace of itself. This immediately implies that all metric spaces are closed, since no sequence in a metric space can be considered to be convergent unless its limit is contained in the space.

An enlightening consequence of this is provided by the metric space consisting of the open interval (a, b) with the natural metric. Like all metric spaces, this one is closed, but it is certainly not closed when considered as a subspace of **R**. To be particular, let the open interval be $(0, 1)$ and consider the sequence $\frac{1}{2}, \frac{1}{3}, \frac{1}{4}, \ldots$. In the metric space consisting of $(0, 1)$ with the natural metric, this is not a convergent sequence (its 'limit' is not in the space). It is however a Cauchy sequence, so this metric space is not complete. Of course, as a sequence in **R**, it has limit 0. This should be looked at carefully in the light of the statement above on closed intervals. That the metric space $(0, 1)$ is neither complete nor closed as a subspace of **R** is a particular case of the following theorem, which is the main result of this section.

Theorem 2.7.3 *A subspace of a complete metric space is complete if and only if it is closed.*

Thus, in a complete metric space, the notions of completeness and closedness of subspaces coincide. To prove the theorem, we suppose that S is a subspace of a complete metric space X, and show first that if S is closed then it is complete. Let $\{x_n\}$ be a Cauchy sequence in S. As S is a subspace of X and X is complete, then $\{x_n\}$, as a sequence in X, must converge. But S is closed, so the limit of the sequence must belong to S. Thus the Cauchy sequence $\{x_n\}$ converges in S, so S is complete. Next, we prove the converse: if S is complete, then it is closed. This time, let $\{x_n\}$ be any sequence in S and suppose that, as a sequence in X, it converges with limit x, say. Then $\{x_n\}$ is a Cauchy sequence in X (Theorem 2.5.6) and hence also in S. But since S is complete, we must have $x \in S$, so S is closed. $\qquad\square$

Having just taken the time to talk carefully of subspaces of metric spaces, we must now foreshadow a loosening of expression. It is common to speak of a subset of a metric space, rather than of a subspace, and we will shortly follow this practice. If we do not do this then the language becomes too confused once we have introduced vector spaces into the discussion. Anticipating a little, a set on which we impose the axioms of a vector space may also be considered as a metric space, and the two notions of subspace (of a vector space and of a metric space) do not coincide. In general, we will later prefer to mean by a subspace the more established idea of a vector subspace.

Speaking of subsets rather than subspaces also allows us conveniently to refer to the empty set as a subset of a metric space. The empty set is in fact always considered to be a closed subset of any metric space.

2.8 Solved problems

(1) Let P be the set of all polynomial functions (of all degrees) defined on $[0, 1]$ and define d by

$$d(x, y) = \max_{0 \leqslant t \leqslant 1} |x(t) - y(t)|, \quad x, y \in P.$$

Prove that (P, d) is a metric space, but that it is not complete. Prove also that (P, d) is a subspace of $C[0, 1]$, but, as a subspace, is not closed.

Solution. Every polynomial function is continuous; d is the restriction to P of the uniform metric of the metric space $C[0, 1]$. These observations prove that (P, d) is a subspace of $C[0, 1]$, so certainly (P, d) is a metric space. (This may also be shown directly.) To prove that (P, d) is not complete, consider (as one of many similar examples) the sequence $\{x_n\}$, where

$$x_n(t) = \sum_{k=0}^{n} \left(\frac{t}{2}\right)^k = 1 + \frac{t}{2} + \frac{t^2}{2^2} + \cdots + \frac{t^n}{2^n}, \quad 0 \leqslant t \leqslant 1.$$

As desired, $x_n \in P$ for each $n \in \mathbf{N}$. This sequence is a Cauchy sequence in (X, d), for, taking $m < n$,

$$\begin{aligned} d(x_n, x_m) &= \max_{0 \leqslant t \leqslant 1} \left| \sum_{k=0}^{n} \left(\frac{t}{2}\right)^k - \sum_{k=0}^{m} \left(\frac{t}{2}\right)^k \right| \\ &= \max_{0 \leqslant t \leqslant 1} \sum_{k=m+1}^{n} \left(\frac{t}{2}\right)^k \\ &= \sum_{k=m+1}^{n} \frac{1}{2^k} = \frac{1}{2^m} - \frac{1}{2^n}, \end{aligned}$$

and this is arbitrarily small for large enough m, n. However, the sequence does not converge in P, because the only candidate for $\lim x_n$ is the function given by $2/(2 - t)$, $0 \leqslant t \leqslant 1$ (using the formula for the limiting

sum of a geometric series), and this is not a polynomial function. Hence
(P, d) is not complete. As $C[0, 1]$ is complete, it immediately follows
from Theorem 2.7.3 that, as a subspace of $C[0, 1]$, (P, d) is not closed.
□

For the second of these solved problems, we will need the following
definition.

Definition 2.8.1

(a) Let (X, d) be a metric space, and let S be a nonempty subset
of X. The number $\delta(S)$ defined by

$$\delta(S) = \sup\{d(x, y) : x, y \in S\}$$

is called the *diameter* of the set S.

(b) A subset of a metric space is said to be *bounded* if it is empty or
if it has a finite diameter.

(2) Show that any Cauchy sequence in a metric space is bounded.

Solution. More precisely, we are to show that the range of any Cauchy
sequence is a bounded set. Let $\{x_n\}$ be a Cauchy sequence in a metric
space (X, d). Then, given any $\epsilon > 0$, there exists a positive integer N
such that $d(x_n, x_m) < \epsilon$ whenever $m, n > N$. In particular, choosing
$m = N + 1$, $d(x_n, x_{N+1}) < \epsilon$ whenever $n > N$. For those $n \leqslant N$, there
being only a finite number of them, the set of distances $d(x_n, x_{N+1})$ is
bounded (in the ordinary sense). Write

$$K = \max\{d(x_n, x_{N+1}) : n = 1,\ 2,\ \ldots,\ N\}.$$

Then surely, for all $n \in \mathbf{N}$, we have $d(x_n, x_{N+1}) < K + \epsilon$. By the triangle
inequality, we then have, for any $n, p \in \mathbf{N}$,

$$d(x_n, x_p) \leqslant d(x_n, x_{N+1}) + d(x_p, x_{N+1}) < 2(K + \epsilon).$$

This provides an upper bound for the set of all distances $d(x_n, x_p)$, so
its least upper bound exists (Theorem 1.5.7). But this means that the
diameter of the subset of X given by the terms of the sequence $\{x_n\}$ is
finite. That is, the Cauchy sequence $\{x_n\}$ is bounded.   □

2.9 Exercises

(1) Use the inequality of Exercise 2.4(1) to prove that if $\{x_n\}$ and $\{y_n\}$ are convergent sequences in a metric space and $\lim x_n = x$, $\lim y_n = y$, then $d(x_n, y_n) \to d(x, y)$, where d is the metric for the space.

(2) Let $\{x_n\}$ and $\{y_n\}$ be two Cauchy sequences in a complete metric space, with metric d. Prove that they have the same limit if and only if $d(x_n, y_n) \to 0$.

(3) Refer to Example 2.6(6). Show that the sequence $\{x_n\}$ in that example is not a Cauchy sequence in $C[a, b]$ (where $a < 0$ and $1 < b$).

(4) Show that any convergent sequence in a metric space is bounded. (More precisely, show that the range of any convergent sequence is a bounded set.)

(5) Let X be any nonempty set and impose on it the discrete metric. (See Example 2.2(14).) Determine whether the resulting metric space is complete.

(6) Prove that the metric space $\mathbf{C}^n$ is complete.

(7) Show that the metric space $(\mathbf{R}^n, d_\infty)$ is complete, d_∞ being the metric of Example 2.2(6).

(8) Prove that the metric space l_1 (Example 2.2(9)) is complete.

(9) Let X be the set of all bounded real-valued sequences. Define a mapping d on $X \times X$ by $d(x, y) = \sup\{|x_k - y_k| : k \in \mathbf{N}\}$ where $x = (x_1, x_2, \dots)$, $y = (y_1, y_2, \dots)$ are elements of X. Prove that (X, d) is a metric space and that it is complete. This space is commonly denoted by m.

(10) If $\{z_n\}$ is a complex-valued sequence, and $z_n \to z$, prove that $|z_n| \to |z|$. Hence show that the subset $\{w : w \in \mathbf{C}, |w| \leqslant c\}$ of $\mathbf{C}$ is closed for any positive number c.

(11) Let Y be the set of all complex-valued sequences $(y_1, y_2, \dots)$ for which $|y_k| \leqslant 1/k$, $k \in \mathbf{N}$. Define d by

$$d(x, y) = \sqrt{\sum_{k=1}^{\infty} |x_k - y_k|^2}, \quad x, y \in Y.$$

Prove that (Y, d) is a subspace of l_2, and that it is closed.

(12) Why does the counterexample in Example 2.6(6) fail when $a = 0$?

. . . .

(13) Show that the metric space s (Exercise 2.4(9)) is complete. Show also that convergence in s is equivalent to convergence by components.

(14) In a semimetric space (Exercise 2.4(12)), convergence of a sequence is defined as it is in a metric space. Let (X, d) be the semimetric space of Exercise 2.4(12)(b), with $a = 0$, $b = \frac{1}{2}$. Show that the sequence $\{x_n\}$, where $x_n(t) = t^n$ $(0 \leqslant t \leqslant \frac{1}{2})$ is convergent and that any constant function on $[0, \frac{1}{2}]$ serves as its limit. (Hence, convergent sequences in semimetric spaces need not have unique limits.)

(15) An *ultrametric space* (X, d) is a nonempty set X together with a 'distance' function $d \colon X \times X \to \mathbf{R}_+$ satisfying (M1), (M2) and, in place of (M3),

$$d(x, z) \leqslant \max\{d(x, y), d(y, z)\} \quad \text{for every } x, y, z \in X.$$

Show that

(a) an ultrametric space is a metric space;

(b) if $d(x, y) \neq d(y, z)$, then $d(x, z) = \max\{d(x, y), d(y, z)\}$, $x, y, z \in X$;

(c) a sequence $\{x_n\}$ in X is a Cauchy sequence (defined as in metric spaces) if and only if $d(x_n, x_{n+1}) \to 0$.

3

The Fixed Point Theorem and its Applications

3.1 Mappings between metric spaces

Let (X, d) and (Y, d') be metric spaces. The definition of a mapping $A: (X, d) \to (Y, d')$ involves nothing more than the definition already given of a mapping $A: X \to Y$ (Definition 1.3.1(a)). The fact that the sets X and Y now have metrics associated with them does not alter the basic notion that to each element $x \in X$ the mapping A assigns a unique element $y \in Y$. The other parts of Definition 1.3.1 are also still used in the context of metric spaces. There are however certain changes of notation which have become established.

We denote the image y of $x \in X$ by $y = Ax$, no longer using parentheses, as in the familiar $y = f(x)$, unless they are necessary to avoid ambiguity. The composition of two mappings (see Definition 1.3.3) is also denoted differently. If $A: X \to Y$ and $B: Y \to Z$ are two mappings between metric spaces, then the composition of A with B is denoted simply by BA. The order of the letters here is important, and natural: if $x \in X$, then

$$(BA)x = B(Ax).$$

As $Ax \in Y$, we have $B(Ax) \in Z$ so BA is a mapping from X into Z, as it should be. The mapping BA is often also called a *product* of A and B.

When A maps a metric space (or simply a set) into itself, it is possible to form the product of A with A, obtaining the mapping AA, which for natural reasons is denoted by A^2. We can then form the product $A(A^2)$, denoted by A^3, and in general may speak of the mapping $A^n: X \to X$ defined inductively by

$$A^n x = A(A^{n-1} x), \quad x \in X, \ n = 2, \ 3, \ 4, \ \ldots .$$

113

By A^1 we of course mean the mapping A itself. It is often useful to use A^0 for the identity mapping I on X defined by $Ix = x$, $x \in X$.

Let $A\colon X \to Y$, $B\colon Y \to Z$, $C\colon Z \to W$ be mappings, where X, Y, Z, W are any sets. Then the products $C(BA)$ and $(CB)A$ both exist, and

$$C(BA) = (CB)A.$$

That is, the associative law is obeyed. To prove this, we let $x \in X$ be arbitrary and then the result follows from:

$$(C(BA))x = C((BA)x) = C(B(Ax)) = (CB)(Ax) = ((CB)A)x.$$

For now, we will indicate just two examples of mappings on metric spaces. The first is the mapping $A\colon C[a,b] \to \mathbf{R}$ defined by $Ax = y$, where $x \in C[a,b]$ and

$$y = \int_a^b x(t)\,dt.$$

Here, A maps each continuous function defined on $[a,b]$ onto the unique real number which is its integral over $[a,b]$. Since the domain of A is the set $C[a,b]$, we are assured that every x in the set does indeed have an image $y \in \mathbf{R}$: this is only a restatement of the fact that every continuous function over a closed interval is integrable over that interval. The second example concerns the Euclidean space $\mathbf{R}^n$. Its elements are n-tuples of real numbers. If the n-tuples are written as columns, then they can be considered as column vectors, or $n \times 1$ matrices. The mapping in mind is $B\colon \mathbf{R}^n \to \mathbf{R}^m$ defined by the equation $Bx = y$, where $x = (x_1, x_2, \ldots, x_n)^T \in \mathbf{R}^n$ and $B = (b_{jk})$ is an $m \times n$ matrix whose elements b_{jk} are real. Then indeed $y = (y_1, y_2, \ldots, y_m)^T$ is an element of $\mathbf{R}^m$ and

$$y_j = \sum_{k=1}^n b_{jk} x_k, \quad j = 1,\, 2,\, \ldots,\, m.$$

It is standard, and we have followed the practice, that in this example the mapping and the matrix by which the mapping is defined are indicated by the same letter.

In these examples, one mapping works on continuous functions, the other on n-tuples of real numbers. What do they have in common? Only this: the domain of each mapping is a complete metric space. Hence if we can conclude anything in general terms about mappings on complete metric spaces, then we will immediately have concrete applications provided by these (and other) examples.

Since we are dealing now with metric spaces, it should be clear that there is no difficulty in coming up with an adequate definition of continuity for a mapping. As models for such a definition, we have a choice between Definition 1.9.1 and Theorem 1.9.2. We choose the latter because of its emphasis on sequences.

Definition 3.1.1 Let X and Y be metric spaces. We say a mapping $A: X \to Y$ is *continuous at* $x \in X$ if, whenever $\{x_n\}$ is a convergent sequence in X with limit x, $\{Ax_n\}$ is a convergent sequence in Y with limit Ax. The mapping A is said to be *continuous on* X if it is continuous at every point of X.

3.2 The fixed point theorem

Probably the most common problem in mathematics, in all branches and at all levels, is: given the mapping A and the image y, solve for x the equation $Ax = y$. In the example above of the mapping $B: \mathbf{R}^n \to \mathbf{R}^m$, this problem is that of solving a set of simultaneous linear equations. Another instance is the need to solve an equation of the form $f(x) = 0$, where f is an ordinary real-valued function. This equation is easy to solve when f is a linear or quadratic function, but for most other functions some method of approximating the roots of the equation is usually employed.

Newton's method provides a means for doing this under certain conditions. We suppose x_0 to be an approximation to the root and then calculate $x_1 = x_0 - f(x_0)/f'(x_0)$. Then x_1 will be a better approximation, and we may repeat the process with x_1 replacing x_0 to obtain a still better approximation x_2, and so on. Such a process is said to be *iterative*. Desirable features of any iterative process are that the successive iterates (x_0, x_1, x_2, ... here) indeed converge to the desired point (a root of $f(x) = 0$ here) and that they converge rapidly in the sense that not too many iterates need to be computed before sufficient accuracy is obtained.

In Application 3.3(1), we will see an alternative approach to the problem of solving an equation of the form $f(x) = 0$, using a different iterative process. This will be just one of many examples arising from the fixed point theorem, which under fairly broad conditions allows us to find or estimate the solution x of an equation of the form $Ax = x$. We are concerned in this section only with mappings from a metric space X into itself. Thus if $x \in X$ and $Ax = y$, then also $y \in X$. The following definitions are pertinent.

Definition 3.2.1 Let A be a mapping from a metric space (X, d) into itself.

(a) A point $x \in X$ such that $Ax = x$ is called a *fixed point* of the mapping A.

(b) If there is a number α, with $0 < \alpha < 1$, such that for every pair of points $x, y \in X$ we have

$$d(Ax, Ay) \leqslant \alpha d(x, y),$$

   then A is called a *contraction mapping*, or simply a *contraction*. The number α is called a *contraction constant* for A.

The reason for calling A a contraction in (b) is clear: since $\alpha < 1$, the effect of applying the mapping A is to decrease the distance between any pair of points in X. We see that the problem we indicated, that of solving the equation $Ax = x$, amounts to asking for the fixed points of A. The fixed point theorem below says that there always exists a fixed point of A when A is a contraction and the space X is complete, and that this fixed point is unique. Before stating this more formally, and proving it, we show that any contraction mapping is continuous.

Theorem 3.2.2 *If A is a contraction mapping on a metric space X then A is continuous on X.*

The proof is simple. Suppose $\{x_n\}$ is a sequence in (X, d) converging to x and let α be a contraction constant for A. Then

$$0 \leqslant d(Ax_n, Ax) \leqslant \alpha d(x_n, x) < d(x_n, x),$$

so $Ax_n \to Ax$ because $x_n \to x$. $\square$

The following is the main theorem of this chapter.

Theorem 3.2.3 (Fixed Point Theorem) *Every contraction mapping on a complete metric space has one and only one fixed point.*

To prove this, let A be a contraction mapping, with contraction constant α, on a complete metric space (X, d). Take any point $x_0 \in X$ and let $\{x_n\}$ be the sequence (in X) defined recursively by

$$x_n = Ax_{n-1}, \quad n \in \mathbf{N}.$$

Thus $x_1 = Ax_0$, $x_2 = Ax_1 = A(Ax_0) = A^2x_0$,

$$x_3 = Ax_2 = A(A^2x_0) = A^3x_0,$$

and so on, so that we may write $x_n = A^nx_0$. We will show that $\{x_n\}$ is a Cauchy sequence. Notice that, for any integer $k > 1$,

$$\begin{aligned}
d(x_k, x_{k-1}) &= d(A^kx_0, A^{k-1}x_0) = d(A(A^{k-1}x_0), A(A^{k-2}x_0)) \\
&\leqslant \alpha d(A^{k-1}x_0, A^{k-2}x_0) \leqslant \alpha^2 d(A^{k-2}x_0, A^{k-3}x_0) \\
&\vdots \\
&\leqslant \alpha^{k-1} d(Ax_0, x_0).
\end{aligned}$$

Now, taking $1 \leqslant m < n$ for definiteness,

$$\begin{aligned}
d(x_n, x_m) &= d(A^nx_0, A^mx_0) \\
&\leqslant d(A^nx_0, A^{n-1}x_0) + d(A^{n-1}x_0, A^{n-2}x_0) \\
&\quad + \cdots + d(A^{m+1}x_0, A^mx_0) \\
&\leqslant \alpha^{n-1} d(Ax_0, x_0) + \alpha^{n-2} d(Ax_0, x_0) + \cdots + \alpha^m d(Ax_0, x_0) \\
&= \alpha^m (1 + \alpha + \alpha^2 + \cdots + \alpha^{n-m-1}) d(x_1, x_0) \\
&< \frac{\alpha^m}{1 - \alpha} d(x_1, x_0),
\end{aligned}$$

using the limiting sum of a geometric series, which we may do since $0 < \alpha < 1$. Since $\alpha^m \to 0$ (as $m \to \infty$), we must have $d(x_n, x_m) < \epsilon$ for any $\epsilon > 0$ whenever m and n are sufficiently large. Hence $\{x_n\}$ is a Cauchy sequence. We see now why we insist that X be a complete metric space: the existence of $\lim x_n$ is assured. We set $x = \lim x_n$ and will show that x is a fixed point of A. For this, we note that, for any positive integer n,

$$\begin{aligned}
0 &\leqslant d(Ax, x) \leqslant d(Ax, x_n) + d(x_n, x) \\
&= d(Ax, Ax_{n-1}) + d(x_n, x) \leqslant \alpha d(x, x_{n-1}) + d(x_n, x),
\end{aligned}$$

and so $d(Ax, x) = 0$ since $d(x_n, x) \to 0$ (and $d(x, x_{n-1}) \to 0$). Thus $Ax = x$, so indeed x is a fixed point of A. Finally, to show that it is the only one, we suppose that y is another, so also $Ay = y$. Then

$$d(x, y) = d(Ax, Ay) \leqslant \alpha d(x, y),$$

which, since $\alpha < 1$, can only be true if $d(x, y) = 0$; that is, if $x = y$. Hence there is just one fixed point of A. $\qquad\square$

Notice how simple it is, in theory, to obtain the fixed point. We take any starting point x_0 and then repeatedly apply the mapping A. The sequence $x_0, Ax_0, A^2x_0, \ldots$ will converge to the fixed point. This is the iteration process that we foreshadowed, the points $A^n x_0$ ($n = 0$, 1, 2, ...) being the successive iterates. In practice, there are important questions of where to start the process and when to stop it. That is, how do we choose x_0 and how many iterates must we take to approximate the fixed point with sufficient accuracy? The contraction mapping itself must often be approximated by some other mapping and this again raises questions of accuracy. We will return to this point at the end of the chapter. For reasons which are clear, the fixed point theorem is often referred to as the *method of successive approximations*.

We will soon consider a number of applications. In the fourth of these, a generalisation of the fixed point theorem will be needed and it is convenient to state and prove it at this stage.

Theorem 3.2.4 *Let A be a mapping on a complete metric space and suppose that A is such that A^n is a contraction for some integer $n \in \mathbf{N}$. Then A has a unique fixed point.*

Let the metric space be X. According to the fixed point theorem, the mapping A^n has a unique fixed point $x \in X$, so that $A^n x = x$. Noting that

$$A^n(Ax) = A^{n+1}x = A(A^n x) = Ax,$$

we see that Ax is also a fixed point of A^n. But there can be only one, so $Ax = x$ and thus x is also a fixed point of A. Now, any fixed point y of A is also a fixed point of A^n since

$$A^n y = A^{n-1}(Ay) = A^{n-1}y = \cdots = Ay = y.$$

It follows that x is the only fixed point of A. $\qquad\square$

3.3 Applications

(1) Let f be a function with domain $[a, b]$ and range a subset of $[a, b]$. Suppose there is some positive constant $K < 1$ such that

$$|f(x_1) - f(x_2)| \leqslant K|x_1 - x_2|,$$

for any points $x_1, x_2 \in [a, b]$. (Then f is said to satisfy a *Lipschitz condition*, with *Lipschitz constant* K.) The fixed point theorem assures us that the equation $f(x) = x$ has a unique solution for x in $[a, b]$.

This is because of the following. First, f may be considered as a mapping from the metric space consisting of the closed interval $[a, b]$ with the natural metric into itself, and this metric space is complete because it is a closed subspace of **R** (Theorem 2.7.3). Second, the Lipschitz condition, with $0 < K < 1$, states that this mapping f is a contraction. Hence f has a unique fixed point.

If f is a differentiable function on $[a, b]$, with range a subset of $[a, b]$, and if there is a constant K such that

$$|f'(x)| \leqslant K < 1,$$

for all x in $[a, b]$, then again the equation $f(x) = x$ has a unique solution for x in $[a, b]$. This is a simpler test than the preceding one, but applies only to differentiable functions. Its truth is a consequence of the mean value theorem of differential calculus: for any $x_1, x_2 \in [a, b]$, with $x_1 < x_2$, there is at least one point c, $x_1 < c < x_2$, such that

$$|f(x_1) - f(x_2)| = |f'(c)(x_1 - x_2)| = |f'(c)|\,|x_1 - x_2| \leqslant K|x_1 - x_2|,$$

and so f satisfies the Lipschitz condition with constant $K < 1$.

As an example, we show that the equation

$$4x^5 - 2x^2 - 4x + 1 = 0$$

has precisely one root in the interval $[0, \frac{1}{2}]$. Introduce the function f, where

$$f(x) = x^5 - \tfrac{1}{2}x^2 + \tfrac{1}{4}, \quad 0 \leqslant x \leqslant \tfrac{1}{2}.$$

The given equation is equivalent to the equation $f(x) = x$, so we seek information about the fixed points, if any, of f. The domain of f is $[0, \frac{1}{2}]$. Its range is shown as follows to be a subset of $[0, \frac{1}{2}]$. We have, for $0 < x < \frac{1}{2}$, $f'(x) = 5x^4 - x$. This is 0 when $x = 1/\sqrt[3]{5}$, and we calculate that $f(0)$, $f(1/\sqrt[3]{5})$ and $f(\frac{1}{2})$ all lie in $[0, \frac{1}{2}]$. Also

$$|f'(x)| = |5x^4 - x| \leqslant 5x^4 + x \leqslant \tfrac{5}{16} + \tfrac{1}{2} < 1$$

for all x in $[0, \frac{1}{2}]$. All the required conditions are met, so f has a single fixed point, and this is the required root of the original equation.

To find the root, we can take $x_0 = 0$. The first three iterates are $x_1 = f(0) = 0.25$, $x_2 = f(0.25) \doteq 0.2197$, $x_3 = f(0.2197) \doteq 0.2264$, and the next three are $x_4 \doteq 0.2250$, $x_5 \doteq 0.2253$, $x_6 \doteq 0.2252$. (The use of the symbol $\doteq$ implies that the result is given correct to the number of decimal places appearing on the right of the symbol.) To three decimal places, the root is 0.225.

(2) Consider the system

$$\sum_{k=1}^{n} a_{jk}x_k = b_j, \quad j = 1, 2, \ldots, n,$$

of n linear equations in n unknowns $x_1, x_2, \ldots, x_n$, where a_{jk} and b_j are real numbers for each j and k. Introducing the $n \times n$ matrix $A = (a_{jk})$ and the column vectors $x = (x_1, x_2, \ldots, x_n)^T$, $b = (b_1, b_2, \ldots, b_n)^T$, the system can be written in matrix form as $Ax = b$, and must be solved for x. Letting $C = (c_{jk})$ be the matrix $I - A$, where I is the $n \times n$ identity matrix, this equation may be written $(I - C)x = b$, or

$$Cx + b = x.$$

Considering the elements of $\mathbf{R}^n$ to be column vectors, we define a mapping $M\colon \mathbf{R}^n \to \mathbf{R}^n$ by

$$Mx = Cx + b,$$

so that our matrix equation is replaced by the equation

$$Mx = x.$$

Hence the solutions of the original system are related to the fixed points of the mapping M. Since $\mathbf{R}^n$ is a complete metric space, there will be just one solution if M is a contraction mapping.

Let $y = (y_1, y_2, \ldots, y_n)^T$ and $z = (z_1, z_2, \ldots, z_n)^T$ be two points of $\mathbf{R}^n$ and let d denote the Euclidean metric:

$$d(y, z) = \sqrt{\sum_{j=1}^{n}(y_j - z_j)^2}.$$

Since My is the vector $Cy + b$, with jth component $\sum_{k=1}^{n} c_{jk}y_k + b_j$ ($j = 1, 2, \ldots, n$), and similarly for Mz, we have

$$d(My, Mz) = \sqrt{\sum_{j=1}^{n}\left(\left(\sum_{k=1}^{n} c_{jk}y_k + b_j\right) - \left(\sum_{k=1}^{n} c_{jk}z_k + b_j\right)\right)^2}$$

$$= \sqrt{\sum_{j=1}^{n}\left(\sum_{k=1}^{n} c_{jk}(y_k - z_k)\right)^2}$$

$$\leqslant \sqrt{\sum_{j=1}^{n}\left(\left(\sum_{k=1}^{n} c_{jk}^2\right)\left(\sum_{k=1}^{n}(y_k - z_k)^2\right)\right)},$$

333333333

3333333333

by the Cauchy–Schwarz inequality, Theorem 2.2.1. Thus we have

$$d(My, Mz) \leqslant \sqrt{\sum_{j=1}^{n}\sum_{k=1}^{n} c_{jk}^2} \cdot d(y,z),$$

so certainly M will be a contraction if

$$0 < \sum_{j=1}^{n}\sum_{k=1}^{n} c_{jk}^2 < 1.$$

In terms of the original matrix A, this condition requires that a_{jk} be near 0 when $j \neq k$ and near 1 when $j = k$.

Different sufficient conditions for M to be a contraction can be obtained by choosing different metrics on the set $\mathbf{R}^n$, as long as the resulting metric space is complete. We are totally free to take whichever metric best serves our purpose. For instance, with the metric d_∞, where

$$d_\infty(y,z) = \max_{1\leqslant k\leqslant n} |y_k - z_k|,$$

we know that $(\mathbf{R}^n, d_\infty)$ is complete (Exercise 2.9(7)), and

$$d_\infty(My, Mz) = \max_{1\leqslant j\leqslant n}\left|\left(\sum_{k=1}^{n} c_{jk}y_k + b_j\right) - \left(\sum_{k=1}^{n} c_{jk}z_k + b_j\right)\right|$$

$$= \max_{1\leqslant j\leqslant n}\left|\sum_{k=1}^{n} c_{jk}(y_k - z_k)\right|$$

$$\leqslant \max_{1\leqslant j\leqslant n}\sum_{k=1}^{n} |c_{jk}|\,|y_k - z_k|$$

$$\leqslant \max_{1\leqslant j\leqslant n}\sum_{k=1}^{n} |c_{jk}| \cdot \max_{1\leqslant k\leqslant n} |y_k - z_k|$$

$$= \max_{1\leqslant j\leqslant n}\sum_{k=1}^{n} |c_{jk}| \cdot d_\infty(y,z),$$

so that M will be a contraction under this metric if

$$0 < \max_{1\leqslant j\leqslant n}\sum_{k=1}^{n} |c_{jk}| < 1,$$

that is, if the sums of the absolute values of the elements in the rows of C are all less than 1 (and C has at least one nonzero element).

A third condition is obtained in Exercise 3.5(3). It only takes one of the conditions to be satisfied to ensure the existence of the unique fixed point.

Once M is known to be a contraction, its fixed point can be found, at least approximately, by iteration. If x_0 is any column vector, then we have successively

$$x_1 = Mx_0 = Cx_0 + b,$$
$$x_2 = Mx_1 = Cx_1 + b = C(Cx_0 + b) + b = C^2x_0 + Cb + b,$$
$$x_3 = Mx_2 = Cx_2 + b = C^3x_0 + C^2b + Cb + b,$$

and so on, the sequence $\{x_n\}$ converging to the unique solution x of $Ax = b$, where $A = I - C$. There are of course other tests for whether a system of linear equations has solutions, and other methods of finding them. However, the above is very simple. The tests essentially require only the operation of addition on the elements of C or their squares, and, if either condition is satisfied, the solution may be obtained to any desired degree of accuracy (subject to computational precision) in terms of powers of C. There is no need to determine the rank, determinant or inverse of any matrix . It must be realised, though, that we have only obtained sufficient conditions: if none of the conditions is met, solutions may still exist.

As a simple example, consider the system of equations

$$16x - 3y + 4z = \ \ 7,$$
$$6x + 7y - 4z = \ \ 4,$$
$$y + 4z = 15.$$

Dividing the equations respectively by 16, 8 and 4 gives the equivalent system

$$x - \tfrac{3}{16}y + \tfrac{1}{4}z = \tfrac{7}{16},$$
$$\tfrac{3}{4}x + \tfrac{7}{8}y - \tfrac{1}{2}z = \tfrac{1}{2},$$
$$\tfrac{1}{4}y + \ \ z = \tfrac{15}{4}.$$

In the notation above, we have

$$A = \begin{pmatrix} 1 & -\tfrac{3}{16} & \tfrac{1}{4} \\ \tfrac{3}{4} & \tfrac{7}{8} & -\tfrac{1}{2} \\ 0 & \tfrac{1}{4} & 1 \end{pmatrix}, \quad C = I - A = \begin{pmatrix} 0 & \tfrac{3}{16} & -\tfrac{1}{4} \\ -\tfrac{3}{4} & \tfrac{1}{8} & \tfrac{1}{2} \\ 0 & -\tfrac{1}{4} & 0 \end{pmatrix},$$

and we find that the sum of the squares of the elements of C is $\tfrac{253}{256}$, less than 1, so our system possesses a unique solution which may be found by iteration. Notice that, in this example, the sum of the absolute values of the elements in the second row of C is $\frac{3}{4} + \frac{1}{8} + \frac{1}{2} = \frac{11}{8}$, so our second condition is of no use here, though the first is. However, other examples can be constructed where the reverse is true.

(3)   Our third application of the fixed point theorem is to prove an important theorem on the existence of a solution to the first-order differential equation

$$\frac{dy}{dx} = f(x, y)$$

with initial condition $y = y_0$ when $x = x_0$. The result is a form of *Picard's theorem*.

Two conditions are imposed on f: first, f is continuous in some rectangle $\{(x, y) : |x - x_0| \leqslant a, \ |y - y_0| \leqslant b\}$; second, f satisfies a Lipschitz condition on y, uniformly in x, in the rectangle. The latter means that there is a positive constant K such that

$$|f(x, y_1) - f(x, y_2)| \leqslant K|y_1 - y_2|$$

for any x in $[x_0 - a, x_0 + a]$ and any y_1, y_2 in $[y_0 - b, y_0 + b]$. Since f is continuous in the rectangle, it must be bounded there (see Section 1.9), so there is a positive constant M such that $|f(x, y)| \leqslant M$.

Under these conditions, we will prove that there is a positive number h such that in $[x_0 - h, x_0 + h]$ there is a unique solution to the differential equation.

Write the differential equation equivalently in integral form as

$$y(x) = y_0 + \int_{x_0}^{x} f(t, y) \, dt,$$

incorporating the initial condition. Let h be a number satisfying

$$h > 0, \quad h < \frac{1}{K}, \quad h \leqslant a, \quad h \leqslant \frac{b}{M}.$$

Denote by J the closed interval $[x_0 - h, x_0 + h]$ and write $C[J]$ for $C[x_0 - h, x_0 + h]$. Let F be the subset of $C[J]$ consisting of continuous functions defined on J for which

$$|y(x) - y_0| \leqslant b, \quad x \in J, \ y \in C[J].$$

Referring to Figure 9, F is the set of all continuous functions with graphs in the shaded rectangle. Impose the uniform metric on F, so that F becomes a subspace of the complete metric space $C[J]$.

We will show that F is a closed subspace, so that, by Theorem 2.7.3,

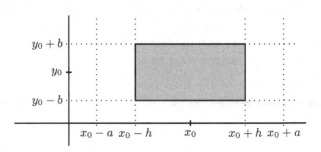

Figure 9

F is a complete metric space. Let $\{y_n\}$ be a sequence of functions in F which, as a sequence in $C[J]$, converges. Write $y = \lim y_n$ (so $y \in C[J]$). By definition of the uniform metric, given $\epsilon > 0$ we can find a positive integer N such that

$$\max_{x \in J} |y_n(x) - y(x)| < \epsilon, \quad n > N.$$

Also, for each $x \in J$ and each $n \in \mathbf{N}$,

$$|y_n(x) - y_0| \leqslant b.$$

Hence, for each $x \in J$, and $n > N$,

$$|y(x) - y_0| \leqslant |y(x) - y_n(x)| + |y_n(x) - y_0| < \epsilon + b.$$

But ϵ is arbitrary, so we must have

$$|y(x) - y_0| \leqslant b$$

for all $x \in J$. This shows that $y \in F$, so F is a closed subspace of $C[J]$.

Now define a mapping A on F by the equation $Ay = z$, where $y \in F$ and

$$z(x) = y_0 + \int_{x_0}^{x} f(t, y(t)) \, dt, \quad x \in J.$$

We will show that $z \in F$ and that A is a contraction mapping. Then the fixed point theorem will imply that A has a unique fixed point. That is, we will have shown the existence of a unique function $y \in F$ such that $Ay = y$, which means

$$y(x) = y_0 + \int_{x_0}^{x} f(t, y(t)) \, dt, \quad x \in J.$$

This will complete the proof of the existence on J of a unique solution of our differential equation.

To show that $z \in F$, we see that, for $x \in J$,

$$|z(x) - y_0| = \left| \int_{x_0}^{x} f(t, y) \, dt \right|$$

$$\leqslant \left| \int_{x_0}^{x} |f(t, y)| \, dt \right| \leqslant M|x - x_0| \leqslant Mh \leqslant b.$$

Thus $z \in F$ (so A maps F into itself). To show that A is a contraction, take $y, \widetilde{y} \in F$. Set $z = Ay$, $\widetilde{z} = A\widetilde{y}$. Let d denote the uniform metric. Then, for $x \in J$,

$$|z(x) - \widetilde{z}(x)| = \left| \int_{x_0}^{x} (f(t, y) - f(t, \widetilde{y})) \, dt \right|$$

$$\leqslant \left| \int_{x_0}^{x} |f(t, y) - f(t, \widetilde{y})| \, dt \right|$$

$$\leqslant K \left| \int_{x_0}^{x} |y(t) - \widetilde{y}(t)| \, dt \right|$$

$$\leqslant K \cdot \max_{x \in J} |y(x) - \widetilde{y}(x)| \cdot |x - x_0| \leqslant Khd(y, \widetilde{y}).$$

We then have

$$d(z, \widetilde{z}) = \max_{x \in J} |z(x) - \widetilde{z}(x)| \leqslant Khd(y, \widetilde{y}).$$

That is, $d(Ay, A\widetilde{y}) \leqslant \alpha d(y, \widetilde{y})$, where $\alpha = Kh$. But $0 < \alpha < 1$ and so A is a contraction.

It is easy to check that this result may be applied successfully to, for example, the linear first-order differential equation

$$\frac{dy}{dx} + P(x)y = Q(x), \quad y(x_0) = y_0,$$

to ensure a unique solution in some interval about x_0, provided the functions P and Q are continuous.

An example of a differential equation where it cannot be applied is the equation

$$\frac{dy}{dx} = 2|y|^{1/2}, \quad y(0) = 0.$$

It is impossible to satisfy the Lipschitz condition for small values of $|y|$: the inequality $||y_1|^{1/2} - |y_2|^{1/2}| \leqslant K|y_1 - y_2|$ cannot hold for any constant K if we take $y_2 = 0$ and $|y_1| < 1/K^2$. In fact, this equation has at least two solutions for x in any interval containing 0. These are the functions defined by the equations

$$y = 0 \quad \text{and} \quad y = \begin{cases} x^2, & x \geqslant 0, \\ -x^2, & x < 0. \end{cases}$$

(4) The differential equation in (3) was considered by first transforming it into an integral equation. We intend now to study two standard types of integral equations, in each case obtaining conditions which ensure a unique solution.

(a) Any equation of the form

$$x(s) = \lambda \int_a^b k(s,t)x(t)\,dt + f(s), \quad a \leqslant s \leqslant b,$$

involving two given functions k (of two variables) and f, an unknown function x, and a nonzero constant λ, is called a *Fredholm integral equation* (of the second kind).

Suppose f is continuous on the interval $[a,b]$, and k is continuous on the square $[a,b] \times [a,b]$. Then k is bounded: there exists a positive constant M so that, in the square, $|k(s,t)| \leqslant M$.

Take any continuous function x on $[a,b]$ and define a mapping A on $C[a,b]$ by $y = Ax$, where

$$y(s) = \lambda \int_a^b k(s,t)x(t)\,dt + f(s).$$

We will obtain a condition for A to be a contraction. Note in the first place that, since k, x and f are continuous, so is y, and so indeed A maps the complete metric space $C[a,b]$ into itself. Now, if d denotes the uniform metric of $C[a,b]$, and if $y_1 = Ax_1$, $y_2 = Ax_2$ $(x_1, x_2 \in C[a,b])$, then

$$
\begin{aligned}
d(y_1, y_2) &= \max_{a \leqslant s \leqslant b} |y_1(s) - y_2(s)| \\
&= \max_{a \leqslant s \leqslant b} \left| \lambda \int_a^b k(s,t)(x_1(t) - x_2(t))\,dt \right| \\
&\leqslant |\lambda| \cdot \max_{a \leqslant s \leqslant b} \int_a^b |k(s,t)|\,|x_1(t) - x_2(t)|\,dt \\
&\leqslant |\lambda| M(b-a) \cdot \max_{a \leqslant s \leqslant b} |x_1(s) - x_2(s)| \\
&= |\lambda| M(b-a) d(x_1, x_2),
\end{aligned}
$$

and hence A is a contraction mapping provided

$$|\lambda| < \frac{1}{M(b-a)}.$$

Thus, provided the constant λ satisfies this inequality, we are assured that the original Fredholm integral equation has a unique solution. This solution may be found by iteration, taking any function in $C[a,b]$ as starting point.

As an example, consider the equation

$$x(s) = \frac{1}{2} \int_0^1 stx(t)\,dt + \frac{5s}{6}.$$

In the above notation, $\lambda = \frac{1}{2}$, $a = 0$, $b = 1$, $k(s,t) = st$, $f(s) = 5s/6$. For $s,t \in [0,1]$, we have $|k(s,t)| = st \leqslant 1$, so take $M = 1$. The inequality for λ is satisfied, so a unique solution is assured. To find it, let us take as starting point the function x_0 where $x_0(s) = 1$, $0 \leqslant s \leqslant 1$. Then we obtain

$$x_1(s) = \frac{1}{2} \int_0^1 st\,dt + \frac{5s}{6} = \frac{13s}{12},$$

$$x_2(s) = \frac{1}{2} \int_0^1 st \frac{13t}{12}\,dt + \frac{5s}{6} = \frac{73s}{72},$$

$$x_3(s) = \frac{1}{2} \int_0^1 st \frac{73t}{72}\,dt + \frac{5s}{6} = \frac{433s}{432},$$

and we are led to suggest

$$x_n(s) = \frac{2 \cdot 6^n + 1}{2 \cdot 6^n}\, s, \quad n \in \mathbf{N}.$$

This should be verified by mathematical induction. The solution of the integral equation is $\lim x_n$: the function x, where $x(s) = s$, $0 \leqslant s \leqslant 1$.

(b) An equation of the form

$$x(s) = \lambda \int_a^s k(s,t)x(t)\,dt + f(s), \quad a \leqslant s \leqslant b,$$

where k, f and λ are as before, is called a *Volterra integral equation* (of the second kind). Note that the upper limit on the integral is now variable. We impose the same conditions on k and f, and give M the same meaning, and will show that this time there is a solution for all values of λ, rather than only for sufficiently small values of $|\lambda|$.

Let B be the mapping of $C[a, b]$ into itself defined by $Bx = y$, where $x \in C[a, b]$ and

$$y(s) = \lambda \int_a^s k(s, t)x(t)\, dt + f(s).$$

Again it is clear that the fixed points of B are solutions of Volterra's equation. We will show that a positive integer n exists such that B^n is a contraction. Then, by Theorem 3.2.4, B will have a unique fixed point.

Take any functions $x_1, x_2 \in C[a, b]$. We show by induction that, for $s \in [a, b]$ and $n \in \mathbf{N}$,

$$|(B^n x_1)(s) - (B^n x_2)(s)| \leqslant |\lambda|^n M^n \frac{(s-a)^n}{n!} \max_{a \leqslant t \leqslant b} |x_1(t) - x_2(t)|.$$

Certainly, the statement is true when $n = 1$, for then

$$|(Bx_1)(s) - (Bx_2)(s)| = \left| \lambda \int_a^s k(s, t)(x_1(t) - x_2(t))\, dt \right|$$

$$\leqslant |\lambda| \int_a^s |k(s, t)|\, |x_1(t) - x_2(t)|\, dt$$

$$\leqslant |\lambda| M \cdot \max_{a \leqslant t \leqslant s} |x_1(t) - x_2(t)| \cdot \int_a^s dt$$

$$\leqslant |\lambda| M \cdot \max_{a \leqslant t \leqslant b} |x_1(t) - x_2(t)| \cdot (s - a).$$

Now assume the statement to be true when $n = j$. Then

$$|(B^{j+1} x_1)(s) - (B^{j+1} x_2)(s)|$$

$$= |(B(B^j x_1))(s) - (B(B^j x_2))(s)|$$

$$= \left| \lambda \int_a^s k(s, t)((B^j x_1)(t) - (B^j x_2)(t))\, dt \right|$$

$$\leqslant |\lambda| \int_a^s |k(s, t)|\, |(B^j x_1)(t) - (B^j x_2)(t)|\, dt$$

$$\leqslant |\lambda| M \frac{|\lambda|^j M^j}{j!} \cdot \max_{a \leqslant t \leqslant b} |x_1(t) - x_2(t)| \cdot \int_a^s (t - a)^j\, dt$$

$$= \frac{|\lambda|^{j+1} M^{j+1}}{j!} \cdot \max_{a \leqslant t \leqslant b} |x_1(t) - x_2(t)| \cdot \frac{1}{j+1} (s - a)^{j+1},$$

and thus it is true also when $n = j + 1$. This concludes the induction. We can now infer that

$$|(B^n x_1)(s) - (B^n x_2)(s)| \leqslant |\lambda|^n M^n \frac{(b-a)^n}{n!} \max_{a \leqslant t \leqslant b} |x_1(t) - x_2(t)|,$$

since $s \leqslant b$. Therefore, if d is the uniform metric, we have

$$d(B^n x_1, B^n x_2) \leqslant \frac{(|\lambda| M(b-a))^n}{n!} d(x_1, x_2)$$

for all $n \in \mathbf{N}$. Choose n so large that $(|\lambda| M(b-a))^n < n!$. (This may always be done since the sequence $\{c^n/n!\}$ converges to 0 for any real number c. One way to see this is to note that the series $\sum_{k=0}^{\infty} c^k/k!$ converges for any c (to e^c), and then to apply Theorem 1.8.3.) For such a value of n, B^n is a contraction and hence B has a unique fixed point. Thus, regardless of the value of λ, the Volterra equation always has a unique solution.

As before, this solution can be found by iteration. The successive approximations are $x_0, x_1, x_2, \ldots$, where

$$x_n(s) = \lambda \int_a^s k(s,t) x_{n-1}(t)\, dt + f(s), \quad n \in \mathbf{N},$$

with any function in $C[a,b]$ chosen as x_0.

The fact that the Volterra equation always has a unique solution implies a simple proof of another important existence theorem in the study of differential equations. Let p, q and g be any functions in $C[a,b]$ and let α and β be any real numbers. We will show that there exists exactly one function y defined on $[a,b]$ with a continuous second-order derivative such that

$$y''(x) + p(x)y'(x) + q(x)y(x) = g(x), \quad a \leqslant x \leqslant b,$$

and satisfying $y(a) = \alpha$ and $y'(a) = \beta$.

For the proof, we suppose at first that there is such a function y. Let u be any number in $[a,b]$. Then, defining a function $z \in C[a,b]$ by $z = y''$, we have

$$\int_a^u z(t)\, dt = y'(u) - y'(a) = y'(u) - \beta$$

and

$$\int_a^x \int_a^u z(t)\, dt\, du = \int_a^x (y'(u) - \beta)\, du$$
$$= y(x) - y(a) - \beta(x-a) = y(x) - \alpha - \beta(x-a).$$

But, inverting the order of integration,

$$\int_a^x \int_a^u z(t)\, dt\, du = \int_a^x \int_t^x z(t)\, du\, dt = \int_a^x z(t)(x-t)\, dt,$$

so that

$$y(x) = \alpha + \beta(x - a) + \int_a^x z(t)(x - t)\,dt.$$

Since y is assumed to satisfy the original differential equation, we have, substituting back y, y' and y'',

$$z(x) + p(x)\left(\beta + \int_a^x z(t)\,dt\right)$$
$$+ q(x)\left(\alpha + \beta(x - a) + \int_a^x z(t)(x - t)\,dt\right) = g(x).$$

This can be written

$$z(x) = \int_a^x (-p(x) - q(x)(x - t))z(t)\,dt$$
$$+ g(x) - \beta p(x) - q(x)(\alpha + \beta(x - a)),$$

which has the form

$$z(x) = \int_a^x k(x, t)z(t)\,dt + f(x),$$

a Volterra equation of the type we have considered. Now, working backwards, if z is the unique solution of this Volterra equation then the function y, where

$$y(x) = \alpha + \beta(x - a) + \int_a^x z(t)(x - t)\,dt,$$

is the required unique solution of the differential equation with its given initial conditions.

3.4 Perturbation mappings

Often in applications it is necessary to approximate a mapping in some way. For example, as we will see, a desirable property of mappings (on vector spaces) is linearity, so that it is common to approximate a nonlinear mapping by a linear one. We will now investigate the errors that can arise when a contraction mapping is uniformly approximated by another mapping in the following sense. Let (X, d) be a metric space and let A be a mapping from X into itself. We call a mapping $\widetilde{A}$ on X a *perturbation*, or *uniform approximation*, of the mapping A if there is some number $\epsilon > 0$ such that

$$d(\widetilde{A}w, Aw) \leqslant \epsilon$$

for all $w \in X$.

Suppose A is a contraction with contraction constant α (so $0 < \alpha < 1$). Let (X, d) be complete, so that, by the fixed point theorem, A has a unique fixed point, x say. Choose any point $x_0 \in X$, set $\tilde{x}_0 = x_0$, and define sequences $\{x_n\}$ and $\{\tilde{x}_n\}$ in X by $x_n = Ax_{n-1}$, $\tilde{x}_n = \tilde{A}\tilde{x}_{n-1}$, where $\tilde{A}$ is the above perturbation of A. Write $\tilde{d}_n = d(\tilde{x}_n, \tilde{x}_{n+1})$ for $n = 0, 1, 2, \ldots$, and set $\delta = \epsilon/(1 - \alpha)$. This sets up the notation for the following result, which has a number of uses in the field of numerical analysis.

Theorem 3.4.1 *In the notation above,*

(a) $d(x_n, \tilde{x}_n) < \delta$ *for* $n = 0, 1, 2, \ldots,$

(b) $d(x, \tilde{x}_1) \leqslant 2\delta + (3 - \alpha)\tilde{d}_0/(1 - \alpha)$,

(c) *for any number* $c > 0$, *we can find a positive integer* N *so that* $d(x, \tilde{x}_n) \leqslant \delta + c$ *when* $n \geqslant N$.

Each of these should be interpreted 'in words'. For instance, (c) says that the sequence of iterates under $\tilde{A}$ can be brought to within a distance δ, in effect, of the fixed point of A by continuing long enough. Note that the starting point of the iteration for $\tilde{A}$ is still arbitrary and that we say nothing at all about the existence of $\lim \tilde{x}_n$. The proofs use little beyond the triangle inequality. We give them in turn.

(a) We use induction to prove that

$$d(x_n, \tilde{x}_n) \leqslant (1 + \alpha + \cdots + \alpha^{n-1})\epsilon, \quad n \in \mathbf{N}.$$

The statement is certainly true when $n = 1$, for

$$d(x_1, \tilde{x}_1) = d(Ax_0, \tilde{A}\tilde{x}_0) = d(Ax_0, \tilde{A}x_0) \leqslant \epsilon,$$

since $x_0 = \tilde{x}_0$. Now suppose it is true when $n = k$. Then

$$
\begin{aligned}
d(x_{k+1}, \tilde{x}_{k+1}) &= d(Ax_k, \tilde{A}\tilde{x}_k) \leqslant d(Ax_k, A\tilde{x}_k) + d(A\tilde{x}_k, \tilde{A}\tilde{x}_k) \\
&\leqslant \alpha d(x_k, \tilde{x}_k) + \epsilon \leqslant \alpha(1 + \alpha + \cdots + \alpha^{k-1})\epsilon + \epsilon \\
&= (1 + \alpha + \alpha^2 + \cdots + \alpha^k)\epsilon,
\end{aligned}
$$

so the inequality holds also when $n = k + 1$. Hence it is true for all positive integers n. But then, since $0 < \alpha < 1$,

$$d(x_n, \tilde{x}_n) < (1 + \alpha + \alpha^2 + \cdots)\epsilon = \frac{\epsilon}{1 - \alpha} = \delta,$$

for all such n, and $d(x_0, \tilde{x}_0) = 0 < \delta$.

(b) Using a result in the proof of the fixed point theorem (Theorem 3.2.3), we have

$$d(x_k, x_{k-1}) \leqslant \alpha^{k-1} d(Ax_0, x_0) = \alpha^{k-1} d(x_1, x_0) \leqslant \alpha^{k-1}(\tilde{d}_0 + \epsilon)$$

for any integer $k > 1$, and this is true also when $k = 1$ since

$$d(x_0, x_1) \leqslant d(x_0, \tilde{x}_1) + d(\tilde{x}_1, x_1)$$
$$= d(\tilde{x}_0, \tilde{x}_1) + d(\tilde{A}x_0, Ax_0) \leqslant \tilde{d}_0 + \epsilon.$$

In the proof of the fixed point theorem, we also deduced that

$$d(x_n, x_m) < \frac{\alpha^m}{1 - \alpha} d(x_1, x_0),$$

where m and n are any positive integers, with $m < n$. For fixed m, the real-valued sequence $\{d(x_n, x_m)\}_{n=1}^{\infty}$ converges to $d(x, x_m)$, since $x_n \to x$, and using Exercise 2.9(1). Hence, making use of Theorem 1.7.7,

$$d(x, x_m) \leqslant \frac{\alpha^m}{1 - \alpha} d(x_1, x_0) \leqslant \frac{\alpha^m}{1 - \alpha} (\tilde{d}_0 + \epsilon)$$

for any integer $m \in \mathbf{N}$. Now,

$$d(x, \tilde{x}_1) \leqslant d(x, x_m) + d(x_m, x_{m-1}) + \cdots + d(x_1, x_0) + d(x_0, \tilde{x}_1)$$
$$\leqslant \frac{\alpha^m}{1 - \alpha} (\tilde{d}_0 + \epsilon) + (\alpha^{m-1} + \alpha^{m-2} + \cdots + \alpha + 1)(\tilde{d}_0 + \epsilon) + \tilde{d}_0$$
$$< \frac{1}{1 - \alpha} (\tilde{d}_0 + \epsilon) + \frac{1}{1 - \alpha} (\tilde{d}_0 + \epsilon) + \tilde{d}_0$$
$$= \frac{2}{1 - \alpha} (\tilde{d}_0 + \epsilon) + \tilde{d}_0 = \frac{3 - \alpha}{1 - \alpha} \tilde{d}_0 + 2\delta.$$

(c) Using (a) and a result from the proof of (b), we have, for $n = 0$, 1, 2, ...,

$$d(x, \tilde{x}_n) \leqslant d(x, x_n) + d(x_n, \tilde{x}_n) \leqslant \frac{\alpha^n}{1 - \alpha} (\tilde{d}_0 + \epsilon) + \delta.$$

Then, no matter how small c is, we may choose n so large that

$$\frac{\tilde{d}_0 + \epsilon}{1 - \alpha} \alpha^n \leqslant c,$$

since $\alpha < 1$.

This ends the proof.   □

In problems involving perturbation mappings it is generally convenient to arrange matters so that the mapping works in a closed proper subspace of a complete metric space. Such a subspace is complete (Theorem 2.7.3) so the fixed point theorem and the preceding results are still true when applied to the elements of the subspace. Doing this allows further estimation of the size of those elements.

To illustrate the use of a perturbation mapping, we will show how a certain type of nonlinear integral equation can be solved approximately by relating it to a Fredholm integral equation.

The nonlinear integral equation that we will consider is

$$x(s) = \lambda \int_a^b k(s, t, x(t)) \, dt + \mu f(s), \quad a \leqslant s \leqslant b,$$

where λ and μ are real constants, with $0 < |\mu| < 1$ and λ as yet unqualified, f is a continuous function on $[a, b]$ with $|f(t)| \leqslant H$ for some number H and all t in $[a, b]$, and k is a continuous function of three variables satisfying a Lipschitz condition in the third variable, uniformly in the others:

$$|k(s, t, u_1) - k(s, t, u_2)| \leqslant M|u_1 - u_2|$$

for some number M, all $s, t \in [a, b]$ and any $u_1, u_2 \in [-H, H]$. We will suppose that the function k has the special form

$$k(s, t, u) = (g(s, t) + \theta(s, t, u))u, \quad |\theta(s, t, u)| < \eta,$$

where g and θ are continuous, and η is some (small) positive number. Notice that u is not an independent variable, but rather $u = x(t)$, where x is the unknown function, and $t \in [a, b]$ is independent. The solution x of the integral equation is required to satisfy $|x(t)| \leqslant H$ for all t in $[a, b]$. As in Application 3.3(1), the Lipschitz condition will be satisfied when $\partial k / \partial u$ exists and $|\partial k(s, t, u)/\partial u| \leqslant M$ for all $s, t \in [a, b]$ and $u \in [-H, H]$.

The above suggests that we work in the complete metric space $C[a, b]$, but restrict ourselves to the subspace F of $C[a, b]$ consisting of continuous functions x for which $|x(t)| \leqslant H$, $a \leqslant t \leqslant b$. Exactly as in Application 3.3(3), F can be shown to be a closed subspace of $C[a, b]$, so F is a complete metric space.

Define the mapping A on F by $Ax = y$ ($x \in F$) where

$$y(s) = \lambda \int_a^b k(s, t, x(t)) \, dt + \mu f(s).$$

The fixed points of A, if any, are our required solutions of the integral equation.

We prove first that, if

$$|\lambda| \leqslant \frac{1 - |\mu|}{M(b-a)},$$

then A maps F into itself. To see this, note in the first place that, since $k(s,t,u) = (g(s,t) + \theta(s,t,u))u$, we have $k(s,t,0) = 0$ for $s,t \in [a,b]$. Then, by the Lipschitz condition (with $u_1 = u$ and $u_2 = 0$), we have $|k(s,t,u)| \leqslant M|u|$ for all s, t, u. Now, take $x \in F$ and put $y = Ax$. Then, for all $s \in [a,b]$, we have

$$
\begin{aligned}
|y(s)| &= \left| \lambda \int_a^b k(s,t,x(t))\, dt + \mu f(s) \right| \\
&\leqslant |\lambda| \int_a^b |k(s,t,x(t))|\, dt + |\mu f(s)| \\
&\leqslant |\lambda| M \int_a^b |x(t)|\, dt + |\mu|\,|f(s)| \\
&\leqslant |\lambda| M H(b-a) + |\mu| H \\
&= (|\lambda| M(b-a) + |\mu|) H \leqslant H
\end{aligned}
$$

if $|\lambda| M(b-a) + |\mu| \leqslant 1$, as stated.

We next prove that this condition on λ implies further that A is a contraction mapping. Let d denote the uniform metric of F. If $x_1, x_2 \in F$ and $Ax_1 = y_1$, $Ax_2 = y_2$, then, for any $s \in [a,b]$,

$$
\begin{aligned}
|y_1(s) - y_2(s)| &= \left| \lambda \int_a^b (k(s,t,x_1(t)) - k(s,t,x_2(t)))\, dt \right| \\
&\leqslant |\lambda| \int_a^b |k(s,t,x_1(t)) - k(s,t,x_2(t))|\, dt \\
&\leqslant |\lambda| M \int_a^b |x_1(t) - x_2(t)|\, dt \\
&\leqslant |\lambda| M \cdot \max_{a \leqslant t \leqslant b} |x_1(t) - x_2(t)| \cdot (b-a)
\end{aligned}
$$

and in particular we have

$$d(y_1, y_2) = d(Ax_1, Ax_2) \leqslant |\lambda| M(b-a) d(x_1, x_2).$$

But $|\lambda| M(b-a) \leqslant 1 - |\mu| < 1$, and we may assume that $\lambda \neq 0$, so A is a contraction.

The fixed point theorem now implies that there is a unique solution in F of our nonlinear integral equation, provided $0 < |\mu| < 1$ and $|\lambda| \leqslant (1 - |\mu|)/M(b-a)$. This solution may be found by iteration, but conceivably this could be very difficult. Fredholm equations are much easier to handle, and so we introduce the mapping $\widetilde{A}$ by $\widetilde{A}x = y$ where $x \in F$ and

$$y(s) = \lambda \int_a^b g(s,t)x(t)\,dt + \mu f(s),$$

with a, b, λ, μ, f and g as above. By our definition, $\widetilde{A}$ is a perturbation of A, for, if $x \in F$,

$$d(Ax, \widetilde{A}x) = \max_{a \leqslant s \leqslant b} \left| \lambda \int_a^b (k(s,t,x(t)) - g(s,t)x(t))\,dt \right|$$

$$\leqslant |\lambda| \cdot \max_{a \leqslant s \leqslant b} \int_a^b |\theta(s,t,x(t))|\,|x(t)|\,dt$$

$$\leqslant |\lambda|\eta H(b-a).$$

Hence in Theorem 3.4.1 we take $\epsilon = |\lambda|\eta H(b-a)$ and $\alpha = |\lambda|M(b-a)$, so that

$$\delta = \frac{\epsilon}{1-\alpha} = \frac{|\lambda|\eta H(b-a)}{1 - |\lambda|M(b-a)}.$$

We notice of course that δ is small if η is. Thus we can solve a nonlinear integral equation which is 'almost' a Fredholm equation by solving that Fredholm equation. This stands to reason. But we have done more. We have a precise estimate of the errors involved in the process. One interesting point remains to be stressed. There is nothing in the above that says that $\widetilde{A}$ is a contraction mapping, so that although we use the iterates under $\widetilde{A}$ to approximate the fixed point of A, there may in fact be no fixed points of $\widetilde{A}$, or there may be many!

3.5 Exercises

(1) Refer to Application 3.3(1). The figure below shows the graphs of two differentiable functions defined on an interval $[a,b]$ and having ranges in $[a,b]$. In (a), the function f is such that, for some constant K and all $x \in [a,b]$, $0 < f'(x) \leqslant K < 1$; in (b), f is such that $-1 < -K \leqslant f'(x) < 0$ in $[a,b]$. Reproduce the diagrams.

Set $x_0 = a$ and in each case sketch a scheme by which the iterates $x_1 = f(x_0)$, $x_2 = f(x_1)$, $x_3 = f(x_2)$, ... may be seen to approach the fixed point of f (which is the x-coordinate of the point of intersection of $y = x$ and $y = f(x)$). Sketch other figures to show the possible nonexistence of a fixed point when the range of f is not a subset of $[a, b]$, or the possible existence of many fixed points when the condition $|f'(x)| \leqslant K < 1$ is violated.

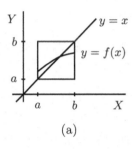

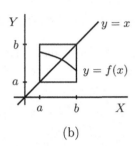

(a)   (b)

(2) In the following, show that the given equation has a unique root in the given interval.

(a) $x^4 + 8x^3 + 32x - 32 = 0$, $[0, 1]$

(b) $\sin x + 2 \sinh x - 8x + 2 = 0$, $[0, \frac{1}{2}\pi]$

In (b), use a calculator to approximate the first four iterates to the root, starting with $x_0 = 0$ and using the method of successive approximations.

(3) For the set $\mathbf{R}^n$ of n-tuples of real numbers, let the metric be $d_1(x, y) = \sum_{k=1}^{n} |x_k - y_k|$, where $x = (x_1, x_2, \ldots, x_n)$ and $y = (y_1, y_2, \ldots, y_n)$ are points of $\mathbf{R}^n$. (See Example 2.2(5).) Show that this defines a complete metric space $\mathbf{R}_1$, say. Define a mapping M from $\mathbf{R}_1$ into $\mathbf{R}_1$ by $y = Mx$, where $x \in \mathbf{R}_1$ and

$$ y_j = \sum_{k=1}^{n} c_{jk} x_k + b_j, \quad j = 1, \, 2, \, \ldots, \, n, $$

with all $c_{jk}, b_j \in \mathbf{R}$. Prove that M is a contraction mapping on $\mathbf{R}_1$ if

$$ 0 < \max_{1 \leqslant k \leqslant n} \sum_{j=1}^{n} |c_{jk}| < 1. $$

(Refer to Application 3.3(2), and compare the above result with the sufficient conditions obtained there for the solution of $Ax = b$ to exist uniquely.)

(4) Use the fixed point theorem to show that the following systems of equations have unique solutions. (No adjustment of the co-efficients, by dividing an equation through by some number, for

example, should be necessary.)

(a) $\frac{3}{4}x - \frac{1}{2}y + \frac{1}{8}z = 2$ (b) $\frac{1}{4}x - \frac{1}{6}y + \frac{2}{5}z = x - 3$

$\frac{1}{6}x + \frac{1}{3}y \qquad = -1$ $\frac{1}{2}x + \frac{2}{3}y - \frac{1}{4}z = y + 1$

$-\frac{2}{5}x + \frac{1}{4}y + \frac{5}{4}z = 1$ $-\frac{1}{8}x \qquad - \frac{1}{4}z = z + 5$

(5) Show that the integral equation

$$x(s) = \frac{1}{3} \int_0^1 stx(t)\, dt + e^s - \frac{s}{3}$$

may be solved by the method of successive approximations. Starting with $x_0(s) = 1$, find the first few iterates and show that

$$x_n(s) = e^s - \frac{s}{2 \cdot 3^{2n-1}}, \quad n \in \mathbf{N}.$$

Hence find $x(s)$.

(6) Solve the Volterra integral equation

$$x(s) = \frac{1}{2} \int_0^s \frac{t}{s^2} x(t)\, dt + \frac{7s^2}{8}$$

by an iterative process, beginning with $x_0(s) = s$. (Hint: Show that the iterates can be given as $x_n(s) = (1/6^n)s + ((8^n - 1)/8^n)s^2$, $n \in \mathbf{N}$.)

(7) It is worth noting that integral equations can often be solved by more direct methods.

(a) Solve

$$x(s) = \frac{1}{4} \int_0^1 st^2 x(t)\, dt + s$$

by first reasoning that any solution x must have the form $x(s) = cs$ for some constant c.

(b) Solve

$$x(s) = \frac{1}{3} \int_0^2 x(t)\, dt + s^2$$

by first integrating the equation with respect to s over $[0, 2]$, and also by adapting the method of (a).

. . . .

(8) Let A be a mapping from a complete metric space (X, d) into itself. Prove that if the contraction condition is weakened to

$$d(Ax, Ay) < d(x, y)$$

(for all $x, y \in X$, $x \neq y$) then the existence of a fixed point of A is no longer assured.

(9) Show how the fixed point theorem may be used to find the unique root of the equation $F(x) = 0$ when F is a differentiable function on $[a, b]$ such that $F(a) < 0$, $F(b) > 0$ and $0 < K_1 \leqslant F'(x) \leqslant K_2$ for some constants K_1, K_2 and all x in $[a, b]$. (Hint: Introduce the function f, where $f(x) = x - \lambda F(x)$, $a \leqslant x \leqslant b$, and choose λ so that f has a unique fixed point. Show that this point is the required root of $F(x) = 0$.)

Apply this technique to the equation of Exercise (2)(a).

(10) Let c be the set of all convergent complex-valued sequences and define a mapping $d \colon c \times c \to \mathbf{R}_+$ by

$$d(x, y) = \sup_{1 \leqslant k} |x_k - y_k|,$$

where $x = (x_1, x_2, \dots)$ and $y = (y_1, y_2, \dots)$ are elements of c.

(a) Prove that (c, d) is a metric space and that it is complete. (The set c was introduced in Section 1.11. The above metric is the one usually associated with c, so c is also commonly used to denote this metric space.)

(b) Define a mapping A on c by

$$A(x_1, x_2, x_3, \dots) = \left(\tfrac{1}{2}x_2, \tfrac{1}{3}x_3, \tfrac{1}{4}x_4, \dots\right).$$

Prove that A is a contraction and hence that A has a unique fixed point (immediately obtained by inspection). Suppose this point is to be obtained by iteration and let $x^{(0)}$, $x^{(1)}$, $x^{(2)}$, $\dots$ denote the successive iterates. Taking $x^{(0)} = \left(1, \tfrac{1}{2}, \tfrac{1}{3}, \tfrac{1}{4}, \dots\right)$, show that $x^{(n)}$ has kth component

$$\frac{k!}{(n + k - 1)!(n + k)^2}, \quad n = 0, 1, 2, \dots, \ k = 1, 2, \dots.$$

(c) Define a mapping B on c by

$$B(x_1, x_2, x_3, \dots)$$
$$= \left(1 + \tfrac{1}{2}x_2 + \tfrac{1}{3}x_3, 1 + \tfrac{1}{2}x_3 + \tfrac{1}{3}x_4, 1 + \tfrac{1}{2}x_4 + \tfrac{1}{3}x_5, \dots\right).$$

Prove that B is a contraction. Find the fixed point of B (by any means).

(11) In the notation of Section 3.4, prove that, for any $n \in \mathbf{N}$,

$$\widetilde{d}_n \leqslant 2\epsilon + \alpha \widetilde{d}_{n-1},$$

and hence that $\widetilde{d}_n < \widetilde{d}_{n-1}$ if $\widetilde{d}_{n-1} > 2\delta$.

4

Compactness

4.1 Compact sets

Before introducing the main ideas of this chapter, we will establish a simple result concerning subsequences of sequences in metric spaces.

The definition of a subsequence of a sequence in Definition 1.7.2 is equally valid for a sequence in a metric space. We have remarked before that subsequences of convergent (real-valued) sequences are themselves convergent and have the same limit as the original sequence, and the proof of the corresponding statement in metric spaces generally is asked for in Exercise 4.5(1). The example $\frac{1}{2}, 2, \frac{1}{3}, 3, \frac{1}{4}, 4, \ldots$ shows that a sequence having a convergent subsequence certainly need not itself converge, for this sequence clearly diverges but has $\frac{1}{2}, \frac{1}{3}, \frac{1}{4}, \ldots$ as a convergent subsequence. We can however say the following.

Theorem 4.1.1 *In a metric space, any Cauchy sequence having a convergent subsequence is itself convergent, with the same limit.*

To prove this, let $\{x_n\}$ be a Cauchy sequence in a metric space (X, d) and let $\{x_{n_k}\}$ be a convergent subsequence of $\{x_n\}$. Set $x = \lim_{k \to \infty} x_{n_k}$. Then, given $\epsilon > 0$, we know there exists a positive integer K such that $d(x_{n_k}, x) < \frac{1}{2}\epsilon$ when $k > K$. As $\{x_n\}$ is a Cauchy sequence, we also know that a positive integer N exists such that $d(x_n, x_m) < \frac{1}{2}\epsilon$ when $m, n > N$. We may assume that $K > N$. If $k > K$, then $n_k \geqslant k > K > N$ and

$$d(x_n, x) \leqslant d(x_n, x_{n_k}) + d(x_{n_k}, x) < \tfrac{1}{2}\epsilon + \tfrac{1}{2}\epsilon = \epsilon$$

whenever $n > N$. Hence indeed the sequence $\{x_n\}$ converges, with limit x. □

Completeness was introduced because of a need to categorise those metric spaces having a property corresponding to the Cauchy convergence criterion for real numbers. It is another classical property of real numbers that leads us to the notion of compactness. If a metric space is complete, then the convergence of any sequence in the space is assured once it can be shown to be a Cauchy sequence. If the space is not complete then there is no such assurance. It would be useful in the latter case to have a criterion which ensures at least the existence of some convergent sequences in the space, whether or not their actual determination is possible. Since this does not impose as much on us, we look to the real number system for something earlier in our treatment of the real number system than the Cauchy convergence criterion. The answer is supplied by the Bolzano–Weierstrass theorem for sequences (Theorem 1.7.11). This says that there exists a convergent subsequence of any (real-valued) sequence, as long as that sequence is bounded, and this prompts our definition of *compactness*.

Definition 4.1.2 A subset of a metric space is called (*sequentially*) *compact* if every sequence in the subset has a convergent subsequence.

Some remarks are necessary. First, we will generally speak of compact sets (or subsets) rather than using the more correct term 'subspace'. This is in line with the comment at the end of Section 2.7.

Secondly, we must comment on the use of the word 'compact', which we have seen before, in Section 1.6. The definition there for point sets does not seem to be too close to that above, which is why this version is referred to more strictly as *sequential* compactness. For the moment, in this chapter, we will use 'compact' as defined in Definition 4.1.2, and we will also use 'closed' as defined in Definition 2.7.2. Some of the discussion that follows, and some of the results, look very similar to the work of Section 1.6. All of this will be brought together and explained in considerable detail in the next chapter.

It should be noted that for a subsequence in some set to be convergent, we require its limit also to belong to the set. Many writers do not make this demand of compact sets and speak additionally of a set as being *relatively compact* or *compact in itself* when referring to what we have simply called a compact set. Notice finally that Definition 4.1.2 can be applied to the metric space itself, so a metric space is compact if every sequence in it has a convergent subsequence.

We remark that the empty set is considered to be a compact subset of any metric space.

There are some immediate consequences of the above definition.

Theorem 4.1.3 *If a metric space is compact, then it is complete.*

This follows from Theorem 4.1.1, for any Cauchy sequence in a compact metric space has a convergent subsequence, by definition of compactness, and hence itself converges.   □

Another way of putting the theorem gives a better emphasis: if a metric space is not complete, then it is not compact. However, it is possible for a metric space to be complete and not compact: the metric space $\mathbf{R}$ is complete but the sequence $1, 2, 3, \dots$ in $\mathbf{R}$ has no convergent subsequence, so $\mathbf{R}$ is not compact.

Theorem 4.1.4 *Every compact set in a metric space is closed.*

In the terminology of other authors, just mentioned, this result would be stated as: a set is relatively compact if and only if it is closed and compact. For us, however, it is little more than our insistence on compact sets containing the limits of their convergent subsequences.   □

Again, the metric space $\mathbf{R}$ provides a counterexample to the converse: $\mathbf{R}$ is closed, but not compact.

The next theorem provides more insight into what compact sets look like. We recall first, from Definition 2.8.1, that a bounded subspace (S, d) of a metric space is one for which the diameter $\sup_{x,y \in S} d(x, y)$ is finite.

Theorem 4.1.5 *Every compact set in a metric space is bounded.*

Again, the converse of the theorem is false, but $\mathbf{R}$ no longer serves to show this since $\mathbf{R}$ is not bounded. For a counterexample, we may take any open interval: Theorem 4.1.4 implies that such an interval is not a compact subset of $\mathbf{R}$, although it is bounded.

The question arises as to which subsets of $\mathbf{R}$ are (sequentially) compact. We know that any such subsets must be both closed and bounded, and a little thought shows that the converse is also true. This is implied by the Bolzano–Weierstrass theorem for sequences. So the compact subsets of $\mathbf{R}$ are therefore fully identified. The more general question (What are the compact subsets of $\mathbf{R}^n$?) will be looked at shortly. We must first give a proof of Theorem 4.1.5, and this requires a little effort for which drawing pictures is helpful. The proof is by contradiction.

The result is clear for the empty set. Let S be a nonempty compact set in a metric space (X, d) and suppose that S is not bounded. Choose any element $x_1 \in S$. We cannot have $d(x, x_1) < 1$ for all $x \in S$, for then we would have $\delta(S) \leqslant 2$, where $\delta(S)$ is the diameter of S. So there is a point $x_2 \in S$ such that $d(x_2, x_1) \geqslant 1$. We write $\lambda_1 = 1$ and $\lambda_2 = \lambda_1 + d(x_2, x_1) = 1 + d(x_2, x_1)$. We cannot have $d(x, x_1) < \lambda_2$ for all $x \in S$, for then we would have $\delta(S) \leqslant 2\lambda_2$. So there is a point $x_3 \in S$ such that $d(x_3, x_1) \geqslant \lambda_2$. Write $\lambda_3 = \lambda_1 + d(x_3, x_1) = 1 + d(x_3, x_1)$. This process can be continued indefinitely: we obtain a sequence $\{x_n\}$ of points of S and an increasing sequence $\{\lambda_n\}$ of numbers such that

$$d(x_n, x_1) = \lambda_n - 1 \geqslant \lambda_{n-1}, \quad n = 2, \ 3, \ \ldots .$$

Then, for any integers m and n, with $n > m \geqslant 2$,

$$\lambda_m \leqslant \lambda_{n-1} \leqslant d(x_n, x_1)$$
$$\leqslant d(x_n, x_m) + d(x_m, x_1) = d(x_n, x_m) + \lambda_m - 1,$$

so that $d(x_n, x_m) \geqslant 1$. It follows from this that the sequence $\{x_n\}$ cannot have a convergent subsequence, and this contradicts the statement that S is a compact set. Hence, S is bounded. $\square$

Another instructive counterexample to the converse of this theorem is provided by a certain subset of the metric space l_2. We let S be the subset of points $e_1 = (1, 0, 0, \ldots)$, $e_2 = (0, 1, 0, 0, \ldots)$, $e_3 = (0, 0, 1, 0, 0, \ldots)$, $\ldots$. It is clear that, if d is the metric of l_2, $d(e_m, e_n) = \sqrt{2}$ whenever $m \neq n$, so that $\delta(S) = \sqrt{2}$, and S is bounded. But by the same token, no sequence in S (other than those with a finite range) can have a convergent subsequence. So S is not compact. Notice that this subset of l_2 is also closed: the only convergent sequences in l_2 consisting of points of S must be those having a finite range and the limit of any such sequence is certainly again an element of S.

We stated that the only compact subsets of $\mathbf{R}$ are those that are both closed and bounded, although, as we have just seen, subsets of l_2 that are closed and bounded need not be compact. The general question of determining which subsets of a metric space are compact is an important one with many uses, for example in approximation theory, as we will see. We will answer the question now for $\mathbf{R}^n$ (leaving $\mathbf{C}^n$ as an exercise) and later will look to the space $C[a, b]$. Compact subsets of l_2 have been identified, but we will not go into this more difficult problem.

Theorem 4.1.6 *A subset of $\mathbf{R}^n$ is compact if and only if it is both closed and bounded.*

This is a direct generalisation of the result when $n = 1$. The two preceding theorems show that closedness and boundedness are necessary conditions for a set in a metric space to be compact. In particular this applies to the space $\mathbf{R}^n$. We must show further that together they are sufficient in $\mathbf{R}^n$.

To this end, we let S be a closed, bounded subset of $\mathbf{R}^n$, and we may assume that S is nonempty. Let $\{x_m\}_{m=1}^\infty$ be any sequence in S. We show that $\{x_m\}$ has a convergent subsequence, and this will prove that S is compact. Let Δ be the diameter of S. Since S is bounded, Δ is finite, and, by definition of the metric in $\mathbf{R}^n$,

$$\sqrt{\sum_{k=1}^n (y_k - z_k)^2} \leqslant \Delta$$

whatever the points $(y_1, y_2, \ldots, y_n)$ and $(z_1, z_2, \ldots, z_n)$ in S. Let the latter be some particular point in S. Then

$$\sqrt{\sum_{k=1}^n y_k^2} \leqslant \sqrt{\sum_{k=1}^n (y_k - z_k)^2} + \sqrt{\sum_{k=1}^n z_k^2} \leqslant \Delta + \sqrt{\sum_{k=1}^n z_k^2}.$$

(Set $a_k = y_k - z_k$ and $b_k = z_k$ in Theorem 2.2.2.) Put

$$M = \Delta + \sqrt{\sum_{k=1}^n z_k^2}.$$

Since

$$|y_k| \leqslant \sqrt{\sum_{k=1}^n y_k^2} \leqslant M$$

for each k, we see that any point of S has bounded components (using 'bounded' here in the old sense of point set theory). In the sequence $\{x_m\}$, write $x_m = (x_{m1}, x_{m2}, \ldots, x_{mn})$ for $m \in \mathbf{N}$. Each x_{mk} is a real number and $\{x_{m1}\}$ (that is, the sequence of first components of the points of the sequence $\{x_m\}$) is a sequence in $\mathbf{R}$. For each m, we know that $|x_{m1}| \leqslant M$, so the sequence $\{x_{m1}\}$ has a convergent subsequence $\{x_{m_k 1}\}$, by the Bolzano–Weierstrass theorem for sequences (Theorem 1.7.11). Form the sequence $\{(x_{m_k 1}, x_{m_k 2}, \ldots, x_{m_k n})\}$ in $\mathbf{R}^n$ (by choosing from $\{x_m\}$ those terms whose first components belong to the subsequence $\{x_{m_k 1}\}$ of $\{x_{m1}\}$). This is a subsequence of $\{x_m\}$ with the property that its sequence of first components converges. From this subsequence we take the sequence in $\mathbf{R}$ of its second components and, as above, obtain a convergent subsequence of it. This allows us to form a new sequence in $\mathbf{R}^n$ which is a subsequence of $\{x_m\}$ with the property that its sequences of first and second components separately converge. (The new first components form a subsequence of the preceding first components. This is a subsequence of a convergent sequence, so it is itself convergent.) This sifting process may be continued through to the nth components, and we finally emerge with a subsequence of $\{x_m\}$ having the property that each of the n sequences of components separately converges. Since convergence in $\mathbf{R}^n$ is equivalent to convergence by components (Theorem 2.5.3(a)), and since S is closed, this last subsequence must converge to some point in S. Thus we have shown the existence of a convergent subsequence of $\{x_m\}$, so S is compact. $\qquad\square$

4.2 Ascoli's theorem

We turn next to the problem of identifying the compact subsets of $C[a, b]$. This will also require a sifting process similar to that just used in $\mathbf{R}^n$ in order to obtain a convergent subsequence, but the criteria that we impose on the sets are more complicated. We need the following definitions.

Definition 4.2.1 Let F be a family (or set) of functions, each with domain D.

(a) We say the family F is *uniformly bounded* on D if there is a positive number M such that $|f(x)| \leqslant M$ for all $f \in F$ and all $x \in D$.

(b) We say F is *equicontinuous* on D if, given any number $\epsilon > 0$, there exists a number $\delta > 0$ such that, for any $f \in F$,

$$|f(x') - f(x'')| < \epsilon \quad \text{whenever } x', x'' \in D \text{ and } |x' - x''| < \delta.$$

Uniform boundedness of a family of functions is a property well described by its name: each function in the family must be bounded and the same bound (M in the definition) must serve for the whole family. It is a property not dependent on any metric that may be defined on F, whereas the notion of a bounded set in Definition 2.8.1 does depend on the metric for the space. However, if F is a subset of $C[a, b]$, with its uniform metric, then the concepts of boundedness and uniform boundedness coincide. This is not true for subsets of the metric space $C_1[a, b]$, for example. The proofs of these statements are left as an exercise.

To understand the definition of equicontinuity, recall Definition 1.9.1 on the continuity of a function at a point: a function f is continuous at a point x_0 in its domain if for any number $\epsilon > 0$ there is a number δ such that, whenever x is in the domain of f and $|x - x_0| < \delta$, we have $|f(x) - f(x_0)| < \epsilon$. If, in that definition, the same δ will do for all points in the domain, then the function is called *uniformly continuous*. Going further, when we have a family of functions and all can be shown to be uniformly continuous on the domain and still only one value of δ is needed, then the family is equicontinuous.

The set of functions $\{x, x^2, x^3, \ldots\}$ on $[0, 1]$ is an example of a family which is not equicontinuous. Like uniform boundedness, equicontinuity is a property of a family F of functions which is independent of any metric defined on F. But if the functions of the equicontinuous family F have domain $[a, b]$ and F is given the uniform metric, then it is clear that F is a subspace of $C[a, b]$.

The criteria for compactness of a subset of $C[a, b]$ are given in the following theorem.

Theorem 4.2.2 (Ascoli's Theorem) *A subset F of the metric space $C[a, b]$ is compact if F is closed, uniformly bounded and equicontinuous.*

Let $\{f_n\}$ be a sequence in F. The proof shows explicitly the existence of a convergent subsequence of $\{f_n\}$ and consists of six main steps.

(a) It follows from Theorem 1.4.3(a) that the set of rational numbers in the interval $[a, b]$ is countable. Suppose that $\{x_1, x_2, \ldots\}$ is a listing of those rational numbers.

(b) Since F is uniformly bounded, there exists a number $M > 0$ such that, for all $x \in [a, b]$ and all $n \in \mathbf{N}$, we have $|f_n(x)| \leqslant M$.

In particular, then $|f_n(x_1)| \leqslant M$ for all n, so the sequence $\{f_n(x_1)\}$ in $\mathbf{R}$ is bounded. By the Bolzano–Weierstrass theorem for sequences, there exists a convergent subsequence $\{f_{n_k}(x_1)\}$ of $\{f_n(x_1)\}$ and this picks out from the sequence $\{f_n\}$ a subsequence $\{f_{n_k}\}$ converging pointwise at x_1.

Write this subsequence as $\left\{f_n^{(1)}\right\}_{n=1}^{\infty}$, rather than $\{f_{n_k}\}_{k=1}^{\infty}$, and apply similar reasoning to $\left\{f_n^{(1)}\right\}$: this time, $|f_n^{(1)}(x_2)| \leqslant M$ for all n, so the sequence $\left\{f_n^{(1)}(x_2)\right\}$ in $\mathbf{R}$ has a convergent subsequence $\left\{f_{n_k}^{(1)}(x_2)\right\}$ which allows us to pick out from $\left\{f_n^{(1)}\right\}$ a subsequence $\left\{f_{n_k}^{(1)}\right\}$, to be written $\left\{f_n^{(2)}\right\}$, with the property of pointwise convergence at x_1 and x_2. This process can be continued indefinitely, producing sequences $\left\{f_n^{(m)}\right\}$,

$m \in \mathbf{N}$, and for each m the sequence is a subsequence of $\{f_n\}$ converging pointwise at $x_1, x_2, \ldots, x_m$. Further, each sequence is a subsequence of the one before it.

(c) We have described the formation of sequences

$$f_1^{(1)}, f_2^{(1)}, f_3^{(1)}, \ldots,$$
$$f_1^{(2)}, f_2^{(2)}, f_3^{(2)}, \ldots,$$
$$f_1^{(3)}, f_2^{(3)}, f_3^{(3)}, \ldots,$$

$$\vdots$$

Consider the diagonal sequence $f_1^{(1)}, f_2^{(2)}, f_3^{(3)}, \ldots$, that is, $\{f_n^{(n)}\}$, which we will write as $\{f^n\}$. (The superscript is an index, not a power. We write the sequence this way to distinguish it from $\{f_n\}$, of which it is a subsequence.) For each $L \in \mathbf{N}$, the sequence $f^L, f^{L+1}, \ldots$ is a subsequence of $\{f_n^{(L)}\}$, so $f^L, f^{L+1}, \ldots$ converges pointwise at $x_1, x_2, \ldots, x_L$. Adding terms at the beginning of a sequence does not change the nature of its convergence, so the sequence $\{f^n\}$ converges also at $x_1, x_2, \ldots, x_L$. Since this is true for all L, we conclude that the sequence $\{f^n\}$ converges at all points $x_1, x_2, \ldots$.

(d) To conclude the proof of the compactness of F, we will show that the sequence $\{f^n\}$ is convergent (rather than simply pointwise convergent at all rational points in $[a, b]$, which is what we have just shown). Take any number $\epsilon > 0$. Since the functions of $\{f^n\}$ are a subset of F, the equicontinuity condition may be applied: there exists a number $\delta > 0$ such that, for any $n \in \mathbf{N}$,

$$|f^n(x') - f^n(x'')| < \tfrac{1}{3}\epsilon$$

whenever $|x'-x''| < \delta$ and x' and x'' are in $[a, b]$. Knowing this number δ, we can choose, say, K rational points in $[a, b]$, where K depends on ϵ, so that any point of $[a, b]$ is within δ of one of those rational points. By renumbering if necessary, we can let those rational points be $x_1, x_2, \ldots, x_K$, so $|x-x_i| < \delta$ for any $x \in [a, b]$ and at least one $i \in \{1, 2, \ldots, K\}$.

(e) Since $\{f^n(x_i)\}$ converges for each $i = 1, 2, \ldots, K$, there exists a positive integer N (also depending on ϵ) such that

$$|f^n(x_i) - f^m(x_i)| < \tfrac{1}{3}\epsilon$$

for all $i = 1, 2, \ldots, K$, when $m, n > N$.

(f) Let x be any point in $[a, b]$ and, as in (d), choose a point x_i from $\{x_1, x_2, \ldots, x_K\}$ such that $|x - x_i| < \delta$. Then

$$|f^n(x) - f^n(x_i)| < \tfrac{1}{3}\epsilon,$$

for all $n \in \mathbf{N}$, and

$$|f^n(x) - f^m(x)| \leqslant |f^n(x) - f^n(x_i)| + |f^n(x_i) - f^m(x_i)|$$
$$+ |f^m(x_i) - f^m(x)|$$
$$< \tfrac{1}{3}\epsilon + \tfrac{1}{3}\epsilon + \tfrac{1}{3}\epsilon = \epsilon,$$

provided $m, n > N$. It follows that

$$\max_{a \leqslant x \leqslant b} |f^n(x) - f^m(x)| < \epsilon$$

when $m, n > N$, so $\{f^n\}$ is a Cauchy sequence in F. But F is a closed subset of the complete metric space $C[a, b]$, so, by Theorem 2.7.3, F is complete. Hence the Cauchy sequence $\{f^n\}$ converges, as required, so F is compact. $\qquad\square$

The converse of Ascoli's theorem is also true: any compact subset of $C[a, b]$ is uniformly bounded and equicontinuous. We will not need the implication in this direction. (Some writers include the converse, due to Arzelá, in the statement of Ascoli's theorem.)

An alternative statement of Ascoli's theorem, having no direct reference to the metric of $C[a, b]$, is the following. From any uniformly bounded, equicontinuous sequence of functions defined on a closed interval may be chosen a subsequence which converges uniformly on the interval. The truth of this is evident from the above proof. It needs only to be noted that convergence in $C[a, b]$ is equivalent to uniform convergence over $[a, b]$ (Theorem 2.5.3(c)).

A simple sufficient condition for a family of functions to be equicontinuous is that all functions of the family satisfy a Lipschitz condition with the same Lipschitz constant. Precisely: a family F of functions defined on an interval $[a, b]$ is equicontinuous if, for all $f \in F$ and any points $x', x'' \in [a, b]$, there is a number K such that

$$|f(x') - f(x'')| \leqslant K|x' - x''|.$$

To see this, we take an arbitrary $\epsilon > 0$ and put $\delta = \epsilon/K$. Then, if $|x' - x''| < \delta$, we have

$$|f(x') - f(x'')| \leqslant K|x' - x''| < K\delta = \epsilon,$$

for all $f \in F$, so that F is indeed an equicontinuous family. Moreover, if the functions of F are all differentiable on $[a, b]$ then there is an even simpler test: the family F is equicontinuous if there is a positive constant K such that $|f'(x)| \leqslant K$ for all $f \in F$ and all $x \in [a, b]$. This follows from the mean value theorem, as in Application 3.3(1).

4.3 Application to approximation theory

One of the most important theorems of classical analysis finds a natural generalisation in the context of compact sets in a metric space. In its turn, the generalised result also assumes considerable importance and has numerous applications. The theorem in question is Theorem 1.9.6, which asserts that a function defined on a closed interval and continuous there actually attains its maximum and minimum values at some points of the interval. As might be anticipated, the clue to the generalisation lies in our insistence on a closed interval as the domain of the function. Closed intervals are compact subsets of $\mathbf{R}$, so we consider in general the effect of a continuous mapping on a compact set in a metric space.

Theorem 4.3.1 *Let $A: X \to Y$ be a continuous mapping between metric spaces X and Y, and let S be a nonempty compact subset of X. Then the image $A(S)$ is a compact subset of Y.*

Briefly, this says that the image under a continuous mapping of a compact set is again a compact set. We will later set $Y = \mathbf{R}$ to obtain the generalisation mentioned above. Let $\{y_n\}$ be a sequence in $A(S)$. For each $n \in \mathbf{N}$, there is at least one point $w \in S$ such that $Aw = y_n$. Choose one and call it x_n. Then $\{x_n\}$ is a sequence in S and $Ax_n = y_n$. Since S is compact, $\{x_n\}$ has a convergent subsequence $\{x_{n_k}\}$, with limit x, say. Then $x \in S$, so $Ax \in A(S)$. Now, $Ax_{n_k} = y_{n_k}$ and $Ax_{n_k} \to Ax$ since A is continuous, so $\{y_{n_k}\}$ is a convergent subsequence of $\{y_n\}$. Hence, $A(S)$ is compact, as required. $\square$

Now take $Y = \mathbf{R}$ in this theorem. Then $A(S)$ is a compact subset of $\mathbf{R}$, and so $A(S)$ is closed and bounded. We know, using Theorems 1.5.7 and 1.5.10, that such a subset of $\mathbf{R}$ contains as members its least upper bound and greatest lower bound. If these numbers are y_M and y_m, respectively, then we have shown the existence of points x_M and x_m in S such that $Ax_M = y_M$ and $Ax_m = y_m$. We have proved the following.

Theorem 4.3.2 *If f is a real-valued continuous mapping on a metric space X and S is any nonempty compact set in X, then there exist points x_M and x_m in S such that*

$$f(x_M) = \max_{x \in S} f(x) \quad and \quad f(x_m) = \min_{x \in S} f(x).$$

We can now prove a basic existence theorem on best approximations in a metric space.

Theorem 4.3.3 *Given a nonempty compact subset S of a metric space (X, d) and a point $x \in X$, there exists a point $p \in S$ such that $d(p, x)$ is a minimum.*

We need to prove the existence of some point $p \in S$ which is such that $d(p, x) \leqslant d(w, x)$ for all $w \in S$. Put differently, p must satisfy

$$d(p, x) = \min_{w \in S} d(w, x).$$

But this is an immediate consequence of Theorem 4.3.2, for in that theorem we let f be the mapping from X into $\mathbf{R}$ defined by $f(y) = d(y, x)$ ($y \in X$) and need only check that f is continuous on X. If $\{y_n\}$ is a sequence in X and $y_n \to y$, then

$$|f(y_n) - f(y)| = |d(y_n, x) - d(y, x)| \leqslant d(y_n, y),$$

by Solved Problem 2.3(1), so $f(y_n) \to f(y)$ since $d(y_n, y) \to 0$. Hence indeed f is continuous on X, and this completes the proof. □

The point p in this theorem is called a *best approximation* in S of the point x in X. There is nothing in the theorem to describe how such a point may be obtained in any practical situation, and there is no suggestion that p is the only point with the given property. These are serious drawbacks in terms of applications. Later, when we have imposed more structure on our sets, we will reconsider the problem of best approximation, including the above difficulties. For now, we can only say they are inherent in the small amount of structure we have allowed ourselves.

The following example in $\mathbf{R}^2$ illustrates the possible non-uniqueness of a best approximation. Let S be the set $\{(y_1, y_2) : 0 < a^2 \leqslant y_1^2 + y_2^2 \leqslant b^2\}$ in $\mathbf{R}^2$ (see Figure 10). It is easy to see that S is both closed and bounded, so it is compact (Theorem 4.1.6). The points x_1 and x_2 clearly have unique best approximations in S, namely p_1 and p_2, respectively. The point x_3 however, at the centre of the circles, has any number of best

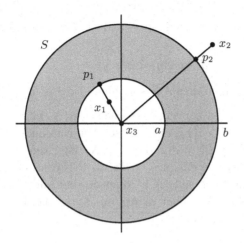

Figure 10

approximations, namely any point on the inner boundary of S. Notice that there are no best approximations of x_1 and x_2, for example, in the set $\{(y_1, y_2) : 0 < a^2 < y_1^2 + y_2^2 < b^2\}$: p_1 and p_2 are excluded from consideration since the new set does not include the boundaries of S. Of course, the new set is not closed, so it is not compact, and Theorem 4.3.3 does not apply.

The following is an application of Theorem 4.3.3 which makes use of Ascoli's theorem (Theorem 4.2.2).

Suppose a, b, c, d are any numbers chosen from a closed interval $[-M, M]$. The family F of functions f of the form

$$f(x) = a \sin bx + c \cos dx, \qquad 0 \leqslant x \leqslant \pi,$$

is uniformly bounded and equicontinuous, since $|f(x)| \leqslant |a| + |c| \leqslant 2M$ and $|f'(x)| \leqslant |ab| + |cd| \leqslant 2M^2$ for all $x \in [0, \pi]$ and any $f \in F$. Since F may be considered as a (closed) subset of $C[0, \pi]$, Ascoli's theorem now implies that it is compact in $C[0, \pi]$. Hence, by Theorem 4.3.3, for any continuous function g defined on $[0, \pi]$, there exist values of a, b, c and d in $[-M, M]$ such that

$$\max_{0 \leqslant x \leqslant \pi} |g(x) - (a \sin bx + c \cos dx)|$$

is a minimum. For an obvious reason, a function $f \in F$ with such values of a, b, c and d is called a *minimax* approximation of g. As discussed above, it is not necessarily unique.

4.4 Solved problems

(1) Let (X, d) be a compact metric space and let A be a mapping from X into itself such that

$$d(Ax, Ay) < d(x, y)$$

for $x, y \in X$, $x \neq y$. Prove that A has a unique fixed point in X.

Solution. Let $\{x_n\}$ be a convergent sequence in X with $\lim x_n = x$. Then $0 \leqslant d(Ax_n, Ax) < d(x_n, x) \to 0$, so A is continuous. Define a mapping $B: X \to \mathbf{R}$ by $Bx = d(x, Ax)$, $x \in X$. By Exercise 2.9(1), we have $Bx_n = d(x_n, Ax_n) \to d(x, Ax) = Bx$, so B is continuous. It now follows from Theorem 4.3.2, since X is compact, that $\min_{x \in X} Bx$ exists, and equals By, say $(y \in X)$. That is, $d(y, Ay) \leqslant d(x, Ax)$ for all $x \in X$. Suppose $d(y, Ay) > 0$. In that case,

$$B(Ay) = d(Ay, A(Ay)) < d(y, Ay) = By,$$

and this contradicts the minimal property of y. Hence $d(y, Ay) = 0$, so $Ay = y$ and y is a fixed point of A. It is the only one, for if $z \in X$ were another then we would have $d(y, z) = d(Ay, Az) < d(y, z)$, which is absurd. Hence A has a unique fixed point in X. $\qquad\square$

The result proved above should be considered in conjunction with the fixed point theorem. See also Exercise 3.5(8).)

For the second of these solved problems, we will need the following definition.

Definition 4.4.1 Let (X, d) be a metric space, S be a nonempty subset of X and $\epsilon > 0$ be a given number. A subset Z of X is called an ϵ-*net* for S if, for any $x \in S$, there exists $z \in Z$ such that $d(x, z) < \epsilon$.

(2) Prove that, whatever the positive number ϵ, a nonempty compact subset of X contains a finite ϵ-net, that is, an ϵ-net consisting of only a finite number of points.

Solution. Let S be a nonempty compact subset of X and suppose S does not contain a finite ϵ-net for some value ϵ_0 of ϵ. Choose any point $x_1 \in S$. There must be a point $x_2 \in S$ such that $d(x_2, x_1) \geqslant \epsilon_0$ (otherwise the set $\{x_1\}$ consisting of the one point x_1 is a finite ϵ_0-net for S). Further, there must be a point $x_3 \in S$ such that $d(x_3, x_1) \geqslant \epsilon_0$, $d(x_3, x_2) \geqslant \epsilon_0$ (otherwise the set $\{x_1, x_2\}$ is a finite ϵ_0-net for S). Continuing in this manner, we find points x_4, x_5, ... in S such that $d(x_{n+1}, x_1) \geqslant \epsilon_0$,

$d(x_{n+1}, x_2) \geqslant \epsilon_0, \ldots, d(x_{n+1}, x_n) \geqslant \epsilon_0$ $(n \in \mathbf{N})$. But this means that we have obtained a sequence $\{x_n\}$ in S such that $d(x_n, x_m) \geqslant \epsilon_0$ for all m, n $(m \neq n)$, so there can be no convergent subsequence, contradicting the compactness of S. Hence S contains a finite ϵ-net, for all $\epsilon > 0$. $\square$

Virtually all of the techniques of numerical analysis, such as the *method of finite differences*, in the end owe their validity to this result, since necessarily those techniques require the division of the domain of interest into only finitely many sub-domains.

4.5 Exercises

(1) Prove that any subsequence of a convergent sequence in a metric space is itself convergent, and has the same limit as the sequence.

(2)   (a) Prove that any finite subset of a metric space is compact.
      (b) Let x be the limit of a convergent sequence $\{x_n\}$ in a metric space. Prove that the set $\{x, x_1, x_2, x_3, \ldots\}$ is compact.

(3) Prove that every closed subset of a compact metric space is compact.

(4) Determine whether the union and intersection of compact subsets of a metric space are compact.

(5) Let X be any nonempty set and impose on X the discrete metric (Example 2.2(14)). Determine whether X is compact, and which subsets of X are compact.

(6) Prove that a subset of $\mathbf{C}^n$ is compact if and only if it is closed and bounded.

(7) Let F be a subset of $C[a, b]$. Prove that F is a uniformly bounded family if and only if it is bounded. Show however that if F is considered as a subset of $C_1[a, b]$ (Example 2.2(12)), then F may be bounded but not uniformly bounded.

(8) Let K and α be given positive numbers and let F be a subset of $C[a, b]$ for which, for all $f \in F$ and any points $x', x'' \in [a, b]$,

$$|f(x') - f(x'')| \leqslant K|x' - x''|^{\alpha}.$$

Show that F is equicontinuous.

(9) Let F be a bounded subset of $C[a, b]$. Prove that the set of all functions g, where

$$g(x) = \int_a^x f(t)\, dt$$

$(f \in F, a \leqslant x \leqslant b)$, is uniformly bounded and equicontinuous.

(10) Prove that Theorem 4.1.5 is a consequence of the result proved in Solved Problem 4.4(2).

$$\dots\dots\dots\dots\dots\dots\dots\dots\dots$$

(11) Let g be a continuous function of two variables satisfying a Lipschitz condition in the second variable. Let $A\colon C[a,b] \to \mathbf{R}$ be a mapping defined by

$$Ax = \int_a^b g(t, x(t))\, dt, \qquad x \in C[a,b].$$

Prove that A is continuous and hence show that, if the domain of A is restricted to a compact subset of $C[a,b]$, then there exists a function x such that $\int_a^b g(t, x(t))\, dt$ is a minimum.

(12) If a subset of a metric space contains a finite ϵ-net for every $\epsilon > 0$, then it is called *totally bounded*.

   (a) Prove that a totally bounded set is bounded.
   (b) Give an example in l_2 of a bounded set that is not totally bounded.

5

Topological Spaces

5.1 Definitions and examples

A topological space is a more basic concept than a metric space. Its building blocks are *open sets*, as suggested by the work for real numbers along the lines of that in Section 1.6.

The abstract idea of a metric space provides a useful and quite visual example of a topological space. Through much of this chapter, we will relate our work to corresponding ideas in metric spaces. In previous chapters, we have spent some time on closed sets and compact sets. These were defined specifically in the context of metric spaces, and each definition made use of the notion of a convergent sequence. The same terms will be used again in this chapter, but they will be redefined in the more general context of topological spaces. To distinguish the different approaches, we will be careful in this chapter to refer to the earlier notions as *sequentially closed* sets and *sequentially compact* sets.

So a set is sequentially closed if convergent sequences in the metric space that belong to the set have their limits in the set, and a set is sequentially compact if every sequence in the set has a convergent subsequence. These are the old definitions; new ones will come soon. It will turn out, and these are two of the important results of this chapter, that the old definitions and the new definitions coincide in metric spaces.

The term 'topology' refers to the work of this chapter in general, but is also used in the technical sense given by the following definition.

Definition 5.1.1 A *topology* on a nonempty set X is a collection $\mathscr{T}$ of subsets of X with the properties

(T1) $X \in \mathscr{T}$ and $\varnothing \in \mathscr{T}$,

(T2) $\bigcup_{T \in \mathscr{S}} T \in \mathscr{T}$ for any subcollection $\mathscr{S}$ of $\mathscr{T}$,

(T3) $T_1 \cap T_2 \in \mathscr{T}$ whenever $T_1, T_2 \in \mathscr{T}$.

The pair $(X, \mathcal{T})$ is called a *topological space*.

The sets $T \in \mathcal{T}$ are called the *open sets* in $(X, \mathcal{T})$. Any subset S of X is said to be *closed* in $(X, \mathcal{T})$ if its complement $\sim\!S$ (that is, $X\backslash S$) is an open set in $(X, \mathcal{T})$.

We often refer to X alone as a topological space, with the understanding that the topology is a certain collection $\mathcal{T}$ of subsets of X. It quickly follows from (T3) that the intersection of any finite number of open sets in X is also an open set in X, while (T2) states that the union of arbitrarily many (perhaps uncountably many) open sets in X is also an open set in X.

Let us remark now that we are not interested in the various unimportant exceptions that arise when X has just one element, so we will always assume that our topological spaces have at least two elements.

In our discussion of the real number system, Theorem 1.6.2 said in other words that the open sets defined then in $\mathbf{R}$ are a topology for $\mathbf{R}$. That is, $(\mathbf{R}, \mathcal{T})$ is a topological space, where $\mathcal{T}$ is the collection of all open sets as given by Definition 1.6.1. This is called the *usual* topology on $\mathbf{R}$, and is always the one we mean when $\mathbf{R}$ is referred to as a topological space. In this space, consider the open intervals $(-1/n, 1/n)$, for $n \in \mathbf{N}$. These are certainly open sets in $\mathbf{R}$. The number 0 belongs to all of them, but no other number does, so $\bigcap_{n=1}^{\infty}(-1/n, 1/n) = \{0\}$. It is easy to see that $\{0\}$ is a closed set in $\mathbf{R}$, so this example suggests why, in (T3), we restrict ourselves to the intersection of only two (or, in effect, finitely many) open sets.

There are two simple topologies that exist for any set X. These are the *discrete* topology, which is the collection of all subsets of X, and the *indiscrete* topology, which is simply $\{\varnothing, X\}$. They are denoted by $\mathcal{T}_{\max}$ and $\mathcal{T}_{\min}$, respectively. It is easy to see that these are indeed topologies for X, and, as the symbols suggest, they are the biggest and smallest possible collections of subsets of X which are topologies.

The following definition sometimes allows us to compare different topologies on the same set.

Definition 5.1.2 If $\mathcal{T}_1$ and $\mathcal{T}_2$ are two topologies on a set X and $\mathcal{T}_1 \subseteq \mathcal{T}_2$, then $\mathcal{T}_1$ is said to be *weaker* than $\mathcal{T}_2$, and $\mathcal{T}_2$ to be *stronger* than $\mathcal{T}_1$.

Then if $\mathcal{T}$ is any topology on X, we must have $\mathcal{T}_{\min} \subseteq \mathcal{T} \subseteq \mathcal{T}_{\max}$, so that, amongst the topologies on a set, the indiscrete topology is the weakest of all and the discrete topology is the strongest of all. Alternative terms for 'weaker' and 'stronger' are *coarser* and *finer*, respectively. Two concepts that are useful in identifying properties of open and closed sets are given next.

Definition 5.1.3 Let X be a topological space.

(a) The *interior* of a subset S of X is the union of all open sets contained in S. It is denoted by $\operatorname{int} S$ or S°.

(b) The *closure* of a subset S of X is the intersection of all closed sets containing S. It is denoted by $\operatorname{cl} S$ or $\overline{S}$.

We think of the interior of a set as the largest open set contained in it, and its closure as the smallest closed set containing it.

The following example illustrates much of the above.

Take $X = \{1,2,3,4,5\}$ and

$$\mathscr{T}_1 = \{\varnothing, \{1\}, \{2\}, \{1,2\}, X\},$$
$$\mathscr{T}_2 = \{\varnothing, \{1\}, \{2\}, \{1,2\}, \{1,2,3\}, \{1,2,3,4\}, X\},$$
$$\mathscr{T}_3 = \{\varnothing, \{1\}, \{1,2\}, \{1,2,3\}, X\},$$
$$\mathscr{T}_4 = \{\varnothing, \{1\}, \{2\}, \{1,2\}, \{2,3,4\}, X\}.$$

We see that $\mathscr{T}_1$ is a topology for X because $\varnothing$ and X are present, the union of any combination of sets in $\mathscr{T}_1$ is also an element of $\mathscr{T}_1$, and the intersection of any two sets in $\mathscr{T}_1$ is an element of $\mathscr{T}_1$ (so (T1), (T2) and (T3) are satisfied). In the same way, $\mathscr{T}_2$ is also a topology for X, and, since $\mathscr{T}_1 \subseteq \mathscr{T}_2$, the topology $\mathscr{T}_1$ is weaker than $\mathscr{T}_2$. We see that $\mathscr{T}_3$ is a third topology for X; it is also weaker than $\mathscr{T}_2$ but is neither weaker nor stronger than $\mathscr{T}_1$. In the topological space $(X, \mathscr{T}_2)$, the closed sets are $X, \{2,3,4,5\}, \{1,3,4,5\}, \{3,4,5\}, \{4,5\}, \{5\}$ and $\varnothing$, while the set $\{2,3\}$, for example, is neither open nor closed; the interior of $\{2,3\}$ is $\{2\}$ and its closure is $\{2,3,4,5\}$. The collection $\mathscr{T}_4$ is not a topology on X since $\{1\} \cup \{2,3,4\} = \{1,2,3,4\} \notin \mathscr{T}_4$ (so (T2) is not satisfied).

Perhaps the most enlightening example is that where X is a metric space. There is a standard way to use the metric on X to define open sets in the metric space, so that every metric space has an associated *metric topology*. At the same time, it should be realised that there are many examples of topological spaces that do not arise this way, such as those in the preceding paragraph.

5 *Topological Spaces*

Definition 5.1.4 Let (X, d) be a metric space.

(a) The set $\{x : x \in X, \ d(x, x_0) < r\}$, where $x_0 \in X$ and $r > 0$, is called an *open ball* in X. Specifically, it is the open ball with centre x_0 and radius r, and is denoted by $b(x_0, r)$.

(b) A subset T of X is *open* if $T = \varnothing$ or if every point in T is the centre of an open ball that is a subset of T.

(c) The *metric topology* for X is the collection of open sets, as just defined. It is denoted by $\mathscr{T}_d$.

Rephrasing (b) when $T \neq \varnothing$, we say T is an open set in X if, for each $x \in T$, there exists an open ball $b(x, r)$ such that $b(x, r) \subseteq T$. The verification that this collection $\mathscr{T}_d$ of open sets does indeed define a topology for the metric space X is left as an exercise. Whenever we refer to a metric space as a topological space, we assume it has the metric topology.

The δ-neighbourhoods that we used in Chapter 1 are examples of open balls in $\mathbf{R}$. It is in $\mathbf{R}^3$ (with the Euclidean metric) that all of this is most familiar. There, the open balls are ordinary three-dimensional spheres of various radii, and the open sets can be thought of as all sorts of bunches of tiny spheres.

In the metric space $C[a, b]$, if x_0 is the function given by $x_0(t) = 1$ for $a \leqslant t \leqslant b$, then $b(x_0, \epsilon)$ is the set of all continuous functions x with $1 - \epsilon < x(t) < 1 + \epsilon$ for $a \leqslant t \leqslant b$. Their graphs all lie in the strip of width 2ϵ lying along the graph of x_0.

5.2 Closed sets

In this section, we will show first that, for any metric space, the closed sets under the metric topology are precisely the sequentially closed sets of Chapter 2. We will follow this with another characterisation of closed sets which looks more like our work on point sets in Section 1.5, and does not rely on a metric.

Theorem 5.2.1 *Let (X, d) be a metric space, and let $\mathscr{T}_d$ be the metric topology on X. A subset of X is closed in $(X, \mathscr{T}_d)$ if and only if it is sequentially closed in (X, d).*

To prove this, suppose first that S is a closed subset of X. Then we must show that it is sequentially closed. So assume $S \neq \varnothing$, let $\{x_n\}$ be a convergent sequence in S, and put $x = \lim x_n$. If $x \in \sim S$, then, since

$\sim S$ is an open set in X, there is an open ball $b(x, \epsilon)$ contained in $\sim S$. Then $d(x_n, x) \geqslant \epsilon$ for all n, and this contradicts the fact that $x_n \to x$. Hence $x \in S$, so S is sequentially closed.

Next, let S be a sequentially closed nonempty subset of X. To show that S is closed, we must show that $\sim S$ is an open set. If this is not true, then there is a point $x \in \sim S$ such that every open ball centred at x contains a point of S. For each $n \in \mathbf{N}$, let x_n be a point of S contained in the ball $b(x, 1/n)$. Then $\{x_n\}$ is a sequence of points in S, and $d(x_n, x) < 1/n$ for all $n \in \mathbf{N}$, so $\lim x_n = x$. Since $x \notin S$, this contradicts the statement that S is sequentially closed. Hence $\sim S$ is open, and S is closed. $\qquad\qquad\square$

So we know now that, in a metric space, sets which are closed in the sense that their complements are open can be described through the idea of convergent sequences in the metric space. The discussion of convergence of sequences given in Section 1.7 made a great deal of use of the earlier Section 1.5, on point sets. We can use the ideas there to give another way of thinking about closed sets.

Definition 5.2.2   Let X be a topological space.

(a) If x is any point in X and U is an open set in X which contains x, then U is called a *neighbourhood* of x.

(b) The point $x \in X$ is called a *cluster point* for a subset S of X if every neighbourhood of x contains a point of S other than x.

(c) The set of all cluster points of a subset S of X is called the *derived set* of S, and is denoted by S'.

Neighbourhoods here are much the same as the δ-neighbourhoods of Section 1.5, but the latter have a certain symmetry which is neither available nor necessary in general. The definition of a cluster point is very much like that in Definition 1.5.2. Other authors now commonly use the term *limit point* for what we have just defined as a cluster point. That would be in conflict with our Definition 1.5.8, so we will stay with the older terminology. Notice, in (b), that x need not be a point of S.

If $\mathcal{T} = \mathcal{T}_{\max}$, so that every subset of X is an open set, then $\{x\}$ is a neighbourhood of $x \in X$ which does not contain any other point of X. Hence no point of X can be a cluster point of any subset of X. Suppose, on the other hand, that $\mathcal{T} = \mathcal{T}_{\min}$. Then every point $x \in X$ is a cluster point of every subset of X, except $\{x\}$ and $\varnothing$, since X is the only neighbourhood of x.

A subset of a topological space is easily identified as closed if its derived set is known.

Theorem 5.2.3 *A set S in a topological space is closed if and only if it contains its cluster points, that is, $S \supseteq S'$.*

To prove this, let X be the topological space and suppose first that S is closed. If $S = X$ then obviously $S \supseteq S'$. Otherwise, $\sim S$ is open and nonempty. If $x \in \sim S$, then $\sim S$ is a neighbourhood of x containing no point of S. So x is not a cluster point for S; that is, $x \notin S'$. Taking the contrapositive, if $x \in S'$ then $x \in S$, so $S' \subseteq S$.

Next, suppose that $S' \subseteq S$. We have to show that $\sim S$ is open. Since $\varnothing$ is an open set, we may assume $\sim S \neq \varnothing$, so take $x \in \sim S$. Then there is a neighbourhood U of x such that $U \subseteq \sim S$. This is so, because otherwise every neighbourhood of x would contain a point of S, which would mean that x is a cluster point for S. That is, $x \in S' \subseteq S$, contradicting the statement that $x \in \sim S$. The union of all such neighbourhoods U for all such points x is a set V, and $V \subseteq \sim S$. Any point of $\sim S$ belongs to some such neighbourhood, and hence to their union V. Thus $\sim S = V$. Since V is a union of open sets, it is itself open, so $\sim S$ is open. $\quad\square$

Exercise 5.7(5), below, gives yet another way of thinking of closed sets, in terms of the closure of a set.

5.3 Compact sets

In any metric space (X, d) containing at least two points x and y, we can always find open balls centred at x and y and not intersecting. For example, take the open balls $b(x, r)$ and $b(y, r)$, with $r < \frac{1}{2}d(x, y)$. Not all topological spaces have this kind of property. It turns out to be the minimal required property to allow us to carry on much of the analysis that we are used to. These spaces have their own name.

Definition 5.3.1   A topological space $(X, \mathscr{T})$ is called a *Hausdorff space*, and $\mathscr{T}$ is called a *Hausdorff topology*, if for every pair of distinct points $x, y \in X$ there is a neighbourhood U_x of x and a neighbourhood U_y of y such that $U_x \cap U_y = \varnothing$.

Briefly, X is a Hausdorff space if distinct points in the space have disjoint neighbourhoods. As we have just shown, every metric space is a Hausdorff space. So is every set with the discrete topology, $\mathscr{T}_{\max}$. However, the indiscrete topology $\mathscr{T}_{\min}$ is not Hausdorff. In the hierarchy of spaces that we have often spoken of, we see that Hausdorff spaces sit between topological spaces in general and metric spaces.

A Hausdorff space is one of a number of types of topological spaces with different levels of 'separation'. We can visualise what this means by comparing Hausdorff spaces with $(X, \mathscr{T}_{min})$, for any X: the points of the latter cannot be separated at all in the sense that every point is contained within the same open set. This in fact is the reasoning behind the term 'indiscrete' for this topology. The discrete topology, on the other hand, has maximal separation of its points, since each point of a discrete topological space is in effect itself an open set.

The Hausdorff separation property is sufficient to allow a generalisation of some of the work on compactness in Section 1.6. Compactness itself is defined much as it was there.

Definition 5.3.2 A subset S of a topological space X is *compact* if any collection of open sets in X whose union contains S has a finite subcollection whose union contains S.

As before, we commonly refer to *open coverings* of S, and say that S is compact if every open covering of S has a finite subcovering. Recall that we are distinguishing compactness, as just defined, from the sequential compactness of Chapter 4. The next theorem is a generalisation of Theorem 1.6.5.

Theorem 5.3.3 *Every compact subset of a Hausdorff space is closed.*

To prove this, let S be a compact subset of a Hausdorff space X. The result is clear if $S = X$, so assume $S \neq X$. We will show that S is closed by showing that $S \supseteq S'$, and employing Theorem 5.2.3. For this, we will suppose that $x \in {\sim}S$ and will show that x is not a cluster point for S. For each point $y \in S$, there are disjoint neighbourhoods U_y of x and V_y of y, as X is Hausdorff. The collection $\{V_y : y \in S\}$ is an open covering of S, so, as S is compact, there is a finite subcollection V_{y_1}, V_{y_2}, ..., V_{y_n}, say, of these that is a covering of S. For the corresponding neighbourhoods $U_{y_1}, U_{y_2}, \ldots, U_{y_n}$ of x, put $U = \bigcap_{k=1}^{n} U_{y_k}$. Then, since U is a finite intersection of open sets, it is itself an open set and is in fact a neighbourhood of x, which is disjoint from $\bigcup_{k=1}^{n} V_{y_k}$ and hence from S. So x is not a cluster point for S. □

In this theorem, the condition that X be a Hausdorff space cannot be dropped. This is shown by the following example, in which open sets are also compact.

Let X be an infinite set and let $\mathscr{T} = \{T : T \subseteq X,\ T = \varnothing$ or $\sim T$ is finite$\}$. Then it is not difficult to see that $\mathscr{T}$ is a topology for X. Take any subset S of X, and let U be one set chosen from an open covering of S. Since $\sim U$ is finite, only finitely many further sets in that open covering would be required to give us, with U, a finite subcovering of S. Thus, every subset of X is compact.

We turn our attention next to proving that, in a metric space, the two notions of compactness and sequential compactness coincide. In order to break up the proof, it is convenient to introduce two further notions. We will say that a metric space X has the *Bolzano–Weierstrass property* if every infinite subset of X has a cluster point. This is obviously a property suggested by Theorem 1.5.3, the Bolzano–Weierstrass theorem. And we will say that X is *countably compact* if every countable open covering of X has a finite subcovering. (By a 'countable open covering', we mean a countable collection of open sets whose union is X.)

Then we can prove the following.

Theorem 5.3.4 *Let (X, d) be a metric space. The following statements are equivalent:*

(a) X *is compact,*

(b) X *is countably compact,*

(c) X *has the Bolzano–Weierstrass property,*

(d) X *is sequentially compact.*

The scheme of the proof is to show that (a) $\Rightarrow$ (b) $\Rightarrow$ (c) $\Rightarrow$ (d) $\Rightarrow$ (b) $\Rightarrow$ (a), where $\Rightarrow$ is read as 'implies'. Then each statement will imply each of the others, so that all four are equivalent.

If X is compact, then in particular X is countably compact, so (a) implies (b).

Suppose X is countably compact, but does not have the Bolzano–Weierstrass property. Then there is an infinite subset, Y say, of X that does not have a cluster point. Let S be any countably infinite subset of Y. Then S also has no cluster point in Y, so each point $x \in S$ has a neighbourhood U_x containing no other point of Y. In a trivial way, by Theorem 5.2.3, S must be a closed set, so $\sim S$ is open. Then the union of all neighbourhoods U_x, with $\sim S$, is a countable open covering of X. But X is countably compact, so we must have a contradiction since no finite subcovering could contain all points of S. Thus (b) implies (c).

Now suppose X has the Bolzano–Weierstrass property, and let $\{x_n\}$ be any sequence in X. If the range of the sequence is finite, then it clearly has a convergent subsequence. Otherwise, the range is infinite and therefore has a cluster point, x say. Every neighbourhood of x contains some term x_n of the sequence, different from x, so the open ball $b(x, 1/k)$ contains a point x_{n_k}, for $k \in \mathbf{N}$, different from x. Since $d(x_{n_k}, x) < 1/k$ for each k, $\{x_{n_k}\}_{k=1}^\infty$ is a convergent subsequence of $\{x_n\}$. So X is sequentially compact, and (c) implies (d).

Let X now be sequentially compact, and suppose there is a countable open covering $\{T_1, T_2, \ldots\}$ of X that has no finite subcovering. Then all of the sets $U_n = \sim \bigcup_{k=1}^n T_k$ ($n \in \mathbf{N}$) are nonempty. For each n, let x_n be a point in U_n, so that $x_n \notin T_k$ for $k = 1, 2, \ldots, n$. We will show that the sequence $\{x_n\}$ has no convergent subsequence, contradicting the statement that X is sequentially compact, and thus showing that (d) implies (b). Suppose there is a subsequence $\{x_{n_k}\}$ which converges, with limit x. We must have $x \in T_N$ for some $N \in \mathbf{N}$, and then $x_{n_k} \in T_N$ for all $k > K$, say. We can assume $K > N$ and then, since $n_k \geqslant k$, we have a contradiction of the statement above that $x_{n_k} \notin T_{n_k}$.

Finally, we must prove that (b) implies (a). We begin by noting that if X is countably compact then it is sequentially compact (as we have already proved) and hence, by the result of Solved Problem 4.4(2), it contains a finite ϵ-net for each $\epsilon > 0$. This means that there exists a set $E(\epsilon) = \{u_1, u_2, \ldots, u_n\} \subseteq X$ such that, if $x \in X$ then $x \in b(u_k, \epsilon)$ for some $k = 1, 2, \ldots, n$. For each $n \in \mathbf{N}$, there is a corresponding finite set $E(1/n)$ and, by Theorem 1.4.2, their union $F = \bigcup_{k=1}^\infty E(1/k)$ is countable and the collection $\mathscr{V} = \{b(u, 1/n) : u \in F, \, n \in \mathbf{N}\}$ of open balls in X is countable. Let x be any point of X, and U any neighbourhood of x. We can clearly find $m \in \mathbf{N}$ such that $b(x, 1/m) \subseteq U$, and then we can find $u \in F$ such that $d(u, x) < 1/2m$. Thus,

$$x \in b\left(u, \frac{1}{2m}\right) \subseteq b\left(x, \frac{1}{m}\right),$$

so we have shown that there exists an open ball $B \in \mathscr{V}$ which is such that $x \in B \subseteq U$.

To complete the proof that (b) implies (a), let $\mathscr{U}$ be an arbitrary open covering of X and let $\mathscr{V}_0$ consist of those $B \in \mathscr{V}$ for which there is an open set $U \in \mathscr{U}$ with $B \subseteq U$. Let U_B be such a set U. The set $\mathscr{U}_0 = \{U_B : B \in \mathscr{V}_0\}$ is a countable subcollection of $\mathscr{U}$. We will show that it is also a covering of X. For this purpose, take any $x \in X$. For some $U \in \mathscr{U}$, we have $x \in U$, and, as above, there exists $B \in \mathscr{V}$ such that $x \in B \subseteq U$. Then, for the corresponding set $U_B \supseteq B$, we have $x \in U_B$. Thus, $\mathscr{U}_0$ is an open covering of X. Since X is countably compact, $\mathcal{U}_0$ has a finite subcovering of X, and hence so too does $\mathcal{U}$.

$\square$

This proof is easily adapted to show that any subset of a metric space is compact if and only if it is sequentially compact.

5.4 Continuity in topological spaces

It is not difficult to define convergence of a sequence in a topological space along the lines of Definition 2.5.1. Then we will use this in a definition of continuity of mappings between topological spaces in the manner of Definition 3.1.1.

Definition 5.4.1

(a) A sequence $\{x_n\}$ in a topological space X is *convergent* to a point $x \in X$ if, given any neighbourhood U of x, there exists a positive integer N such that $x_n \in U$ whenever $n > N$. As usual, we say the sequence has *limit* x, and we write $x_n \to x$.

(b) Let X and Y be topological spaces. A mapping $A: X \to Y$ is said to be *sequentially continuous at* $x \in X$ if, whenever $\{x_n\}$ is a convergent sequence in X with limit x, $\{Ax_n\}$ is a convergent sequence in Y with limit Ax. The mapping A is *sequentially continuous on* X if it is sequentially continuous at every point of X.

We have, from the beginning in Section 1.9, thought of this approach to continuity through convergent sequences as an alternative to the original 'ϵ–δ' version of Definition 1.9.1. Theorem 1.9.2 showed the two approaches to be equivalent in **R**. We will shortly give the generalisation of that original approach to mappings between topological spaces, and it will turn out that it is not equivalent to the sequential continuity which we have just defined. In metric spaces, though, the two are equivalent.

We first need a further concept to do with functions. Let X and Y be any sets, and let $f: X \to Y$ be a function from X into Y. We recall that, when $C \subseteq X$, the set $\{y : y \in Y,\ y = f(x) \text{ for some } x \in C\}$ is called the *image* $f(C)$ of C. Furthermore, if $D \subseteq Y$, then we call the set $\{x : x \in X,\ f(x) \in D\}$ the *inverse image*, or *preimage*, of D. This subset of X is denoted by $f^{-1}(D)$. The notation must not be confused with that for an inverse function. The following theorem lists a number of properties of images and inverse images.

Theorem 5.4.2 *Let* $f\colon X \to Y$ *be a function from a set* X *into a set* Y.
Let C_1, C_2 *and* C *be subsets of* X, *and let* D_1, D_2 *and* D *be subsets of* Y.

(a) $f(C_1) \subseteq f(C_2)$ *if* $C_1 \subseteq C_2$; $f^{-1}(D_1) \subseteq f^{-1}(D_2)$ *if* $D_1 \subseteq D_2$.

(b) $f(C_1 \cup C_2) = f(C_1) \cup f(C_2)$; $f^{-1}(D_1 \cup D_2) = f^{-1}(D_1) \cup f^{-1}(D_2)$.

(c) $f(C_1 \cap C_2) \subseteq f(C_1) \cap f(C_2)$; $f^{-1}(D_1 \cap D_2) = f^{-1}(D_1) \cap f^{-1}(D_2)$.

(d) $f(C_1 \backslash C_2) \subseteq f(C_1)$; $f^{-1}(D_1 \backslash D_2) = f^{-1}(D_1) \backslash f^{-1}(D_2)$.

(e) $C \subseteq f^{-1}(f(C))$; $f(f^{-1}(D)) \subseteq D$.

Results corresponding to those in (b) and (c) are true for unions and intersections of arbitrarily many sets. The second result of (d) may be given in a natural way as $f^{-1}(\sim D) = \sim f^{-1}(D)$, where we have written D for D_2. We will prove just (c), here. The proof of the rest of the theorem is left as an exercise.

Consider the first result in (c). If $f(C_1 \cap C_2) = \varnothing$, the result is clear, so suppose $f(C_1 \cap C_2) \neq \varnothing$ and let $y \in f(C_1 \cap C_2)$. Then $y = f(x)$ for some $x \in C_1 \cap C_2$. Since $x \in C_1$ and $x \in C_2$, then $f(x) \in f(C_1)$ and $f(x) \in f(C_2)$, so $f(x) \in f(C_1) \cap f(C_2)$. Thus $f(C_1 \cap C_2) \subseteq f(C_1) \cap f(C_2)$, and we are done.

For the second result, suppose that $f^{-1}(D_1 \cap D_2) \neq \varnothing$ and take any $x \in f^{-1}(D_1 \cap D_2)$. Then $f(x) \in D_1 \cap D_2$ so $f(x) \in D_1$ and $f(x) \in D_2$. Hence $x \in f^{-1}(D_1)$ and $x \in f^{-1}(D_2)$, so $x \in f^{-1}(D_1) \cap f^{-1}(D_2)$. It follows that $f^{-1}(D_1 \cap D_2) \subseteq f^{-1}(D_1) \cap f^{-1}(D_2)$, and this is true also if $f^{-1}(D_1 \cap D_2) = \varnothing$. Next, suppose $f^{-1}(D_1) \cap f^{-1}(D_2) \neq \varnothing$ and take $x \in f^{-1}(D_1) \cap f^{-1}(D_2)$. Then $x \in f^{-1}(D_1)$ and $x \in f^{-1}(D_2)$, so $f(x) \in D_1$ and $f(x) \in D_2$. Hence $f(x) \in D_1 \cap D_2$, so $x \in f^{-1}(D_1 \cap D_2)$. This time, we conclude that $f^{-1}(D_1) \cap f^{-1}(D_2) \subseteq f^{-1}(D_1 \cap D_2)$, and this is true also if $f^{-1}(D_1) \cap f^{-1}(D_2) = \varnothing$. The result now follows. $\square$

The ϵ–δ definition of continuity of a real-valued function f at x_0 may be viewed as describing a relationship between neighbourhoods. The values of $f(x)$ such that $|f(x) - f(x_0)| < \epsilon$ lie in a certain neighbourhood V of $f(x_0)$, and the values of x such that $|x - x_0| < \delta$ are in a neighbourhood U of x_0. The definition states that f is continuous at x_0 if $f(x) \in V$ whenever $x \in U$; that is, if $x \in f^{-1}(V)$ whenever $x \in U$; that is, if $U \subseteq f^{-1}(V)$. This is how we arrive at our definition of continuity in topological space.

Definition 5.4.3 Let X and Y be topological spaces. We say that a mapping $A\colon X \to Y$ is *continuous at* $x \in X$ if, given any neighbourhood V of Ax, there exists a neighbourhood U of x such that

$U \subseteq A^{-1}(V)$. The mapping A is *continuous on* X if it is continuous at every point of X.

We will give some equivalent formulations of continuity in topological space, and then will relate this to sequential continuity.

Theorem 5.4.4 *Let* $A \colon X \to Y$ *be a mapping between topological spaces. The following statements are equivalent:*

(a) A *is continuous on* X,

(b) $A^{-1}(T)$ *is an open set in* X *for each open set* T *in* Y,

(c) $A^{-1}(S)$ *is a closed set in* X *for each closed set* S *in* Y.

The equivalence of (a) and (b) justifies a common alternative definition of continuity: the mapping is continuous on X if the inverse image of every open set in Y is an open set in X. We will prove the theorem according to the scheme: (a) $\Rightarrow$ (b) $\Rightarrow$ (c) $\Rightarrow$ (b) $\Rightarrow$ (a).

Suppose A is continuous on X and T is open in Y. For each point $y = Ax \in T$, there is a neighbourhood U_x of x in X which is such that $U_x \subseteq A^{-1}(T)$. It follows that $A^{-1}(T)$ is equal to the union of all such open sets U_x, and is consequently itself an open set. So (a) implies (b). To show that (b) implies (c), suppose that S is a closed set in Y, so $\sim S$ is an open set in Y. We are assuming (b) is true, so $A^{-1}(\sim S)$ is open in X. By Theorem 5.4.2 (d), $A^{-1}(S) = \sim A^{-1}(\sim S)$, so $A^{-1}(S)$ is a closed set in X. The same argument, interchanging 'open' and 'closed', shows that (c) implies (b). Finally, to show that (b) implies (a), let $x \in X$ and let V be a neighbourhood of Ax. Then $A^{-1}(V)$ is an open set in X so it is a neighbourhood of x, and itself serves to show that A is continuous at x. $\Box$

Theorem 5.4.5 *Let* X *and* Y *be topological spaces, and let* $A \colon X \to Y$ *be a continuous mapping on* X. *Then* A *is sequentially continuous on* X.

To prove this, take any point $x \in X$ and let $\{x_n\}$ be a sequence in X convergent to x. Let V be a neighbourhood of Ax. By Theorem 5.4.4 and the continuity of A, $A^{-1}(V)$ is a neighbourhood of x, so a positive integer N exists such that $x_n \in A^{-1}(V)$ for $n > N$. Then $Ax_n \in V$ for $n > N$, so $Ax_n \to Ax$. Hence, A is sequentially continuous at x. $\Box$

We will give an example now to show that the converse of this theorem is not true, so continuity and sequential continuity are not equivalent in general. Take the set $\mathbf{R}$, and let $\mathscr{T}$ be the collection of sets consisting of $\varnothing$ and the complements of countable subsets of $\mathbf{R}$. It is easy enough to verify that $\mathscr{T}$ is a topology on $\mathbf{R}$. Let $\{x_n\}$ be a convergent sequence in $(\mathbf{R}, \mathscr{T})$. Let its limit be x and its range be S. Then $S\backslash\{x\}$ is a countable set, so its complement is a neighbourhood of x. In order for this set to contain all terms x_n for large enough n, we must have $x_n = x$ for all $n > N$, say. That is, the range S of $\{x_n\}$ must be finite. Now consider the identity map $I\colon (\mathbf{R}, \mathscr{T}) \to (\mathbf{R}, \mathscr{T}')$, where $\mathscr{T}'$ is the usual topology on $\mathbf{R}$, and consider such a sequence $\{x_n\}$. If $n > N$, then we have $Ix_n = x_n = x = Ix$ so I is certainly sequentially continuous on $\mathbf{R}$. However, choose any nonempty set $T' \in \mathscr{T}'$, for which $\sim T'$ is uncountable. (For example, let T' be any open interval.) Then $I^{-1}(T') = T'$, but $T' \notin \mathscr{T}$ since $T' \neq \varnothing$ and T' is not the complement of a countable subset of $\mathbf{R}$. Hence, I is not continuous on $\mathbf{R}$.

In metric spaces, the two forms of continuity do coincide. That is what we prove next.

Theorem 5.4.6 *Let X and Y be metric spaces. A mapping $A\colon X \to Y$ is continuous on X if and only if A is sequentially continuous on X.*

Following on from the preceding theorem, it is only necessary to show that if A is sequentially continuous at some point $x \in X$ then it is continuous at x. Put $y = Ax$ and let V be a neighbourhood of y. By definition of an open set in the metric topology (Definition 5.1.4), there exists an open ball $b_Y(y, \epsilon)$ in Y with $b_Y(y, \epsilon) \subseteq V$. Suppose A is not continuous at x. Then, since open balls are open sets in the metric topology (to be proved in Exercise 5.6(4)), there is no open ball $b_X(x, \delta)$ in X such that $b_X(x, \delta) \subseteq A^{-1}(b_Y(y, \epsilon))$, whatever the value of δ. (Otherwise, $b_X(x, \delta) \subseteq A^{-1}(V)$, by Theorem 5.4.2(a), and then A is continuous at x.) Then, for each $n \in \mathbf{N}$, there is a point x_n in X such that $x_n \in b_X(x, 1/n)$ and $x_n \notin A^{-1}(b_Y(y, \epsilon))$. This generates a sequence $\{x_n\}$, and clearly $x_n \to x$. So $Ax_n \to y$ since A is sequentially continuous at x. Then for all sufficiently large n we must have $Ax_n \in b_Y(y, \epsilon)$. This is a contradiction, so A must indeed be continuous at x. $\square$

5.5 Homeomorphisms; connectedness

A particular type of continuous mapping that is basic to the further study of topology is given in the following definition.

Definition 5.5.1 Let X and Y be topological spaces. A *homeomorphism* between X and Y is a continuous bijection $A\colon X \to Y$, with the property that A^{-1} is also continuous. If such a homeomorphism exists, then X and Y are said to be *homeomorphic*.

Recall that a bijection is a one-to-one onto mapping, and that a bijection always has an inverse mapping. So there is a one-to-one correspondence between the points of homeomorphic spaces. Furthermore, the property of continuous mappings that inverse images of open sets are open sets, and the fact that a homeomorphism and its inverse are both continuous, mean that there is a one-to-one correspondence between the open sets of homeomorphic spaces. For these reasons, topologically speaking, two homeomorphic spaces are considered to be essentially identical.

A *topological property* is one which is common to homeomorphic topological spaces and is made evident by the homeomorphism. *Topology* itself can be thought of as the study of topological properties. Compactness is one such property. Completeness, in metric space, is not a topological property: examples can be given of homeomorphic metric spaces where one is complete and the other is not. Topology is often known colloquially as 'rubber sheet geometry'. This term comes about by considering a topological space as drawn (in some sense) on a rubber sheet. Homeomorphic images of that space result from stretching and bending the sheet, provided it does not tear. Thus, a circle is topologically identical to any ellipse, or to any rectangle.

It is therefore important to be able to determine whether a mapping is a homeomorphism. One such result in this direction is Theorem 5.5.3, below. The following result is required first. It is the generalisation of Theorem 4.3.1 to topological spaces.

Theorem 5.5.2 *Let $A\colon X \to Y$ be a continuous mapping between topological spaces X and Y, and let S be a compact subset of X. Then $A(S)$ is a compact subset of Y.*

For the proof, let $\mathcal{V}$ be an open covering of $A(S)$. Since A is continuous, $A^{-1}(V)$ is an open set in X, for each $V \in \mathcal{V}$. We will show that $\mathcal{U} = \{A^{-1}(V) : V \in \mathcal{V}\}$ is an open covering of S. If $x \in S$, then $Ax \in A(S)$ so that $Ax \in V$ for some $V \in \mathcal{V}$. Then $x \in A^{-1}(V)$. So indeed $\mathcal{U}$ is an open covering of S. Since S is compact, there is a finite subcovering $\{A^{-1}(V_1), A^{-1}(V_2), \ldots, A^{-1}(V_n)\}$, say, chosen from $\mathcal{U}$. If $y \in A(S)$, then $y = Ax$ for some $x \in S$, and $x \in A^{-1}(V_k)$ for some $k = 1, 2, \ldots, n$. Then $Ax = y \in V_k$. This shows that $\{V_1, V_2, \ldots, V_n\}$ is a finite subcovering of $A(S)$, chosen from $\mathcal{V}$. Hence $A(S)$ is compact. $\qquad\square$

Theorem 5.5.3 *Let X be a compact topological space, Y a Hausdorff space, and $A\colon X \to Y$ a continuous bijection. Then A is a homeomorphism.*

All the conditions for A to be a homeomorphism are present except for the continuity of A^{-1}, so this is all we need to show. Since A^{-1} is a mapping from Y onto X, and $(A^{-1})^{-1} = A$, we must show that the image $A(T)$ of an arbitrary open set T in X is an open set in Y. Theorem 1.6.6 stated that a closed subset of a compact set is compact. That was with reference to $\mathbf{R}$, but there is a direct analogue, proved the same way, for any topological space. So, since $\sim T$ is a closed subset of the compact space X, it is compact. By the preceding theorem and Theorem 5.3.3, $A(\sim T)$ is a compact subset of Y, and is closed. So $\sim A(\sim T)$ is open. By Theorem 5.4.2(d), $\sim A(\sim T) = A(T)$, and so we are finished. $\square$

We will end this chapter with a few comments regarding another important topological property, *connectedness*. This notion may be familiar from complex variable theory, where the domain of an analytic function is typically required to be an open connected set.

The term 'separation' arose earlier in this chapter. We now give it a precise meaning. Connectedness is then defined as a lack of separation.

Definition 5.5.4

(a) A *separation*, or *partition*, of a subset S of a topological space X is a disjoint pair (T_1, T_2) of open sets in X with the properties:

   (i) $T_1 \cap S \neq \varnothing$ and $T_2 \cap S \neq \varnothing$,
   (ii) $S = (T_1 \cap S) \cup (T_2 \cap S)$.

(b) A subset of a topological space is *connected* if it has no separation.

The definition may be more easily visualised in terms of the special case $S = X$: a separation of the topological space X is a disjoint pair of nonempty open sets T_1 and T_2 such that $X = T_1 \cup T_2$. We can say that X is connected if it cannot be expressed as a union of disjoint nonempty open sets. Otherwise, X is *disconnected*. Intuitively, a connected set consists of one piece.

When a topological space X has a separation, we can write $X = T_1 \cup T_2$ for disjoint open sets T_1 and T_2. These sets are then the complements of each other, so they are also both closed. It is easy to see that X is connected if and only if $\varnothing$ and X are the only subsets of X which are both open and closed.

It follows from the definition that the set $\{x\}$ is connected for any point x of a topological space $(X, \mathscr{T})$. When $\mathscr{T} = \mathscr{T}_{\max}$, these are the only connected subsets of X, while if $\mathscr{T} = \mathscr{T}_{\min}$ every subset of X is connected. In $\mathbf{R}$, with the usual topology, the only connected sets are the sets consisting of a single point and all the various intervals described at the beginning of Section 1.5. We will omit the proof of this statement.

Under a continuous mapping, a connected set stays connected. That is the content of the next theorem.

Theorem 5.5.5 *Let $A\colon X \to Y$ be a continuous mapping between topological spaces. If S is a connected subset of X then $A(S)$ is a connected subset of Y.*

To prove this, suppose there exists a separation (T_1, T_2) of $A(S)$. Then we will show that (S_1, S_2), where $S_1 = A^{-1}(T_1)$ and $S_2 = A^{-1}(T_2)$, is a separation of S, contradicting the fact that S is connected. Certainly, S_1 and S_2 are open sets in X, since T_1 and T_2 are open in Y and A is continuous. If $x \in S_1 \cap S_2$, then we easily see that $Ax \in T_1 \cap T_2$. But $T_1 \cap T_2 = \varnothing$, so $S_1 \cap S_2 = \varnothing$. We know that $T_1 \cap A(S) \neq \varnothing$. Take any point $y \in T_1 \cap A(S)$ and say $y = Ax$. Then $x \in A^{-1}(T_1) = S_1$ and $x \in S$, so $S_1 \cap S \neq \varnothing$, and similarly $S_2 \cap S \neq \varnothing$. Finally, suppose $x \in S$, so that $Ax \in A(S) = (T_1 \cap A(S)) \cup (T_2 \cap A(S))$. If $Ax \in T_1 \cap A(S)$ then $x \in A^{-1}(T_1 \cap A(S)) = A^{-1}(T_1) \cap A^{-1}(A(S))$, by Theorem 5.4.2(c). In particular, $x \in A^{-1}(T_1) = S_1$, so $x \in S_1 \cap S$. If $Ax \in T_2 \cap A(S)$, then we proceed similarly, and conclude that $x \in (S_1 \cap S) \cup (S_2 \cap S)$, so that $S \subseteq (S_1 \cap S) \cup (S_2 \cap S)$. The reverse inclusion is obvious, so $S = (S_1 \cap S) \cup (S_2 \cap S)$. We have shown that (S_1, S_2) is a separation of S, as required. $\qquad\square$

5.6 Solved problems

In the first of these solved problems, we will need the following definition.

Definition 5.6.1

(a) An *open base* for a topological space X is a collection $\mathscr{U}$ of open sets in X with the property that every open set in X is the union of sets in $\mathscr{U}$.

(b) A topological space which has a countable open base is said to be *second countable*, or to satisfy the *second axiom of countability*. (We do not need the definition of a first countable space.)

(1) Prove *Lindelöf's Theorem*: Let X be a second countable space and let S be a nonempty subset of X. If $S = \bigcup_{T \in \mathscr{S}} T$ for some collection $\mathscr{S}$ of open sets in X, then S is the union of a countable subcollection of sets in $\mathscr{S}$.

Solution. Let $\mathscr{U}$ be a countable open base for X. Any point $x \in S$ satisfies $x \in T_x$ for some $T_x \in \mathscr{S}$ and thus $x \in U_x \subseteq T_x$ for some $U_x \in \mathscr{U}$, since T_x is a union of sets in $\mathscr{U}$ by definition of an open base. The collection $\{U_x : U_x \in \mathscr{U}, \; x \in S\}$ is certainly countable, and its union is S. The corresponding collection $\{T_x : T_x \in \mathscr{S}, \; U_x \subseteq T_x\}$ then clearly satisfies the requirements of the theorem. $\qquad\square$

(2) Let f and g be two functions from a topological space X into $\mathbf{R}$, with the usual topology. Prove that $f + g$ is continuous on X if f and g are.

Solution. Put $h = f + g$, take any point $x \in X$, and write $y = h(x)$. Let V be a neighbourhood of y. We must show that there is a neighbourhood U of x such that $U \subseteq h^{-1}(V)$. Since V is an open set in $\mathbf{R}$, we can find $\epsilon > 0$ such that $(y - \epsilon, y + \epsilon) \subseteq V$. Let V_1 and V_2 be the intervals

$$V_1 = (f(x) - \tfrac{1}{2}\epsilon, f(x) + \tfrac{1}{2}\epsilon), \quad V_2 = (g(x) - \tfrac{1}{2}\epsilon, g(x) + \tfrac{1}{2}\epsilon).$$

Since f and g are continuous, there are neighbourhoods U_1, U_2 of x such that $U_1 \subseteq f^{-1}(V_1)$ and $U_2 \subseteq g^{-1}(V_2)$. The intersection $U_1 \cap U_2$ is also a neighbourhood of x. For any point $x' \in U_1 \cap U_2$, we have $x' \in U_1$ and $x' \in U_2$, so

$$\begin{aligned}
|y - h(x')| &= |(f(x) + g(x)) - (f(x') + g(x'))| \\
&\leqslant |f(x) - f(x')| + |g(x) - g(x')| < \tfrac{1}{2}\epsilon + \tfrac{1}{2}\epsilon = \epsilon.
\end{aligned}$$

That is, $h(x') \in V$, or $x' \in h^{-1}(V)$. Thus $U_1 \cap U_2 \subseteq h^{-1}(V)$, so we may take $U = U_1 \cap U_2$, showing that $f + g$ is continuous at x, and hence on X. $\qquad\square$

5.7 Exercises

(1) Let $X = \{a, b, c, d\}$,

$$\mathscr{T}_1 = \{\varnothing, \{a\}, \{b\}, \{a, b\}, \{a, b, c\}, \{a, b, d\}, X\}$$

and $\mathscr{T}_2 = \{\varnothing, \{a\}, \{a, b\}, \{a, c\}, \{a, b, c\}, X\}$.

  (a) Verify that $\mathscr{T}_1$ and $\mathscr{T}_2$ are topologies on X.

(b) In $(X, \mathcal{T}_1)$, find the closed sets, and find the interiors and closures of $\{a\}$, $\{c\}$ and $\{a,c\}$.

(c) Do the same in $(X, \mathcal{T}_2)$.

(2) In the topological spaces $(X, \mathcal{T}_{\max})$ and $(X, \mathcal{T}_{\min})$, what are int S and $\overline{S}$ for any subset S of X?

(3) In any topological space X, prove: (a) X and $\varnothing$ are closed sets, (b) arbitrary intersections of closed sets are closed, (c) finite unions of closed sets are closed.

(4) (a) In a metric space, prove that every open ball is an open set. That is, for each x belonging to an open ball $b(x_0, r)$ in a metric space, show that there is an open ball $b(x, \epsilon)$ satisfying $b(x, \epsilon) \subseteq b(x_0, r)$.

(b) Verify that the metric topology $\mathcal{T}_d$ of Definition 5.1.4 defines a topology on every metric space.

(5) (a) Let S be a subset of a topological space. Prove that $\overline{S} = S \cup S'$.

(b) Let S_1, S_2 be subsets of a topological space, with $S_1 \subseteq S_2$. Prove that $\overline{S}_1 \subseteq \overline{S}_2$.

(6) Let X be a topological space, and let S be a subset of X. Prove: (a) $\overline{S}$ is closed, (b) $S \subseteq \overline{S}$, (c) S is closed if and only if $S = \overline{S}$.

(7) Let X be a Hausdorff space. Show that, for each $x \in X$, the subset $\{x\}$ is closed.

(8) Prove that the discrete topology is the only Hausdorff topology on a finite set.

(9) Prove parts (a), (b), (d) and (e) of Theorem 5.4.2.

(10) (a) In the topological space $(X, \mathcal{T}_{\min})$, show that any sequence is convergent, and any point of X is its limit.

(b) Prove that any convergent sequence in a Hausdorff space has a unique limit.

(11) Let $\mathcal{T}_1$ and $\mathcal{T}_2$ be two topologies on a set X. Show that the identity map $I: (X, \mathcal{T}_1) \to (X, \mathcal{T}_2)$, for which $Ix = x$ for all $x \in X$, is continuous if and only if $\mathcal{T}_1$ is stronger than $\mathcal{T}_2$.

(12) Prove that two metric spaces (X_1, d_1) and (X_2, d_2) are homeomorphic if there exists a mapping A of X_1 onto X_2 such that

$$\alpha d_1(x, y) \leqslant d_2(Ax, Ay) \leqslant \beta d_1(x, y)$$

for some positive real constants α and β, and all $x, y \in X_1$.

(13) Let X and Y be connected subsets of a topological space, and suppose $X \cap Y \neq \varnothing$. Prove that $X \cup Y$ is connected.

....

(14) Let f and g be two functions from a topological space X into $\mathbf{R}$, with the usual topology. Prove that fg is continuous on X if f and g are.

(15) A topological space X is called a T_1-*space* if, given any two distinct points of X, each has a neighbourhood which does not contain the other.

    (a) Show that every Hausdorff space is a T_1-space.

    (b) Prove that a topological space X is a T_1-space if and only if, for every $x \in X$, $\{x\}$ is a closed set.

    (c) Show that every finite T_1-space has the discrete topology.

(16) Prove that a collection $\mathscr{U}$ of open sets in a topological space $(X, \mathscr{T})$ is an open base for X if and only if for each $T \in \mathscr{T}$ and each $x \in T$ there exists $U \in \mathscr{U}$ such that $x \in U \subseteq T$.

6

Normed Vector Spaces

6.1 Definition of a normed vector space; examples

In this and the following chapters we will give an indication of the advantages to be gained by superimposing onto vector spaces the ideas we have developed for metric spaces. It is worthwhile spending a few lines now to enlarge on the reasons previously given for wanting to do this.

All the work of Chapters 2, 3 and 4 was developed from the three axioms (M1), (M2) and (M3) for a metric space. The numerous applications that we have given from many fields are a pointer to just how much can be developed in this way. In all of those applications, the metric was defined in a way suggested by our ultimate aim within the application and we then made use of the general theorems deduced earlier. Within each application our knowledge of the subject matter of that application allowed us to carry out the usual manipulations that occur in any piece of mathematics. A common operation was of course the addition of elements. The pertinent point is that this could only be done within applications because, according to the axioms of a metric space X, no meaning is attached to any form of sum of elements of X. Imagine therefore what extra general theorems could be obtained if in the axioms themselves we did incorporate such an operation.

In a vector space we may add elements together. We may also multiply them by scalars. These operations alone give rise to a vast number of applications, as any book on linear algebra will show. When we incorporate the idea of a metric (which allows us to speak of the convergence of sequences of elements in the space), we may combine the two fields of algebra and analysis, and this is a basic feature of modern analysis.

In Section 1.11, we detailed most of what we need to know about vector spaces. Remember that whenever we use vector spaces whose

elements are n-tuples, functions or sequences, the operations of addition
and multiplication by scalars will be as given for the spaces $\mathbf{C}^n$, $C[a,b]$
and c. We remark that, as in metric spaces, we will generally refer to
the elements of a vector space as 'points'.

One vector space we will need which has not previously been mentioned, as a vector space, is the space l_2. This is defined on the same set
as the corresponding metric space, namely the set of all complex-valued
sequences $(x_1, x_2, \dots)$ for which $\sum_{k=1}^{\infty} |x_k|^2$ converges. The convergence
of this series implies that $x_n \to 0$ (Theorem 1.8.3), so the set l_2 is a subset of the set c_0 of all complex-valued sequences that converge with
limit 0. To show that l_2 is indeed a vector space, we may show that
it is a subspace of the vector space c_0. According to Definition 1.11.2,
this follows by showing that $x + y \in l_2$ when $x, y \in l_2$ and that $\alpha x \in l_2$
when $x \in l_2$, $\alpha \in \mathbf{C}$. The latter is easy. For the former, we note that
$(|x_k| - |y_k|)^2 \geqslant 0$, so

$$|x_k + y_k|^2 \leqslant (|x_k| + |y_k|)^2 \leqslant 2(|x_k|^2 + |y_k|^2)$$

for any complex numbers x_k, y_k. Then the convergence of $\sum |x_k|^2$ and
$\sum |y_k|^2$ implies that of $\sum |x_k + y_k|^2$. That l_2 is an infinite-dimensional
vector space may be shown in precisely the same way that c was shown
to be infinite-dimensional following Theorem 1.11.4.

By use of the discrete metric of Example 2.2(14), we have seen that
any set can be made into a metric space. When that set is a vector
space, it soon becomes evident that the use of a metric alone does not
allow us to take full benefit of the vector space properties of the set.
The following illustrates this. Denoting as usual the zero vector of a
vector space X by θ, and imposing a metric d on X, the number $d(\theta, x)$
represents the distance between θ and x. Using only the metric space
axioms, it is impossible to prove the very desirable and natural property

$$d(\theta, 2x) = 2d(\theta, x), \quad x \in X.$$

This equation is in fact false when d is the discrete metric and $x \neq \theta$.
Something further is required to relate the two types of structure to each
other and to allow anything new to be developed.

The quantity $d(\theta, x)$ provides the clue. For ordinary three-dimensional
vectors, the distance between θ and $\mathbf{x}$ is referred to as the length or
magnitude of $\mathbf{x}$. This is the notion that we will abstract. For any
point x in a vector space, we will define a new quantity, called the
norm of x and denoted by $\|x\|$, to generalise the idea of the length of
a vector. We will carefully specify the properties it must have, so that

in particular we will have $\|2x\| = 2\|x\|$, getting around the problem described above. Then $\|x - y\|$ will denote the length of the vector $x - y$, or in other words (thinking again of ordinary three-dimensional vectors) the distance between x and y. But this would be $d(x, y)$, allowing us to retrieve the metric space properties.

In the definition and discussion that follow, do not be dismayed by the blank look of $\|\ \|$. This is simply a symbol (conforming to a long-established convention) for a certain mapping. Its image at a point x is denoted by $\|x\|$ and this allows a visual interpretation as a generalisation of the length $|\mathbf{x}|$ of an ordinary vector $\mathbf{x}$.

Definition 6.1.1 A *normed vector space* is a vector space X together with a mapping $\|\ \|: X \to \mathbf{R}_+$ with the properties

(N1) $\|x\| = 0$ if and only if $x = \theta$ $(x \in X)$,

(N2) $\|\alpha x\| = |\alpha|\, \|x\|$ for all $x \in X$ and every scalar α,

(N3) $\|x + y\| \leqslant \|x\| + \|y\|$ for all $x, y \in X$.

This normed vector space is denoted by $(X, \|\ \|)$ and the mapping $\|\ \|$ is called the *norm* for the space.

It is possible to define different norms for the same vector space X. These may be written for example as $\|\ \|_1, \|\ \|_2, \ldots$ and then $(X, \|\ \|_1)$, $(X, \|\ \|_2), \ldots$ are different normed vector spaces. This notation is analogous to that for metric spaces but here it has a much less satisfying look. There is a correspondingly greater tendency, which we will follow, to denote the normed vector space itself by X and to introduce without prior specification $\|x\|$ for the norm of a point $x \in X$. Only in a few instances in this book (though such instances are common in deeper topics) will we be considering in the same context different norms for the same vector space, so no confusion should arise.

The term 'normed vector space' is commonly abbreviated to *normed space*.

It is left as an exercise to prove that any normed space X can now be given a metric in a natural way by defining

$$d(x, y) = \|x - y\|, \quad x, y \in X,$$

as we anticipated above. In verifying (M3), use will be made of (N3) alone; the latter is also termed a *triangle inequality*. In concert with our preliminary remarks, we now go a step further and specify that the only metric ever to be used in conjunction with a given normed space X will be that defined by $d(x,y) = \|x - y\|$, $x, y \in X$. This is called the metric *associated with*, or *generated by*, or *induced by*, the norm.

In the following examples of normed spaces, the verification of (N1), (N2) and (N3) is easy, the triangle inequality in each case following from the work done for the associated metric.

The vector space $\mathbf{C}^n$ may be normed by defining

$$\|x\| = \sqrt{\sum_{k=1}^{n} |x_k|^2},$$

where $x = (x_1, x_2, \ldots, x_n) \in \mathbf{C}^n$. This is called the *Euclidean norm* for $\mathbf{C}^n$ and is the norm we always mean when we refer to the normed space $\mathbf{C}^n$. The associated metric is of course the Euclidean metric. Other norms can be defined for this vector space; for example,

$$\|x\| = \max_{1 \leqslant k \leqslant n} |x_k|.$$

The real vector space $\mathbf{R}^n$ may be similarly normed. The Euclidean norm is

$$\|x\| = \sqrt{\sum_{k=1}^{n} x_k^2},$$

where $x = (x_1, x_2, \ldots, x_n) \in \mathbf{R}^n$, and is the norm always implied by reference to the normed space $\mathbf{R}^n$.

The vector space l_2 is normed when we define

$$\|x\| = \sqrt{\sum_{k=1}^{\infty} |x_k|^2},$$

where $x = (x_1, x_2, \ldots)$ is any element of l_2. Note that for any $x \in l_2$, $\|x\|$ is finite by the very definition of the space l_2.

By the normed space $C[a, b]$, we will always mean the real vector space $C[a, b]$ with norm given by

$$\|x\| = \max_{a \leqslant t \leqslant b} |x(t)|, \quad x \in C[a, b].$$

This is called the *uniform* norm. Other norms on the same vector space are given by

$$\|x\| = \int_a^b |x(t)|\, dt \quad \text{and} \quad \|x\| = \sqrt{\int_a^b (x(t))^2\, dt},$$

and, by reference to the associated metrics, these normed spaces are denoted by $C_1[a, b]$ and $C_2[a, b]$, respectively.

6.2 Convergence in normed spaces

We have stated that we will consider a given normed space to be a metric space in one way only, that for which the metric is generated by the norm. Then all the notions associated with the convergence of a sequence in a metric space are easily transferred to normed spaces. We summarise these.

A sequence $\{x_n\}$ in a normed space is *convergent* if for any number $\epsilon > 0$ there exists an element $x \in X$ and a positive integer N such that

$$\|x_n - x\| < \epsilon \quad \text{whenever } n > N.$$

Again we write $x_n \to x$, or $\lim x_n = x$, and call x the *limit* of the sequence. The sequence $\{x_n\}$ is a *Cauchy sequence* if, given $\epsilon > 0$, there exists a positive integer N such that

$$\|x_n - x_m\| < \epsilon \quad \text{whenever } m, n > N.$$

When every Cauchy sequence in a normed space converges, the associated metric space is *complete*. This term may be applied to the normed space itself. Complete normed spaces occur so predominantly in all of modern analysis that a special term is used for them.

Definition 6.2.1 A complete normed vector space is called a *Banach space*.

All the theorems of Section 2.5 still hold: the limit of a convergent sequence in a normed space is unique; any convergent sequence in a normed space is a Cauchy sequence. The discussion of convergence in the spaces $\mathbf{R}^n$, $\mathbf{C}^n$, l_2 and $C[a, b]$, given in conjunction with Theorem 2.5.3, also remains valid. All of these are Banach spaces.

The fixed point theorem for normed spaces says that every contraction mapping on a Banach space has a unique fixed point. As you would expect, a contraction mapping on a normed space X is a mapping $A: X \to X$ for which there is a number α, with $0 < \alpha < 1$, such that $\|Ax - Ay\| \leqslant \alpha \|x - y\|$ for any $x, y \in X$.

In the same vein, a subset of a normed space is sequentially compact if every sequence in the subset has a convergent subsequence. (Recall the note at the end of Section 2.7: a compact subset of a normed space X

certainly need not be a subspace of X in the sense of a vector subspace.) Since every normed space is a metric space, there is a metric topology induced by the norm and consequently a normed space may be defined to be compact in the topological sense of Definition 5.3.2. Then, by Theorem 5.3.4, a normed space is compact if and only if it is sequentially compact, so we may always use the simpler term 'compact' in the present context. There are natural analogues for normed spaces of all the theorems of Chapter 4, on compactness.

In a vector space, where we may add elements together, we have available the idea of an infinite series. Once the space is normed it is a simple matter to come up with a definition of convergence of a series entirely analogous to that of Definition 1.8.1 for series of real or complex numbers. Let X be a normed vector space, let $\{x_n\}$ be a sequence in X and let $s_n = \sum_{k=1}^{n} x_k$. Then $\{s_n\}$ is also a sequence in X and, as in Definition 1.8.1, we say the series $\sum_{k=1}^{\infty} x_k$ (or simply $\sum x_k$) *converges* if $\lim s_n$ exists. In that case, we say $\lim s_n$ is the *sum* of the series. It is a natural generalisation of Definition 1.8.4 to call the series $\sum x_k$ *absolutely convergent* if the series $\sum \|x_k\|$ (of real numbers) is convergent.

In the discussion of Figure 3 on page 47, we pointed out in picturesque fashion that the convergence of an absolutely convergent series of real numbers is a consequence of the completeness of the real number system. We will now state and prove the generalisation to Banach spaces of Theorem 1.8.5. We will also prove the converse, that if every absolutely convergent series in a normed space converges then the space must be a Banach space. Applied to **R**, this means that we may finish off the ring of arrows in Figure 3 with an arrow from 1.8.5 to 1.5.4. Hence all the theorems on the outer ring of Figure 3 are actually equivalent, so any one of them could have been taken as an axiom to generate the theory leading to the others.

Theorem 6.2.2 *A normed vector space X is a Banach space if and only if every absolutely convergent series in X is convergent.*

To prove this, suppose first that X is a Banach space and that $\sum x_k$ is an absolutely convergent series in X. Then, by definition, $\sum \|x_k\|$ converges. Let $\epsilon > 0$ be given. Using the triangle inequality (N3) and Theorem 1.8.2,

$$\left\| \sum_{k=m}^{n} x_k \right\| \leqslant \sum_{k=m}^{n} \|x_k\| < \epsilon$$

for all sufficiently large integers m, n $(m \leqslant n)$. This implies that the partial sums $\sum_{k=1}^{n} x_k$ form a Cauchy sequence. Since X is a Banach space (that is, X is complete), it follows that $\sum_{k=1}^{\infty} x_k$ converges, as required.

A deeper argument is required for the converse. We will be calling on Theorem 4.1.1, that a Cauchy sequence with a convergent subsequence is itself convergent. We suppose now that every absolutely convergent series in X is convergent, and let $\{x_n\}$ be a Cauchy sequence in X. Then, for any $\epsilon > 0$, we can find a positive integer N so that

$$\|x_n - x_m\| < \epsilon$$

when $m, n > N$. In particular, we may take $\epsilon = 1/2^k$ with $k = 1, 2, \ldots$ in turn and find the corresponding integers $N_1, N_2, \ldots$. We may assume $N_1 < N_2 < \cdots$. Now choose any integers $n_1, n_2, \ldots$ with $n_k > N_k$ for each k. Then for each k we also have $n_{k+1} > N_k$ and so

$$\|x_{n_{k+1}} - x_{n_k}\| < \frac{1}{2^k}.$$

It follows that

$$\sum_{k=1}^{\infty} \|x_{n_{k+1}} - x_{n_k}\| \leqslant \sum_{k=1}^{\infty} \frac{1}{2^k} = 1,$$

so the series $\sum \|x_{n_{k+1}} - x_{n_k}\|$ is convergent. This means that the series $\sum (x_{n_{k+1}} - x_{n_k})$ is absolutely convergent and by assumption it is therefore also convergent. Thus its sequence $\{s_m\}_{m=1}^{\infty}$ of partial sums converges. But

$$s_m = \sum_{k=1}^{m} \left(x_{n_{k+1}} - x_{n_k} \right)$$
$$= (x_{n_2} - x_{n_1}) + (x_{n_3} - x_{n_2}) + \cdots + \left(x_{n_{m+1}} - x_{n_m} \right)$$
$$= x_{n_{m+1}} - x_{n_1}.$$

Now, x_{n_1} is some fixed term of $\{x_n\}$, so the sequence $\{x_{n_{m+1}}\}_{m=1}^{\infty}$ converges. This is a convergent subsequence of the Cauchy sequence $\{x_n\}$ and so, by Theorem 4.1.1, $\{x_n\}$ itself converges. Hence the normed space X is complete; that is, X is a Banach space. □

A consequence of this theorem is that in every normed space which is not complete there must be an absolutely convergent series that is not convergent. We will give an example to illustrate this strange-looking notion.

The normed space $C_1[a,b]$, for which we define $\|f\| = \int_a^b |f(x)|\,dx$ for any continuous function f defined on $[a,b]$, is not complete. This was shown in Example 2.6(6). In the space $C_1[0,2]$, consider the sequence $\{f_n\}$ where

$$f_n(x) = \begin{cases} 1 - \dfrac{n^2 x}{2}, & 0 \leqslant x < \dfrac{2}{n^2}, \\ 0, & \dfrac{2}{n^2} \leqslant x \leqslant 2. \end{cases}$$

It is easy to check that

$$\|f_n\| = \int_0^2 |f_n(x)|\,dx = \frac{1}{n^2}$$

for $n \in \mathbf{N}$, so $\sum \|f_k\|$ is a convergent series. This means that $\sum f_k$ is an absolutely convergent series in this normed space. If it were also to be a convergent series, then the sequence $\{s_m\}_{m=1}^\infty$, where $s_m = \sum_{k=1}^m f_k$, would have to converge: its limit would again have to be a continuous function defined on $[0,2]$. To show that this is not possible, let g be any function belonging to $C_1[0,2]$ and define a function h, with domain $[0,2]$ but discontinuous at 0, by

$$h(x) = \sum_{j=1}^k \left(1 - \frac{j^2 x}{2}\right), \quad \frac{2}{(k+1)^2} < x \leqslant \frac{2}{k^2},$$

with $h(0)$ fixed but unspecified. It can be seen that for any $x > 0$, we have $h(x) = s_m(x)$ for all m large enough. Now

$$\int_0^2 |g(x) - h(x)|\,dx \leqslant \int_0^2 |g(x) - s_m(x)|\,dx + \int_0^2 |s_m(x) - h(x)|\,dx$$

for any positive integer m. The integral on the left must be positive because g is continuous while h is continuous on $(0,2]$ but unbounded. The final integral must approach 0 as $m \to \infty$. Therefore, we cannot have $\|g - s_m\| \to 0$, no matter what the function g is. Hence the sequence $\{s_m\}$ does not converge.

6.3 Solved problems

(1) Let V be a vector space of dimension n, and let $\{v_1, v_2, \ldots, v_n\}$ be a basis for V. Prove that

(a) we may define a norm for V by $\|x\| = \max_{1 \leqslant k \leqslant n} |\alpha_k|$, where $x = \sum_{k=1}^n \alpha_k v_k \in V$,

(b) if $\{x_m\}_{m=1}^\infty$ is a sequence in V, and $x_m = \sum_{k=1}^n \alpha_{mk} v_k$, then, with the norm of (a), convergence of $\{x_m\}$ is equivalent to the convergence of each sequence $\{\alpha_{mk}\}_{m=1}^\infty$, for $k = 1, 2, \ldots, n$.

Solution. (a) If $x = \theta$, then $\alpha_1 = \alpha_2 = \cdots = \alpha_n = 0$, by the linear independence of $\{v_1, v_2, \ldots, v_n\}$, so $\|x\| = 0$; conversely, if $\|x\| = 0$, then $\alpha_1 = \alpha_2 = \cdots = \alpha_n = 0$, so $x = \theta$. This verifies (N1). For (N2), we have, for any scalar α,

$$\|\alpha x\| = \left\| \sum_{k=1}^n (\alpha \alpha_k) v_k \right\| = \max_{1 \leqslant k \leqslant n} |\alpha \alpha_k| = |\alpha| \max_{1 \leqslant k \leqslant n} |\alpha_k| = |\alpha| \, \|x\|.$$

Finally, let $y = \sum_{k=1}^n \beta_k v_k$ be another vector in V. For each k,

$$|\alpha_k + \beta_k| \leqslant |\alpha_k| + |\beta_k| \leqslant \max_{1 \leqslant k \leqslant n} |\alpha_k| + \max_{1 \leqslant k \leqslant n} |\beta_k| = \|x\| + \|y\|$$

so

$$\|x + y\| = \left\| \sum_{k=1}^n (\alpha_k + \beta_k) v_k \right\| = \max_{1 \leqslant k \leqslant n} |\alpha_k + \beta_k| \leqslant \|x\| + \|y\|,$$

verifying (N3).

(b) Suppose $x_m \to x$, say, and put $x = \sum_{k=1}^n \alpha_k v_k$. Given $\epsilon > 0$, there is a positive integer N such that $\|x_m - x\| = \max_{1 \leqslant k \leqslant n} |\alpha_{mk} - \alpha_k| < \epsilon$ when $m > N$. Then $|\alpha_{mk} - \alpha_k| < \epsilon$ when $m > N$ for each $k = 1, 2, \ldots, n$. This means that each sequence $\{\alpha_{mk}\}_{m=1}^\infty$ in $\mathbf{C}$ (or $\mathbf{R}$, if V is a real vector space) is convergent.

Conversely, suppose each sequence $\{\alpha_{mk}\}_{m=1}^\infty$ in $\mathbf{C}$ (or $\mathbf{R}$) converges, and set $\alpha_k = \lim_{m \to \infty} \alpha_{mk}$, $k = 1, 2, \ldots, n$. Then, given $\epsilon > 0$, for each k there is a positive integer N_k such that $|\alpha_{mk} - \alpha_k| < \epsilon$ when $m > N_k$. If we set $N = \max\{N_1, N_2, \ldots, N_n\}$ and $x = \sum_{k=1}^n \alpha_k v_k$, then $\|x_m - x\| = \max_{1 \leqslant k \leqslant n} |\alpha_{mk} - \alpha_k| < \epsilon$ when $m > N$, so the sequence $\{x_m\}$ converges. This completes the proof. $\square$

(2) Let t be the vector space of complex-valued sequences x for which the series

$$\sum_{k=1}^\infty |x_{k+1} - x_k|$$

converges, where $x = (x_1, x_2, \ldots)$. Show that

(a) $\|x\| = |x_1| + \sum_{k=1}^\infty |x_{k+1} - x_k|$ defines a norm for t,
(b) with this norm, t is a Banach space.

Solution. (a) We must verify (N1), (N2) and (N3) for the definition $\|x\| = |x_1| + \sum_{k=1}^{\infty} |x_{k+1} - x_k|$. Certainly, by definition of t, the expression on the right is always finite. Furthermore, that expression is positive unless $x = \theta$, and $\|\theta\| = 0$, so (N1) is true. It is also quickly seen that (N2) is true. For (N3), if $y = (y_1, y_2, \ldots)$ is another element of t, and n is any positive integer, then

$$\sum_{k=1}^{n} |(x_{k+1} + y_{k+1}) - (x_k + y_k)| = \sum_{k=1}^{n} |(x_{k+1} - x_k) + (y_{k+1} - y_k)|$$

$$\leqslant \sum_{k=1}^{n} |x_{k+1} - x_k| + \sum_{k=1}^{n} |y_{k+1} - y_k|$$

$$\leqslant \sum_{k=1}^{\infty} |x_{k+1} - x_k| + \sum_{k=1}^{\infty} |y_{k+1} - y_k|.$$

Also, $|x_1 + y_1| \leqslant |x_1| + |y_1|$, and it follows that

$$\|x + y\| = |x_1 + y_1| + \sum_{k=1}^{\infty} |(x_{k+1} + y_{k+1}) - (x_k + y_k)|$$

$$\leqslant |x_1| + |y_1| + \sum_{k=1}^{\infty} |x_{k+1} - x_k| + \sum_{k=1}^{\infty} |y_{k+1} - y_k|$$

$$= \|x\| + \|y\|,$$

as required.

(b) We must show that t is complete with this norm. The procedure is the same as in metric spaces. Let $\{x_n\}$ be a Cauchy sequence in t, and write $x_n = (x_{n1}, x_{n2}, \ldots)$, $n \in \mathbf{N}$. Given $\epsilon > 0$, we know there is a positive integer N so that $\|x_n - x_m\| < \epsilon$ when $m, n > N$; that is,

$$|x_{n1} - x_{m1}| + \sum_{k=1}^{\infty} |(x_{n,k+1} - x_{m,k+1}) - (x_{nk} - x_{mk})| < \epsilon$$

when $m, n > N$. Noting that, for any $j = 2, 3, \ldots$,

$$x_{nj} - x_{mj} = x_{n1} - x_{m1} + \sum_{k=1}^{j-1} ((x_{n,k+1} - x_{m,k+1}) - (x_{nk} - x_{mk})),$$

we have

$$|x_{nj} - x_{mj}| \leqslant |x_{n1} - x_{m1}| + \sum_{k=1}^{j-1} |(x_{n,k+1} - x_{m,k+1}) - (x_{nk} - x_{mk})|$$

$$\leqslant |x_{n1} - x_{m1}| + \sum_{k=1}^{\infty} |(x_{n,k+1} - x_{m,k+1}) - (x_{nk} - x_{mk})| < \epsilon$$

when $m, n > N$. Hence $\{x_{nj}\}$ is a Cauchy sequence in $\mathbf{C}$ for each $j = 2, 3, \ldots$, and the same is clearly true when $j = 1$. Since $\mathbf{C}$ is a Banach space, we know $\lim_{n\to\infty} x_{nj}$ exists for all j. Put $x_{\cdot j} = \lim x_{nj}$ and write $x = (x_{\cdot 1}, x_{\cdot 2}, \ldots)$. It remains to show that $x \in t$ and that $x_n \to x$. For any positive integer K, we know that

$$|x_{n1} - x_{m1}| + \sum_{k=1}^{K} |(x_{n,k+1} - x_{m,k+1}) - (x_{nk} - x_{mk})| < \epsilon$$

when $m, n > N$. Fixing n, and using the fact that $\lim_{m\to\infty} x_{mk} = x_{\cdot k}$ for all k, we obtain

$$|x_{n1} - x_{\cdot 1}| + \sum_{k=1}^{K} |(x_{n,k+1} - x_{\cdot,k+1}) - (x_{nk} - x_{\cdot k})| \leqslant \epsilon.$$

Once we know that $x \in t$, this inequality will imply that $\|x_n - x\| \leqslant \epsilon$ when $n > N$; that is, $x_n \to x$. But the last displayed inequality implies that

$$\sum_{k=1}^{K} |(x_{n,k+1} - x_{\cdot,k+1}) - (x_{nk} - x_{\cdot k})| \leqslant \epsilon$$

$(n > N)$, and so

$$\sum_{k=1}^{K} |x_{\cdot,k+1} - x_{\cdot k}|$$
$$= \sum_{k=1}^{K} |x_{\cdot,k+1} - x_{n,k+1} + x_{n,k+1} - x_{nk} + x_{nk} - x_{\cdot k}|$$
$$= \sum_{k=1}^{K} |(x_{n,k+1} - x_{\cdot,k+1}) - (x_{nk} - x_{\cdot k}) + (x_{nk} - x_{n,k+1})|$$
$$\leqslant \sum_{k=1}^{K} |(x_{n,k+1} - x_{\cdot,k+1}) - (x_{nk} - x_{\cdot k})| + \sum_{k=1}^{K} |x_{n,k+1} - x_{nk}|$$
$$\leqslant \epsilon + \sum_{k=1}^{\infty} |x_{n,k+1} - x_{nk}|,$$

when $n > N$. The final expression is finite since $x_n \in t$, so the series $\sum_{k=1}^{\infty} |x_{\cdot,k+1} - x_{\cdot k}|$ converges; that is, $x \in t$. The proof is finished. $\quad\square$

6.4 Exercises

(1) Let X be a normed vector space and let d be the associated metric, given by $d(x, y) = \|x - y\|$ for $x, y \in X$.

  (a) Verify that d is indeed a metric.

  (b) For any $x, y, z \in X$ and any scalar α, prove that

$$d(x + z, y + z) = d(x, y) \quad \text{and} \quad d(\alpha x, \alpha y) = |\alpha| d(x, y).$$

  (Such a metric is called *translation invariant* and *homogeneous*.)

(2) In a normed space, prove that

  (a) $\|x - y\| \geqslant |\, \|x\| - \|y\|\, |$,

  (b) $\|(1/\alpha)x\| = 1$ if $\alpha = \|x\|$, $x \neq \theta$.

(3) (a) Let $\{x_n\}$ and $\{y_n\}$ be convergent sequences in a normed space, with $\lim x_n = x$, $\lim y_n = y$. Prove that $x_n + y_n \to x + y$.

  (b) Let $\{x_n\}$ be a convergent sequence in a normed space, with $\lim x_n = x$, and let $\{\alpha_n\}$ be a convergent sequence of scalars, with $\lim \alpha_n = \alpha$. Prove that $\alpha_n x_n \to \alpha x$.

  (c) Let $\{x_n\}$ be a convergent sequence in a normed space, with $\lim x_n = x$. Prove that $\|x_n\| \to \|x\|$. (Thus, $\| \ \|$ is a continuous mapping on a normed space.)

(4) (a) For the vector space $\mathbf{C}^n$ of n-tuples $x = (x_1, x_2, \ldots, x_n)$ of complex numbers, prove in full that

$$\|x\|_c = \max\{|x_1|, \ldots, |x_n|\}, \quad \|x\|_o = |x_1| + \cdots + |x_n|$$

  are valid definitions of norms.

  (b) The norms $\| \ \|_c$ and $\| \ \|_o$ are sometimes referred to as the *cubic* and *octahedral* norms, respectively, for the vector space $\mathbf{C}^n$. If $\| \ \|$ is the Euclidean norm for $\mathbf{C}^n$, prove that, when $x \in \mathbf{C}^n$,

  (i) $\|x\|_c \leqslant \|x\| \leqslant \sqrt{n}\, \|x\|_c$,

  (ii) $\|x\| \leqslant \|x\|_o \leqslant \sqrt{n}\, \|x\|$,

  (iii) $\dfrac{1}{n}\|x\|_o \leqslant \|x\|_c \leqslant \|x\|_o$.

(5) Prove that a nonempty subset S of a normed space is bounded if and only if there is a positive number M such that $\|x\| \leqslant M$ for all $x \in S$. (Hint: This is to be deduced as a consequence of Definition 2.8.1.)

(6) Let P be the set of all polynomial functions p. Show that P is a normed real vector space when

$$\|p\| = |a_0| + |a_1| + \cdots + |a_n|,$$

where $p(t) = a_0 + a_1 t + \cdots + a_n t^n$ (and $a_0, a_1, \ldots, a_n \in \mathbf{R}$). Show, however, that P is not a Banach space.

(7) Let V be a vector space of dimension n and let $\{v_1, v_2, \ldots, v_n\}$ be a basis for V. Prove that we may define a norm for V by

$$\|x\| = \sum_{k=1}^{n} |\alpha_k|,$$

where $x = \sum_{k=1}^{n} \alpha_k v_k \in V$. Deduce a theorem analogous to that of Solved Problem 6.3(1)(b).

(8) Define a sequence $\{x_n\}$ of functions continuous on $[0, 1]$ by

$$x_n(t) = \begin{cases} nt, & 0 \leqslant t < \dfrac{1}{n}, \\ 1, & \dfrac{1}{n} \leqslant t \leqslant 1. \end{cases}$$

Show that $\{x_n\}$ is convergent (with limit x, where $x(t) = 1$, $0 \leqslant t \leqslant 1$) when considered as a sequence in $C_1[0, 1]$, but not convergent when considered as a sequence in $C[0, 1]$.

. . . .

(9) Let $\{x_n\}$ be a sequence in a normed space and suppose that the series $\sum_{k=1}^{\infty}(x_k - x_{k+1})$ is absolutely convergent. Determine whether $\{x_n\}$ is (a) Cauchy, (b) convergent.

(10) In the normed space l_2, let

$$u_1 = (-1, 1, 0, 0, \ldots),$$
$$u_2 = (0, -1, 1, 0, \ldots),$$
$$u_3 = (0, 0, -1, 1, \ldots), \quad \ldots .$$

Show that the series $\sum_{k=1}^{\infty} u_k/2^k$ is absolutely convergent and deduce that

$$\left\| \sum_{k=1}^{\infty} \frac{u_k}{2^k} \right\| = \frac{1}{\sqrt{3}}, \qquad \sum_{k=1}^{\infty} \left\| \frac{u_k}{2^k} \right\| = \sqrt{2}.$$

(11) Recall that c_0 is the vector space of all sequences $(x_1, x_2, \ldots)$ for which $x_n \to 0$.

(a) Show that c_0 is a Banach space under the norm

$$\|x\| = \max_{k \geqslant 1} |x_k|, \quad x = (x_1, x_2, \dots) \in c_0.$$

(b) Let $e_1 = (1, 0, 0, \dots)$, $e_2 = (0, 1, 0, \dots)$, $e_3 = (0, 0, 1, \dots)$, $\dots$. Prove that the series $\sum_{k=1}^{\infty} e_k / k$ is convergent but not absolutely convergent in c_0. Find the sum of the series.

(12) If X is a vector space, a *seminorm* for X is a mapping $\nu \colon X \to \mathbf{R}_+$ satisfying $\nu(\theta) = 0$, $\nu(\alpha x) = |\alpha| \nu(x)$, $\nu(x + y) \leqslant \nu(x) + \nu(y)$, for $x, y \in X$ and any scalar α. (The second requirement of (N1) is omitted; compare this with the definition of a semimetric in Exercise 2.4(12).)

Let P be the real vector space of all polynomial functions. Prove that

$$\nu(p) = |p(0)| + |p'(0)| + |p''(0)|, \quad p \in P,$$

defines a seminorm for P, but not a norm. Determine all polynomial functions $p \in P$ for which $\nu(p) = 0$ and show that they form a subspace of P.

6.5 Finite-dimensional normed vector spaces

A number of theorems will be proved in this section giving a quite detailed account of completeness and compactness in finite-dimensional normed vector spaces. These will lead to some approximation theory, extending the result of Theorem 4.3.3.

The work was actually begun in Solved Problem 6.3(1) where it was shown in the first place that a norm can always be defined for a finite-dimensional vector space, namely by setting

$$\|x\| = \max_{1 \leqslant k \leqslant n} |\alpha_k|$$

where $x = \sum_{k=1}^{n} \alpha_k v_k$ and $\{v_1, v_2, \dots, v_n\}$ is a basis for the space, and in the second place that under this norm the convergence of a sequence in the space is equivalent to the separate convergence of the sequences of coefficients of the basis vectors. The existence of a second norm for this vector space, namely that given by

$$\|x\| = \sum_{k=1}^{n} |\alpha_k|,$$

is a consequence of Exercise 6.4(7). There are other norms that can

always be defined for a finite-dimensional vector space, but the second
theorem of this section will show that they are all the same in the sense
that sequences convergent under one particular norm will also be convergent under any other. This is not the case for infinite-dimensional vector
spaces. In Exercise 6.4(8), we gave a sequence of continuous functions
defined on $[0, 1]$ which is convergent when considered as a sequence in
$C_1[0, 1]$ but not when considered as a sequence in $C[0, 1]$.

Because we will be comparing different norms for the same vector
space, we will specify now that throughout this section V will be a
vector space of dimension n, the set $\{v_1, v_2, \ldots, v_n\}$ of vectors in V will
be a basis for V, α's with or without subscripts will be scalars, and the
first norm mentioned above will be distinguished by writing it as $\| \ \|_\infty$.
Thus

$$\|x\|_\infty = \max_{1 \leqslant k \leqslant n} |\alpha_k|$$

when $x = \sum_{k=1}^{n} \alpha_k v_k$. All the statements of this section are equally valid
when V is replaced by a real vector space. Only very minor adjustments
would be required to handle this.

We begin with a theorem about compact sets in V, under this special
norm.

Theorem 6.5.1 *The subset $\{x : x \in V, \ \|x\|_\infty \leqslant 1\}$ of $(V, \| \ \|_\infty)$ is
compact.*

The proof uses mathematical induction on the dimension n of the
vector space. Bear in mind below that the norm $\| \ \|_\infty$ depends on n,
but we will not clutter the notation by making this explicit.

Suppose the space has dimension 1 and that the vector v is a basis
for the space. Then any vector x in the space can be written as $x = \alpha v$,
and $\|x\|_\infty = |\alpha|$. Let $Q_1 = \{x : \|x\|_\infty \leqslant 1\}$ be the subset of the theorem
for this vector space of dimension 1 and let $Z = \{\alpha : \alpha \in \mathbf{C}, \ |\alpha| \leqslant 1\}$.
Define a mapping $A \colon Z \to Q_1$ by $A\alpha = \alpha v = x$. The closed disc Z is
compact in $\mathbf{C}$ (Exercise 4.5(6)), and clearly $A(Z) = Q_1$, so that once we
have shown A to be a continuous mapping it will follow from Theorem
4.3.1 that Q_1 is compact. If α is any point in Z and $\{\alpha_m\}$ is any sequence
in Z convergent to α, then the equations

$$\|A\alpha_m - A\alpha\|_\infty = \|\alpha_m v - \alpha v\|_\infty = \|(\alpha_m - \alpha)v\|_\infty = |\alpha_m - \alpha|$$

imply that $A\alpha_m \to A\alpha$, so indeed A is continuous.

Now suppose the theorem to be true when the dimension n of V

satisfies $n = h - 1$ for some integer $h > 1$. We will show that it is then also true when $n = h$. In general, write

$$Q_i = \{x : x \in V, \ \|x\|_\infty \leqslant 1\},$$

when $n = i$, $i \in \mathbf{N}$. We know Q_1 is compact and are assuming that Q_{h-1} is compact. Let $\{x_m\}$ be a sequence in Q_h and write

$$x_m = \sum_{j=1}^{h} \alpha_{mj} v_j = \alpha_{m1} v_1 + \sum_{j=2}^{h} \alpha_{mj} v_j,$$

for $m \in \mathbf{N}$. Now $\{\alpha_{m1} v_1\}$ is a sequence in a vector space of dimension 1 and

$$\|\alpha_{m1} v_1\|_\infty = |\alpha_{m1}| \leqslant \max_{1 \leqslant j \leqslant h} |\alpha_{mj}| = \|x_m\|_\infty \leqslant 1$$

so $\{\alpha_{m1} v_1\}$ is a sequence in Q_1, which is compact. Hence there is a convergent subsequence $\{\alpha_{m_k 1} v_1\}_{k=1}^{\infty}$ of $\{\alpha_{m1} v_1\}$. The sequence $\{x_{m_k}\}_{k=1}^{\infty}$ is therefore a subsequence of $\{x_m\}$ such that the sequence of coefficients of v_1 converges. The sequence $\left\{ \sum_{j=2}^{h} \alpha_{m_k j} v_j \right\}_{k=1}^{\infty}$ belongs to the vector space $\mathrm{Sp}\{v_2, v_3, \ldots, v_h\}$ (defined in Definition 1.11.3(c)), of dimension $h - 1$, and since

$$\left\| \sum_{j=2}^{h} \alpha_{m_k j} v_j \right\|_\infty = \max_{2 \leqslant j \leqslant h} |\alpha_{m_k j}| \leqslant \max_{1 \leqslant j \leqslant h} |\alpha_{m_k j}| = \|x_{m_k}\|_\infty \leqslant 1,$$

for each $k \in \mathbf{N}$, it is a sequence in Q_{h-1}, which is assumed to be compact. It therefore has a convergent subsequence, so that, by applying the result of Solved Problem 6.3(1)(b), we are able to pick out from the original sequence $\{x_m\}$ a convergent subsequence, showing that Q_h is compact. □

Now we can clarify our earlier statement about different norms for a finite-dimensional vector space being all much the same. The relevant definition follows.

Definition 6.5.2 Two norms $\| \ \|_1$ and $\| \ \|_2$ for a vector space X are said to be *equivalent* if there exist positive numbers a and b such that

$$a\|x\|_1 \leqslant \|x\|_2 \leqslant b\|x\|_1$$

for all $x \in X$.

Following on from the definition, we then also have

$$\frac{1}{b}\|x\|_2 \leqslant \|x\|_1 \leqslant \frac{1}{a}\|x\|_2$$

for any $x \in X$, so the definition is quite symmetrical. The normed spaces $(X, \| \ \|_1)$ and $(X, \| \ \|_2)$ are different, but the point is that if the norms are equivalent then any sequence $\{x_n\}$ of points in X which converges in one of the normed spaces converges also in the other: to say the sequence converges in $(X, \| \ \|_1)$ means $\|x_n - x\|_1 \to 0$ for some $x \in X$, but since $0 \leqslant \|x_n - x\|_2 \leqslant b\|x_n - x\|_1$ for all n, we also have $\|x_n - x\|_2 \to 0$, so $\{x_n\}$ converges also in $(X, \| \ \|_2)$. The same can be said of Cauchy sequences: a sequence which is Cauchy in one of the normed spaces will be Cauchy in the other.

A special instance of the next theorem was given in Exercise 6.4(4), in which three different norms for the vector space $\mathbf{C}^n$ were shown to be equivalent in pairs.

Theorem 6.5.3 *Any two norms for a finite-dimensional vector space are equivalent.*

We will only prove that any norm $\| \ \|$ for our vector space V is equivalent to the norm $\| \ \|_\infty$. That is, we will show that there exist positive numbers a and b such that

$$a\|x\|_\infty \leqslant \|x\| \leqslant b\|x\|_\infty$$

for any $x \in V$. This readily implies the theorem, but the details are left as an exercise.

In Theorem 6.5.1, we showed that the subset $Q = \{x : \|x\|_\infty \leqslant 1\}$ of V is compact. It is another simple exercise to use this fact, in conjunction with Exercise 4.5(3), to conclude that the set $Q' = \{x : \|x\|_\infty = 1\}$ in V is also compact. On any normed space, the norm is a continuous mapping (Exercise 6.4(3)(c)) so we may invoke Theorem 4.3.2 to ensure the existence of points x_M and x_m in Q' such that

$$\|x_M\| = \max_{x \in Q'} \|x\|, \quad \|x_m\| = \min_{x \in Q'} \|x\|.$$

Thus $\|x_m\| \leqslant \|x\| \leqslant \|x_M\|$ for all $x \in Q'$. Also, since $\|x_m\|_\infty = 1$, we cannot have $x_m = \theta$, so $\|x_m\| > 0$. For any nonzero vector $x \in V$, we have

$$\left\|\frac{1}{\|x\|_\infty}x\right\|_\infty = \frac{1}{\|x\|_\infty}\|x\|_\infty = 1,$$

so $(1/\|x\|_\infty)x \in Q'$. Hence, for $x \neq \theta$,

$$\|x_m\| \leqslant \left\| \frac{1}{\|x\|_\infty} x \right\| \leqslant \|x_M\|.$$

We thus have

$$\|x_m\| \, \|x\|_\infty \leqslant \|x\| \leqslant \|x_M\| \, \|x\|_\infty$$

or $a\|x\|_\infty \leqslant \|x\| \leqslant b\|x\|_\infty$, where $a = \|x_m\| > 0$ and $b = \|x_M\|$, and this is clearly true also when $x = \theta$. Hence the norms $\| \ \|$ and $\| \ \|_\infty$ are equivalent.   □

To lead into our next theorem, we note that if we can prove that the finite-dimensional vector space V is complete under the norm $\| \ \|_\infty$ then it will quickly follow from the preceding theorem that V is complete regardless of the norm defined for it. This is just another way of putting the earlier comment that a Cauchy sequence in V with one norm is again a Cauchy sequence with any equivalent norm, and similarly for a convergent sequence. We will have proved the following.

Theorem 6.5.4 *Every finite-dimensional normed vector space is a Banach space.*

Hence we prove that the normed space $(V, \| \ \|_\infty)$ is complete. Let $\{x_m\}$ be a Cauchy sequence in this space. Then, given any $\epsilon > 0$, there exists a positive integer N such that

$$\|x_m - x_j\|_\infty = \max_{1 \leqslant k \leqslant n} |\alpha_{mk} - \alpha_{jk}| < \epsilon$$

when $j, m > N$, where we write $x_m = \sum_{k=1}^{n} \alpha_{mk} v_k$, for $m \in \mathbf{N}$. Then $|\alpha_{mk} - \alpha_{jk}| < \epsilon$ when $j, m > N$, for each $k = 1, 2, \ldots, n$, so $\{\alpha_{mk}\}_{m=1}^{\infty}$ is a Cauchy sequence in $\mathbf{C}$ for each k. Since $\mathbf{C}$ is complete, each of these sequences converges so, by the result of Solved Problem 6.3(1)(b), the sequence $\{x_m\}$ converges, and the theorem follows.   □

We will employ a similar technique for the next theorem.

Theorem 6.5.5 *A subset of any finite-dimensional normed vector space is compact if and only if it is both closed and bounded.*

This provides a generalisation of Theorem 4.1.6, in which we determined that the compact subsets of $\mathbf{R}^n$ are precisely those that are closed and bounded. We must prove here the sufficiency of the condition, since we know, by Theorems 4.1.4 and 4.1.5, that any compact subset must be closed and bounded. We do this first for a closed and bounded subset S of V, with norm $\| \ \|_\infty$.

There is little to do. We observe that the proof of Theorem 6.5.1 could have been carried through in the same way to prove that the subset $Q^{(L)} = \{x : \|x\|_\infty \leqslant L\}$ of $(V, \| \ \|_\infty)$ is compact for any positive number L. Since S is bounded, a value of L certainly exists so that S is a subset of $Q^{(L)}$. Since S is closed, we may then use Exercise 4.5(3) to infer that S is compact.

Now let $\| \ \|$ be any other norm for V and let S be a closed, bounded subset of V with respect to this other norm. We leave it as an exercise to show that, because of the equivalence of all norms on V, S is also closed and bounded with respect to $\| \ \|_\infty$. Thus S is a compact subset of $(V, \| \ \|_\infty)$ by what was just said, so any sequence $\{x_m\}$ in S has a subsequence $\{x_{m_k}\}$ which is convergent with respect to $\| \ \|_\infty$. But the equivalence of the norms implies that this subsequence is also convergent with respect to $\| \ \|$, and so the result follows. $\square$

6.6 Some approximation theory

The preceding theorem has far-reaching consequences in approximation theory. In terms of normed spaces, Theorem 4.3.3 stated: given a compact subset S of a normed space X and a point $x \in X$, there exists a point $p \in S$ such that $\|p - x\|$ is a minimum. Proving that S is compact in a given situation may be difficult, but the result we prove next replaces compactness of S by a much more easily tested condition: the same conclusion is true if S is a finite-dimensional subspace of X.

Theorem 6.6.1 *A finite-dimensional subspace of a normed vector space contains at least one point of minimum distance from a given point.*

To prove this, continue with the notation above and take any point $p_0 \in S$. Consider the set

$$Y = \{y : y \in S, \ \|y - x\| \leqslant \|p_0 - x\|\}.$$

If there is a point $p \in S$ such that $\|p - x\|$ is a minimum, then certainly $p \in Y$, so the desired result will follow from Theorem 4.3.3 if we can show that Y is compact. Since Y is a subset of the finite-dimensional space S, the compactness of Y will follow by the preceding theorem once it has been shown to be closed and bounded. This is not difficult. First,

Y is bounded since

$$\|y\| = \|(y - x) + x\| \leqslant \|y - x\| + \|x\| \leqslant \|p_0 - x\| + \|x\|,$$

for any $y \in Y$. Secondly, Y is closed since if $\{y_n\}$ is any sequence of points in Y that is convergent as a sequence in S, and $\lim y_n = y$ say, then for any $\epsilon > 0$ we can find n large enough to ensure that

$$\|y - x\| \leqslant \|y - y_n\| + \|y_n - x\| < \epsilon + \|p_0 - x\|.$$

Then the arbitrariness of ϵ implies that $\|y - x\| \leqslant \|p_0 - x\|$, so $y \in Y$.

$\square$

As an example of this existence theorem, we have the following. In the notation above, let $X = C[0, 1]$ and let S be the set of all polynomial functions on $[0, 1]$ of degree less than some fixed positive integer r. Then S is a vector space of dimension r (the set of functions defined by $\{1, t, t^2, \ldots, t^{r-1}\}$, $0 \leqslant t \leqslant 1$, is a basis for S) and S may be taken as a subspace of $C[0, 1]$. Given a function $f \in C[0, 1]$, the theorem implies that there is a polynomial function $p \in S$ such that

$$\|p - f\| = \max_{0 \leqslant t \leqslant 1} |p(t) - f(t)|$$

is a minimum.

This leads us to Weierstrass' famous approximation theorem: if we are not restricted in the degree of the polynomial functions, then there exists a polynomial function p such that $\|p - f\|$ is as small as we please. We take this up in the next section.

Theorem 6.6.1 has the same drawbacks as the earlier Theorem 4.3.3: there is no suggestion that there is only one best approximation nor any indication of how to find such a point. The theorem assures us only of the existence of at least one best approximation. By imposing more structure on a normed space we can at least give in general terms a sufficient condition for the best approximation to be unique.

Definition 6.6.2 A normed space X is said to be strictly convex if the equation

$$\|x + y\| = \|x\| + \|y\|,$$

where $x, y \in X$, $x \neq \theta$, $y \neq \theta$, holds only when $x = \beta y$ for some (real) positive number β.

The triangle inequality tells us that $\|x + y\| \leqslant \|x\| + \|y\|$ for any $x, y \in X$. If $x = \beta y$ and $\beta > 0$, it is readily checked that $\|x + y\| = \|x\| + \|y\|$.

However equality can hold in the triangle inequality in some normed spaces in cases other than this, as we show below for $C[a, b]$, so such spaces are not strictly convex.

Now we can put a few things together.

Theorem 6.6.3 *If X is a strictly convex normed space, then a finite-dimensional subspace of X contains a unique best approximation of any point in X.*

That is, the best approximation whose existence is implied by the earlier Theorem 6.6.1 is unique when the space is strictly convex. To prove this, let S be a finite-dimensional subspace of X and let x be a given point in X. We suppose $x \notin S$ since otherwise x is obviously its own unique best approximation. By Theorem 6.6.1, there exists at least one point $p \in S$ such that $\|x - p\|$ is a minimum. Suppose that $p' \in S$ shares this property. Set

$$\|x - p\| = \|x - p'\| = d.$$

Now, since S is a vector space, $\frac{1}{2}(p + p') \in S$ and

$$d \leqslant \|x - \tfrac{1}{2}(p + p')\| = \|\tfrac{1}{2}(x - p) + \tfrac{1}{2}(x - p')\|$$
$$\leqslant \tfrac{1}{2}\|x - p\| + \tfrac{1}{2}\|x - p'\| = d.$$

Hence $\|x - \frac{1}{2}(p + p')\| = d$ so $\frac{1}{2}(p + p')$ is also a best approximation of x. (This averaging process can be continued indefinitely to show the existence of infinitely many best approximations in a normed space once there are two different best approximations.) It follows that

$$\|x - \tfrac{1}{2}(p + p')\| = \tfrac{1}{2}\|x - p\| + \tfrac{1}{2}\|x - p'\|,$$

from which, since X is strictly convex,

$$x - p = \beta(x - p')$$

for some number $\beta > 0$. If $\beta \neq 1$, we get

$$x = \frac{1}{1 - \beta}p - \frac{\beta}{1 - \beta}p'.$$

This is impossible since it represents x as belonging to the vector space S, whereas $x \notin S$. So we must have $\beta = 1$. Thus $p = p'$ and we have proved that the best approximation is unique. $\qquad\square$

To see that $C[a, b]$ is not a strictly convex normed space, we take the following simple example. The functions f, g, where

$$f(t) = bt, \quad g(t) = t^2, \quad a \leqslant t \leqslant b,$$

are such that $\|f\| = \|g\| = b^2$ while $\|f + g\| = 2b^2 = \|f\| + \|g\|$. But certainly $f \neq \beta g$ for any number β. (We have assumed here that $|a| < b$.)

However, as is to be shown in Exercise 6.10(5), the normed space $C_2[a, b]$ is strictly convex . It follows then from Theorem 6.6.3 that for any function $f \in C_2[a, b]$ there is a unique polynomial function p of given degree or less such that

$$\|f - p\| = \sqrt{\int_a^b (f(x) - p(x))^2 \, dx}$$

is a minimum. This function p is called the *best least squares polynomial approximation* of f, and will be more fully discussed in Chapter 8.

6.7 Chebyshev theory

Although Theorem 6.6.3 does not apply to the space $C[a, b]$, since it is not strictly convex, it can be shown nonetheless that any function in this space does have a unique best approximation from the set of all polynomial functions of degree less than a given integer. This is a consequence of some work initiated by Chebyshev. We will not go very far into that theory, contenting ourselves mainly with the problem of approximating a polynomial function of degree r by one of smaller degree.

Since we will be working here with the norm of the space $C[a, b]$, the approximations we will obtain are known as *uniform approximations*. They are also called *minimax approximations*, as noted at the end of Chapter 4. In general, the best uniform approximation of a function will not be the same as its best least squares approximation, or its best approximation under many other criteria that may be used. We note that in the context of approximation theory, the uniform norm is often referred to as the *Chebyshev norm*.

Specifically, we will seek in the first place the best uniform approximation of x^r by a polynomial function of the form $a_0 + a_1 x + a_2 x^2 + \cdots + a_{r-1} x^{r-1}$, over the interval $[-1, 1]$. Thus we must determine the numbers $a_0, a_1, \ldots, a_{r-1}$ so that

$$\max_{-1 \leqslant x \leqslant 1} |x^r - (a_0 + a_1 x + a_2 x^2 + \cdots + a_{r-1} x^{r-1})|$$

is a minimum. Write

$$P_a(x) = x^r - a_{r-1}x^{r-1} - \cdots - a_1x - a_0,$$

the subscript a indicating the dependence of P_a on the coefficients. Such a polynomial function, where the coefficient of the term of highest degree (called the *leading coefficient*) is 1, is said to be *monic*. Our immediate problem can be phrased this way: To find the monic polynomial function of degree r which best approximates the zero function over $[-1, 1]$, with the uniform norm.

The set of all monic polynomial functions is not a vector space, but our first formulation of the problem shows, by Theorem 6.6.1, that a solution certainly exists. Thus there exist values for a_0, a_1, $\ldots$, a_{r-1} such that

$$\max_{-1 \leqslant x \leqslant 1} |x^r - (a_0 + a_1x + a_2x^2 + \cdots + a_{r-1}x^{r-1})| = \|P_a\|$$

is minimised. Let this minimum value be m, so for any other monic polynomial function P_b on $[-1, 1]$, of degree r, we have $\|P_b\| \geqslant m$. Consider a function which has alternate maxima and minima, with values m and $-m$, at $r + 1$ points x_0, x_1, $\ldots$, x_{r-1}, x_r, where

$$-1 = x_0 < x_1 < \cdots < x_{r-1} < x_r = 1.$$

(See Figure 11, where we have $r = 6$.) Certainly, this function has norm m. For the moment, we will assume that there is a monic polynomial function of degree r with this property. Under that assumption, we will show there is in fact at most one, and later we will actually create such a function. In the interim, we may continue to use P_a to denote the function.

Suppose P_c is any other monic polynomial function of degree r also satisfying $|P_c(x)| \leqslant m$ on $[-1, 1]$. Then the difference $P_a - P_c$ is a polynomial function of degree at most $r - 1$ satisfying

$$(P_a - P_c)(-1) \geqslant 0 \quad \text{if, say, } P_a(-1) = m, \text{ as in Figure 11,}$$
$$(P_a - P_c)(x_1) \leqslant 0,$$
$$(P_a - P_c)(x_2) \geqslant 0,$$
$$\vdots$$
$$(P_a - P_c)(x_{r-1}) \leqslant 0 \quad \text{if, say, } P_a(1) = m, \text{ as in Figure 11,}$$
$$(P_a - P_c)(1) \geqslant 0.$$

Since $(P_a - P_c)(x)$ is alternately positive and negative, or is 0, at the

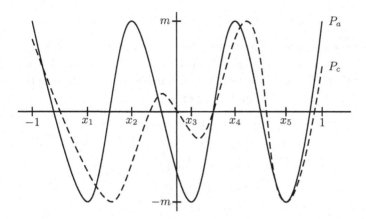

Figure 11

$r+1$ points -1, x_1, x_2, ..., x_{r-1}, 1, it must equal 0 at at least r points. Being a polynomial function of degree at most $r-1$, this is impossible unless in fact $P_c = P_a$.

Under the assumption that there is such a monic polynomial function P_a, this proves its uniqueness. To actually find it, we observe that trigonometric functions, the sine and cosine in particular, possess an oscillatory property like that described above. (This is termed the *equal-ripple property* in approximation theory.) In fact, $\cos r\theta$ is alternately 1 and -1 for $r+1$ values of θ in the interval $[0, \pi]$, including the endpoints. If we set

$$x = \cos\theta \quad \text{and} \quad T_r(x) = \cos r\theta,$$

then T_r has domain $[-1, 1]$ and has the desired equal-ripple property. We will show that T_r is a polynomial function of degree r, so that, once we divide by its leading coefficient to make it monic, we will have the required function P_a.

Now,

$$T_0(x) = 1, \quad T_1(x) = x, \qquad -1 \leqslant x \leqslant 1,$$

and, since

$$\cos(r+1)\theta = 2\cos\theta\cos r\theta - \cos(r-1)\theta,$$

we have

$$T_{r+1}(x) = 2xT_r(x) - T_{r-1}(x), \quad r = 1, 2, \ldots, \quad -1 \leqslant x \leqslant 1.$$

It follows by mathematical induction that T_r is a polynomial function of degree r. It also follows that T_r has leading coefficient 2^{r-1} (for $r \geqslant 1$) and that $\|T_r\| = 1$. These polynomial functions are known as the *Chebyshev polynomials*. The next few are

$$T_2(x) = 2x^2 - 1,$$
$$T_3(x) = 4x^3 - 3x,$$
$$T_4(x) = 8x^4 - 8x^2 + 1,$$
$$T_5(x) = 16x^5 - 20x^3 + 5x,$$

all for $-1 \leqslant x \leqslant 1$.

The monic polynomial function $P_a = 2^{1-r}T_r$ is thus the one satisfying our initial problem. It has maximum modulus $m = 2^{1-r}$ in the interval $[-1, 1]$ and takes on the values m and $-m$ alternately at the $r+1$ points

$$x = \cos\frac{k\pi}{r}, \quad k = 0, 1, \ldots, r,$$

with zeros between them at the points

$$x = \cos\frac{(2k+1)\pi}{r}, \quad k = 0, 1, \ldots, r-1.$$

Our original problem here was to find the best approximation of a polynomial function of degree r from the polynomial functions of degree less than r, under the uniform norm. We will answer this on $[-1, 1]$. Let p_r be the given function. Then we require a polynomial function q, of degree less than r, such that $\|p_r - q\|$ is a minimum. Put another way, we require q so that $p_r - q$ is the best approximation of the zero function on $[-1, 1]$. But this is λT_r where the number λ is chosen so that λT_r and $p_r - q$ have the same leading coefficient. If p_r has leading coefficient a_r, then $2^{r-1}\lambda = a_r$ so that $\lambda = 2^{1-r}a_r$. The required polynomial is thus $p_r - 2^{1-r}a_rT_r$.

As an example, suppose we wish to approximate the polynomial function $p_3(x) = x^3 - 2x^2 + 2$ on $[-1, 1]$ by one of lower degree. Here, $r = 3$ and $a_r = 1$ so the required function $p_r - 2^{1-r}a_rT_r$ is given by

$$x^3 - 2x^2 + 2 - 2^{-2}(4x^3 - 3x) = -2x^2 + \tfrac{3}{4}x + 2.$$

It follows that solving the equation $2x^2 - \tfrac{3}{4}x - 2 = 0$, which we may do easily to high accuracy, will give information on the roots of the equation

$x^3 - 2x^2 + 2 = 0$, which is not so easy to obtain. The quadratic equation has the root -0.830, to three decimal places, in $[-1, 1]$. The cubic can be shown to have the root -0.839, to the same degree of accuracy. This idea certainly provides at least a method of finding a decent starting point for an iterative solution of the cubic equation. Notice that the other root of the quadratic equation has no relevance since it lies outside the interval $[-1, 1]$.

6.8 The Weierstrass approximation theorem

We gave the gist of the Weierstrass theorem following Theorem 6.6.1. Before stating it more formally, we will prepare some preliminary results.

The notion of uniform continuity of a function was mentioned briefly in Section 4.2. Though our main application will require the elementary form already given, we will take the opportunity here to present the ideas in a more general setting.

Definition 6.8.1 Let X and Y be normed vector spaces. A mapping $A \colon S \to Y$ is said to be uniformly continuous on a subset S of X if, for any number $\epsilon > 0$, there exists a number $\delta > 0$ such that

$$\|Ax' - Ax''\| < \epsilon \quad \text{whenever } x', x'' \in S \text{ and } \|x' - x''\| < \delta.$$

Suppose $S = X$ here and that δ is the number stated in the definition. If x is any point of X and $\{x_n\}$ is any sequence in X convergent to x, then there exists a positive integer N so that $\|x_n - x\| < \delta$ when $n > N$. It follows immediately that $\|Ax_n - Ax\| < \epsilon$ when $n > N$. Hence $Ax_n \to Ax$ so that the mapping A is also continuous on X. Of more interest is that we can give a partial converse of this result.

Theorem 6.8.2 *Suppose X and Y are normed vector spaces and that $A \colon S \to Y$ is a mapping continuous on a nonempty compact subset S of X. Then A is uniformly continuous on S.*

To prove this, we will suppose that A is not uniformly continuous on S. This means that that there is some number $\epsilon > 0$ such that, regardless of the value of δ, there are points $x', x'' \in S$ with $\|x' - x''\| < \delta$ but for which $\|Ax' - Ax''\| \geqslant \epsilon$. Take $\delta = 1/n$, for $n = 1, 2, \ldots$ in turn, and for each n let x'_n, x''_n be points in S (known to exist by our supposition) such that

$$\|x'_n - x''_n\| < \frac{1}{n} \quad \text{and} \quad \|Ax'_n - Ax''_n\| \geqslant \epsilon.$$

As S is compact, the sequence $\{x'_n\}$ has a convergent subsequence $\{x'_{n_k}\}$, with limit x, say. Take any number $\eta > 0$. There exists a positive integer K such that $\|x'_{n_k} - x\| < \frac{1}{2}\eta$ when $k > K$. We may suppose $K > 2/\eta$. For such k, $n_k \geqslant k > 2/\eta$ and

$$\|x''_{n_k} - x\| \leqslant \|x''_{n_k} - x'_{n_k}\| + \|x'_{n_k} - x\| < \frac{1}{n_k} + \frac{\eta}{2} < \eta,$$

so that $\{x''_{n_k}\}$ is a convergent subsequence of $\{x''_n\}$, also with limit x. Further, the sequence $x'_{n_1}, x''_{n_1}, x'_{n_2}, x''_{n_2}, \ldots$ must then have limit x and so, since A is continuous on S, the sequence $Ax'_{n_1}, Ax''_{n_1}, Ax'_{n_2}, Ax''_{n_2}, \ldots$ in Y must converge with limit Ax. Hence there is an integer N such that, when $k > N$,

$$\|Ax'_{n_k} - Ax''_{n_k}\| \leqslant \|Ax'_{n_k} - Ax\| + \|Ax''_{n_k} - Ax\| < \tfrac{1}{2}\epsilon + \tfrac{1}{2}\epsilon = \epsilon,$$

and this gives us a contradiction. Thus A is indeed uniformly continuous on S.   □

It follows in particular that a real-valued function that is continuous on a closed interval is also uniformly continuous.

An interesting property of uniform continuity, whose proof is asked for in Exercise 6.10(12), is the following: If $\{x_n\}$ is a Cauchy sequence in X, and $A\colon X \to Y$ is uniformly continuous, then $\{Ax_n\}$ is a Cauchy sequence in Y. This is not true of mappings that are only continuous. The function f, where

$$f(x) = \frac{x}{1-x}, \quad 0 \leqslant x < 1,$$

is continuous, and $\{1 - 1/n\}$ is a Cauchy sequence in its domain. However, $f(1 - 1/n) = n - 1$ and $\{n - 1\}$ is certainly not a Cauchy sequence in $\mathbf{R}$. Of course, f is not uniformly continuous.

In an unexpected and clever way, the binomial theorem, reviewed at the end of Section 1.8, enters our proof of the Weierstrass theorem. This is via the following identities:

(a) $\displaystyle\sum_{k=0}^{n} \binom{n}{k} x^k (1-x)^{n-k} = 1,$

(b) $\displaystyle\sum_{k=0}^{n} (k - nx)^2 \binom{n}{k} x^k (1-x)^{n-k} = nx(1-x).$

Here, n is any positive integer and x is any real number.

We stated the binomial theorem in the form

$$(a + b)^n = \sum_{k=0}^{n} \binom{n}{k} a^{n-k} b^k$$

in Section 1.8, so we need only set $a = 1 - x$ and $b = x$ to obtain (a). To prove (b), note that it is certainly true when $x = 0$ or 1 and assume henceforth that $x \neq 0$, $x \neq 1$. Differentiate the identity (a) with respect to x:

$$\sum_{k=0}^{n} \binom{n}{k} \left(kx^{k-1}(1-x)^{n-k} - (n-k)x^k(1-x)^{n-k-1} \right) = 0,$$

so

$$\frac{1}{x} \sum_{k=0}^{n} k \binom{n}{k} x^k(1-x)^{n-k} = \frac{1}{1-x} \sum_{k=0}^{n} (n-k) \binom{n}{k} x^k(1-x)^{n-k}.$$

Then

$$\left(\frac{1}{x} + \frac{1}{1-x} \right) \sum_{k=0}^{n} k \binom{n}{k} x^k(1-x)^{n-k} = \frac{n}{1-x} \sum_{k=0}^{n} \binom{n}{k} x^k(1-x)^{n-k},$$

and, using (a), we obtain

$$\text{(c)} \quad \sum_{k=0}^{n} k \binom{n}{k} x^k(1-x)^{n-k} = nx.$$

Now differentiate this identity with respect to x:

$$\sum_{k=0}^{n} k \binom{n}{k} \left(kx^{k-1}(1-x)^{n-k} - (n-k)x^k(1-x)^{n-k-1} \right) = n,$$

so

$$\frac{1}{x} \sum_{k=0}^{n} k^2 \binom{n}{k} x^k(1-x)^{n-k}$$

$$= n + \frac{1}{1-x} \sum_{k=0}^{n} k(n-k) \binom{n}{k} x^k(1-x)^{n-k}.$$

Then

$$\left(\frac{1}{x} + \frac{1}{1-x} \right) \sum_{k=0}^{n} k^2 \binom{n}{k} x^k(1-x)^{n-k}$$

$$= n + \frac{n}{1-x} \sum_{k=0}^{n} k \binom{n}{k} x^k(1-x)^{n-k},$$

and, using (c),

(d) $\displaystyle\sum_{k=0}^{n} k^2 \binom{n}{k} x^k (1-x)^{n-k} = nx(1-x) + n^2 x^2.$

From (a), (c) and (d), we now have

$$\sum_{k=0}^{n} (k - nx)^2 \binom{n}{k} x^k (1-x)^{n-k}$$

$$= nx(1-x) + n^2 x^2 - 2nx \cdot nx + n^2 x^2 \cdot 1 = nx(1-x),$$

and (b) is proved.

We come to the main theorem.

Theorem 6.8.3 (Weierstrass Approximation Theorem) *Given any function $f \in C[0,1]$ and any number $\epsilon > 0$, there exists a polynomial function p such that $\|p - f\| < \epsilon$.*

It is not too difficult to extend this to obtain a similar result for functions in $C[a,b]$, and we will leave the details as an exercise.

Let $\{p_n\}$ be the sequence of polynomial functions on $[0,1]$ defined by

$$p_n(x) = \sum_{k=0}^{n} f\left(\frac{k}{n}\right) \binom{n}{k} x^k (1-x)^{n-k},$$

where f is the given function in $C[0,1]$. These are known as the *Bernstein polynomials* for f. The first three are

$p_1(x) = f(0)(1-x) + f(1)x,$

$p_2(x) = f(0)(1-x)^2 + 2f(\frac{1}{2})x(1-x) + f(1)x^2,$

$p_3(x) = f(0)(1-x)^3 + 3f(\frac{1}{3})x(1-x)^2 + 3f(\frac{2}{3})x^2(1-x) + f(1)x^3.$

We have, using (a),

$$|f(x) - p_n(x)| = \left| \sum_{k=0}^{n} \left(f(x) - f\left(\frac{k}{n}\right) \right) \binom{n}{k} x^k (1-x)^{n-k} \right|$$

$$\leqslant \sum_{k=0}^{n} \left| f(x) - f\left(\frac{k}{n}\right) \right| \binom{n}{k} x^k (1-x)^{n-k}.$$

Since f is continuous on a closed interval, it is uniformly continuous. Take any number $\epsilon > 0$. Then there is a number $\delta > 0$ such that, for any points x', x'' in $[0,1]$ satisfying $|x'-x''| < \delta$, we have $|f(x')-f(x'')| < \frac{1}{2}\epsilon$. We choose such a δ and maintain it through the following. Let x_0 be a fixed point in $[0,1]$ and partition the set $S = \{0,1,\ldots,n\}$ into two disjoint parts:

$$S_1 = \left\{ k : k \in S, \left| \frac{k}{n} - x_0 \right| < \delta \right\},$$

$$S_2 = \left\{ k : k \in S, \left| \frac{k}{n} - x_0 \right| \geqslant \delta \right\}.$$

Using an obvious abbreviation, we have

$$|f(x_0) - p_n(x_0)| \leqslant \sum_{k \in S_1} + \sum_{k \in S_2}$$

$$< \frac{\epsilon}{2} \sum_{k \in S_1} \binom{n}{k} x_0^k (1 - x_0)^{n-k}$$

$$+ 2\|f\| \sum_{k \in S_2} \binom{n}{k} x_0^k (1 - x_0)^{n-k},$$

since $|f(x_0) - f(k/n)| < \epsilon/2$ when $k \in S_1$, and since

$$\left| f(x_0) - f\left(\frac{k}{n}\right) \right| \leqslant |f(x_0)| + \left| f\left(\frac{k}{n}\right) \right| \leqslant 2 \max_{0 \leqslant x \leqslant 1} |f(x)| = 2\|f\|.$$

Now, using (a),

$$\sum_{k \in S_1} \binom{n}{k} x_0^k (1 - x_0)^{n-k} \leqslant \sum_{k=0}^{n} \binom{n}{k} x_0^k (1 - x_0)^{n-k} = 1,$$

and, using (b),

$$nx_0(1 - x_0) = \sum_{k=0}^{n} (k - nx_0)^2 \binom{n}{k} x_0^k (1 - x_0)^{n-k}$$

$$= n^2 \sum_{k=0}^{n} \binom{n}{k} \left(\frac{k}{n} - x_0 \right)^2 x_0^k (1 - x_0)^{n-k}$$

$$\geqslant n^2 \sum_{k \in S_2} \binom{n}{k} \left(\frac{k}{n} - x_0 \right)^2 x_0^k (1 - x_0)^{n-k}$$

$$\geqslant n^2 \delta^2 \sum_{k \in S_2} \binom{n}{k} x_0^k (1 - x_0)^{n-k}.$$

Then

$$\sum_{k \in S_2} \binom{n}{k} x_0^k (1 - x_0)^{n-k} \leqslant \frac{x_0(1 - x_0)}{n\delta^2} \leqslant \frac{1}{4n\delta^2},$$

as $x_0(1 - x_0) = \frac{1}{4} - (x_0 - \frac{1}{2})^2 \leqslant \frac{1}{4}$.

We thus obtain

$$|f(x_0) - p_n(x_0)| < \frac{\epsilon}{2} + \frac{\|f\|}{2n\delta^2}$$

for all x_0 in $[0, 1]$ and all n. Choose $n > \|f\|/\epsilon\delta^2$, so that $\|f\|/n\delta^2 < \epsilon$, and we have

$$|f(x_0) - p_n(x_0)| < \frac{\epsilon}{2} + \frac{\epsilon}{2} = \epsilon.$$

As this is true for all x_0 in $[0, 1]$, it follows that

$$\max_{0 \leqslant x \leqslant 1} |f(x) - p_n(x)| = \|f - p_n\| < \epsilon.$$

We have thus exhibited a polynomial function p such that $\|f - p\| < \epsilon$.

$\square$

There is a simple application of the Weierstrass theorem to a problem in statistics. For any function f, continuous on $[0, 1]$, the *moments* of f are the numbers $\int_0^1 x^n f(x)\,dx$, for $n = 0, 1, 2, \ldots$. (When f is the probability density function of a continuous random variable, then this is precisely the definition of the moments of the random variable.) We will prove that if all the moments of f are 0, then f must be the zero function on $[0, 1]$. It follows that any continuous function on $[0, 1]$ is uniquely determined by its moments: if two such functions both had the same moments then all the moments of their difference would be 0, and so the difference would be the zero function. In statistics, the moments of a continuous random variable uniquely determine its probability density function.

To prove the result, take any number $\epsilon > 0$ and let M be such that $|f(x)| \leqslant M$ for all x in $[0, 1]$. By the Weierstrass theorem, there exists a polynomial function p such that

$$|f(x) - p(x)| < \frac{\epsilon}{M},$$

for all x in $[0, 1]$. Since all moments of f are zero, and since p is a polynomial function, we have $\int_0^1 f(x)p(x)\,dx = 0$. Hence

$$0 \leqslant \int_0^1 (f(x))^2\,dx = \int_0^1 f(x)(f(x) - p(x))\,dx$$

$$\leqslant \int_0^1 |f(x)|\,|f(x) - p(x)|\,dx < M \cdot \frac{\epsilon}{M} = \epsilon.$$

But ϵ is arbitrary, so we must have $\int_0^1 (f(x))^2\,dx = 0$. It follows now, since f is continuous, that $f(x) = 0$ for all x in $[0,1]$, as we set out to prove.

We end this section with a slightly more specialised form of the Weierstrass theorem. It will be called on in Chapter 9.

Theorem 6.8.4 *Given any function $f \in C[0,1]$ and any number $\epsilon > 0$, there exists a polynomial function p, all of whose coefficients are rational numbers, such that $\|p - f\| < \epsilon$.*

Certainly, by the Weierstrass theorem itself, there exists a polynomial function q such that $\|q - f\| < \frac{1}{2}\epsilon$. Suppose q has degree r, and

$$q(x) = \sum_{k=0}^{r} a_k x^k, \quad 0 \leqslant x \leqslant 1,$$

where some or all of the coefficients a_0, a_1, ..., a_r may be irrational. For each coefficient a_k we can find a rational number b_k so that

$$|b_k - a_k| \leqslant \frac{\epsilon}{2(r+1)}.$$

Let p be the polynomial function given by

$$p(x) = \sum_{k=0}^{r} b_k x^k, \quad 0 \leqslant x \leqslant 1.$$

Then, for all x in $[0,1]$,

$$|p(x) - q(x)| = \left| \sum_{k=0}^{r} (b_k - a_k) x^k \right|$$

$$\leqslant \sum_{k=0}^{r} |b_k - a_k| \, |x|^k \leqslant \sum_{k=0}^{r} \frac{\epsilon}{2(r+1)} = \frac{\epsilon}{2},$$

so $\|p - q\| \leqslant \frac{1}{2}\epsilon$. Hence

$$\|p - f\| \leqslant \|p - q\| + \|q - f\| < \epsilon,$$

and this proves the theorem. $\qquad \square$

6.9 Solved problems

(1) Determine a cubic function which is an approximation of the function sin over the interval $[-1, 1]$ with error less than 0.001.

Solution. We know that

$$\sin x = x - \frac{x^3}{3!} + \frac{x^5}{5!} - \frac{x^7}{7!} + \cdots,$$

the series converging for any value of $x \in \mathbf{R}$. An immediate suggestion for a cubic function approximating sin is that given by $x - \frac{1}{6}x^3$. However, $\sin 1 - (1 - \frac{1}{6}) > 0.008$ so this cubic function is not within the given error bound on $[-1, 1]$. The series does provide a quintic function which approximates sin with acceptable accuracy on $[-1, 1]$, since elementary considerations show that

$$\left| \sin x - \left(x - \frac{x^3}{6} + \frac{x^5}{120} \right) \right| \leqslant \frac{|x|^7}{7!} \leqslant \frac{1}{7!} < 0.0002$$

when $|x| \leqslant 1$.

We obtain an expression for this quintic function in terms of the Chebyshev polynomials:

$$x - \frac{x^3}{6} + \frac{x^5}{120} = T_1(x) - \frac{1}{6} \cdot \frac{1}{4}(3T_1(x) + T_3(x))$$

$$+ \frac{1}{120} \cdot \frac{1}{16}(10T_1(x) + 5T_3(x) + T_5(x))$$

$$= \frac{169}{192}T_1(x) - \frac{5}{128}T_3(x) + \frac{1}{1920}T_5(x).$$

Since $|T_5(x)| \leqslant 1$ when $|x| \leqslant 1$, omitting the term $\frac{1}{1920}T_5(x)$ will admit a further error of at most $\frac{1}{1920} < 0.0006$ which gives a total error less than 0.0008, still within the given bound. Now,

$$\frac{169}{192}T_1(x) - \frac{5}{128}T_3(x) = \frac{169}{192}x - \frac{5}{128}(4x^3 - 3x) = \frac{383}{384}x - \frac{5}{32}x^3,$$

and the cubic function we end with has the desired property.   □

This solution demonstrates the technique known as *economisation of power series*, used in numerical analysis.

(2) Find the linear function (polynomial function of degree 1) that is the best uniform approximation of the function sin on the interval $[0, \frac{1}{2}\pi]$.

Solution. Theorem 6.6.1 implies the existence of such a linear function and the work of Chebyshev, referred to in Section 6.7, shows its uniqueness. That the approximation we derive below is unique is clear enough in this particular example, but we will not go into a full justification of this fact.

Define the error function E by

$$E(x) = (a + bx) - \sin x, \quad 0 \leqslant x \leqslant \tfrac{1}{2}\pi.$$

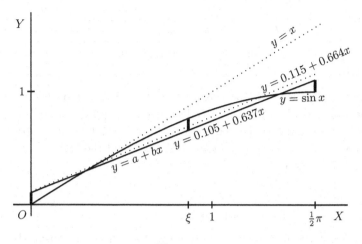

Figure 12

We wish to determine values for the constants a and b so that

$$\|E\| = \max_{0 \leqslant x \leqslant \pi/2} |E(x)|$$

is a minimum. Look at Figure 12. It is clear that varying the values of a and b allows two of the three indicated values of $|E|$ (the lengths of the heavy vertical line segments at $x = 0$, $x = \frac{1}{2}\pi$ and $x = \xi$ for some ξ in $(0, \frac{1}{2}\pi)$) to be decreased but at the expense of increasing the third. The best values of a and b are those for which these three values of $|E|$ are equal. Two further unknowns are introduced: the value of ξ where this occurs and the common value E_M of $|E|$ at the three points.

We have $|E(x)| = E_M$ at $x = 0$, $x = \xi$ and $x = \frac{1}{2}\pi$; that is,

$$a = E_M,$$
$$(a + b\xi) - \sin\xi = -E_M,$$
$$(a + \tfrac{1}{2}b\pi) - 1 = E_M.$$

A fourth equation, allowing the determination of the four unknowns, follows by noting that the function $-E$ has a minimum value when $x = \xi$ so, setting the derivative of $-E$ equal to 0 at $x = \xi$, we have

$$\cos\xi - b = 0.$$

From the first and third equations, $b = 2/\pi$, so $\xi = \cos^{-1}(2/\pi)$, and adding the first two equations leads to

$$a = \frac{1}{2}\left(\sin\left(\cos^{-1}\frac{2}{\pi}\right) - \frac{2}{\pi}\cos^{-1}\frac{2}{\pi}\right) = \frac{1}{2\pi}\sqrt{\pi^2 - 4} - \frac{1}{\pi}\cos^{-1}\frac{2}{\pi}.$$

Using three places of decimals, the required linear function is given by $0.105 + 0.637x$. □

The error E_M in using the approximation above is less than 0.106. For comparison, we note that the best least squares approximation of $\sin$ over $[0, \frac{1}{2}\pi]$ is the function given by $8(\pi - 3)/\pi^2 + 24(4 - \pi)x/\pi^3$, or $0.115 + 0.664x$ to three decimal places. This line, and the line $y = x$, are also shown in Figure 12.

6.10 Exercises

(1) Show that the fact that any two norms for a finite-dimensional vector space are equivalent follows from the fact that any norm for the space is equivalent to $\| \; \|_\infty$. (See Theorem 6.5.3.)

(2) Prove that the subset $\{x : \|x\|_\infty = 1\}$ of a finite-dimensional vector space is compact. (See the proof of Theorem 6.5.3.)

(3) Prove that a subset of a finite-dimensional vector space that is closed and bounded with respect to some norm for the space is closed and bounded also with respect to any other norm for the space.

(4) Prove that, whatever the norm for a finite-dimensional vector space, convergence of a sequence in the space is equivalent to convergence of the sequences of coefficients.

(5) Show that $C_2[a, b]$ is a strictly convex normed space. (Hint: See Exercise 2.4(6).)

(6) Find $T_6(x)$, $T_7(x)$, $T_8(x)$. Obtain x^6, x^7, x^8 as linear combinations of the Chebyshev polynomials.

(7) Prove that $T_r(-x) = (-1)^r T_r(x)$, for $r = 0, 1, 2, \ldots$.

(8) Use Chebyshev polynomials to obtain the fourth-degree polynomial $\frac{191}{192} - \frac{29}{32}x^2 + \frac{1}{4}x^4$ as an approximation for e^{-x^2}, having uniform error less than 0.05 for x in $[-1, 1]$.

(9) Show that the best uniform approximation of $8x^4 - 38x^3 + 11x^2 - 3x - 27$ over $[-1, 1]$ by a cubic polynomial is $-38x^3 + 19x^2 - 3x - 28$. Obtain the best uniform quadratic approximation of this cubic function and find the zero of the quadratic function in $[-1, 1]$. (This could serve as a first trial in an iterative solution for the

zero in $[-1, 1]$ of the original quartic function. This zero is -0.75, to two decimal places.)

(10) Find the linear function that is the best uniform approximation of the function given by $\sqrt{x}$ on the interval (a) $[0, 1]$, (b) $[1, 4]$. Set $x = \frac{5}{4}$ in (b) to show that $\sqrt{5} = 2.25$, approximately. Estimate the error using the maximum error found in (b).

(11) Find the linear function that is the best uniform approximation of the function given by $1/(1 + x)$ on the interval $[0, 1]$.

(12) Prove that the sequence of images of the terms of a Cauchy sequence under a uniformly continuous mapping is again a Cauchy sequence.

(13) Prove that a contraction mapping is uniformly continuous.

(14) Generalise the Weierstrass theorem to show that, given $\epsilon > 0$, for any function $f \in C[a, b]$ there is a polynomial function p so that $\|p - f\| < \epsilon$. (Hint: Define a function g by $g(y) = f(a + (b-a)y)$. Then $g \in C[0, 1]$ so there is a polynomial function q such that $\|q - g\| < \epsilon$. Set $p(x) = q((x - a)/(b - a))$, so p is a polynomial function with the desired property.)

(15) Let $f \in C^{(1)}[a, b]$, which is the space of all differentiable functions defined on $[a, b]$, with the uniform norm. Show that, if $\epsilon > 0$ is given and p is a polynomial function such that $\|p - f'\| < \epsilon$, then $\|q - f\| < \epsilon(b - a)$, where q is the polynomial function defined by $q(x) = \int_a^x p(t)\, dt + f(a)$.

(16) Find the best uniform quadratic approximations for the functions indicated by (a) $1/(1 + x^2)$, (b) $|x|$, both on $[-1, 1]$.

. . . .

(17)   (a) Suppose X is a strictly convex normed vector space. Show that $\|\frac{1}{2}(x + y)\| < 1$ if $x, y \in X$ and $\|x\| = \|y\| = 1$, $x \neq y$.
        (b) Prove the converse of the result in (a).

(18) Verify that the Chebyshev polynomial T_r is a solution of the differential equation

$$(1 - x^2)\frac{d^2 y}{dx^2} - x\frac{dy}{dx} + r^2 y = 0.$$

7

Mappings on Normed Spaces

7.1 Bounded linear mappings

In this chapter, we are concerned with mappings between normed vector spaces. We will see applications to numerical analysis, the theory of integral equations, and quantum mechanics.

There is nothing new in the notion of a mapping $A\colon X \to Y$ when X and Y are normed spaces beyond what we have described for mappings between metric spaces. We have already used such mappings, for example in the discussion of uniform continuity. However, the fact that X and Y are vector spaces for which norms have been defined allows us to distinguish more easily different types of mappings and therefore to develop more precise theories for those different types.

The simplest class of mappings between vector spaces, taking fullest advantage of the vector space properties, turns out to be the most important in practice. These are the linear maps.

Definition 7.1.1 A mapping $A\colon X \to Y$, where X and Y are vector spaces, is said to be *linear* when

$$A(\alpha_1 x_1 + \alpha_2 x_2) = \alpha_1 A x_1 + \alpha_2 A x_2$$

for any points $x_1, x_2 \in X$ and any scalars α_1, α_2.

It is not required here that X and Y be normed. When we insist on normed spaces we are able, in a certain sense, to measure the size of mappings. This leads to our second class.

Definition 7.1.2 A mapping $A\colon X \to Y$, where X and Y are normed vector spaces, is said to be *bounded* when there exists a constant $K > 0$

210

such that

$$\|Ax\| \leqslant K\|x\|$$

for all $x \in X$.

The magnitude of the number K here gives the size of the mapping A in a way to be made precise in the next section. We remark that this definition uses the notation $\|\ \|$ for the norms of both X and Y, though these may well be different. This is a common practice which we have already used in discussing uniform continuity, and will continue to follow. Notice also that the use of the word 'bounded' is quite different from earlier uses of that word. In particular, we must carefully distinguish the earlier idea of a bounded function: a function f for which there is a constant $K > 0$ such that $|f(x)| \leqslant K$ for all x in the domain of f.

Bounded mappings need not be linear, but it is the class of mappings that are both bounded and linear on which we will spend most of our time. So much so, that we give such mappings a special name.

Definition 7.1.3 A bounded linear mapping between normed vector spaces is called an *operator*.

Thus, we emphasise, whenever we refer to an operator we mean a mapping that is both linear and bounded. It turns out that operators are always continuous. To show this, we need first the following result, which is a surprising one at first glance.

Theorem 7.1.4 *A linear mapping that is continuous at any given point of a normed vector space X is continuous on X.*

Suppose A is a linear mapping continuous at a point $x_0 \in X$. Then we must show that A is continuous at any other point $x \in X$. Let $\{x_n\}$ be a convergent sequence in X, with limit x; that is, $x_n \to x$. But then $x_n - x + x_0 \to x_0$ and, since A is continuous at x_0, we have $A(x_n - x + x_0) \to Ax_0$. Since A is linear, we have $Ax_n - Ax + Ax_0 \to Ax_0$ and hence $Ax_n \to Ax$. That is, A is continuous at x. $\qquad\square$

Now we prove the result mentioned above, and its converse.

Theorem 7.1.5 *Let A be a linear mapping on a normed space X. Then A is continuous on X if and only if it is bounded.*

To prove this, suppose first that A is bounded: $\|Ax\| \leqslant K\|x\|$ for some number $K > 0$ and all $x \in X$. Take any point $x \in X$ and let $\{x_n\}$ be a sequence in X with $\lim x_n = x$. Let ϵ be any positive number. For all n large enough, we have $\|x_n - x\| < \epsilon/K$. But then, using in turn the linearity and boundedness of A,

$$\|Ax_n - Ax\| = \|A(x_n - x)\| \leqslant K\|x_n - x\| < \epsilon,$$

for such n. Thus A is continuous at x and, by the preceding theorem, A is continuous on X.

For the converse, suppose that A is continuous on X but that A is not bounded. We will obtain a contradiction. Since A is not bounded, for each positive integer n there is a point $x_n \in X$ so that

$$\|Ax_n\| > n\|x_n\|.$$

Notice that we cannot have $x_n = \theta$, because $A\theta = \theta$ for any linear map (to be proved as an exercise). Hence $\|x_n\| \neq 0$ for any n. Define a sequence $\{y_n\}$ in X by

$$y_n = \frac{1}{n\|x_n\|} x_n.$$

Then, for all n,

$$\|Ay_n\| = \left\|A\left(\frac{1}{n\|x_n\|} x_n\right)\right\| = \left\|\frac{1}{n\|x_n\|} Ax_n\right\| = \frac{1}{n\|x_n\|} \|Ax_n\| > 1.$$

But

$$\|y_n\| = \frac{1}{n\|x_n\|} \|x_n\| = \frac{1}{n}$$

so $\|y_n\| < \epsilon$ for any $\epsilon > 0$ if n is large enough. Hence the sequence $\{y_n\}$ converges to the zero vector θ in X. Since A is continuous on X, it is in particular continuous at θ so $Ay_n \to A\theta = \theta$. This is contradicted by the fact that $\|Ay_n\| > 1$ for all n, so A must indeed be bounded. $\qquad\square$

As a result of this theorem, we may use the words 'bounded' and 'continuous' interchangeably when referring to a linear mapping on a normed space.

One example of an operator on a normed space X is the mapping A defined by

$$Ax = \beta x, \quad x \in X,$$

for some fixed scalar β. It is indeed linear, since

$$A(\alpha_1 x_1 + \alpha_2 x_2) = \beta(\alpha_1 x_1 + \alpha_2 x_2)$$
$$= \alpha_1(\beta x_1) + \alpha_2(\beta x_2) = \alpha_1 A x_1 + \alpha_2 A x_2$$

($x_1, x_2 \in X$, scalars α_1, α_2). And it is bounded, since

$$\|Ax\| = \|\beta x\| = |\beta|\,\|x\|$$

so $\|Ax\| \leqslant K\|x\|$ for some constant K (such as $|\beta|$ or any larger number). If $\beta = 1$, A is the *identity operator* or *unit operator* on X and is denoted by I. Thus I maps every element of X into itself. If $\beta = 0$, A is called the *zero operator* on X and maps every element of X into θ.

For a second example, we take the mapping $A \colon C[a, b] \to C[a, b]$ defined by the equation $Ax = y$ where

$$y(s) = \lambda \int_a^b k(s, t) x(t)\, dt, \quad x \in C[a, b],\ a \leqslant s \leqslant b.$$

Here, k is a function of two variables, which is continuous in the square $[a, b] \times [a, b]$, and λ is a given nonzero real number. The mapping A is linear, since, for $x_1, x_2 \in C[a, b]$, scalars α_1, α_2, and any s in $[a, b]$,

$$(A(\alpha_1 x_1 + \alpha_2 x_2))(s) = \lambda \int_a^b k(s, t)(\alpha_1 x_1(t) + \alpha_2 x_2(t))\, dt$$
$$= \alpha_1 \lambda \int_a^b k(s, t) x_1(t)\, dt + \alpha_2 \lambda \int_a^b k(s, t) x_2(t)\, dt$$
$$= (\alpha_1 A x_1)(s) + (\alpha_2 A x_2)(s);$$

that is, $A(\alpha_1 x_1 + \alpha_2 x_2) = \alpha_1 A x_1 + \alpha_2 A x_2$. Also, A is bounded. To see this, let M be a positive constant such that $|k(s, t)| \leqslant M$ for (s, t) in the square. Then

$$\|Ax\| = \|y\| = \max_{a \leqslant s \leqslant b} |y(s)| = \max_{a \leqslant s \leqslant b} \left| \lambda \int_a^b k(s, t) x(t)\, dt \right|$$
$$\leqslant |\lambda| \max_{a \leqslant s \leqslant b} \int_a^b |k(s, t)|\,|x(t)|\, dt$$
$$\leqslant |\lambda| M \max_{a \leqslant t \leqslant b} |x(t)| \cdot (b - a)$$
$$= |\lambda| M (b - a) \|x\|.$$

Thus, for $K = |\lambda| M (b - a)$, say, we have $\|Ax\| \leqslant K\|x\|$ for all $x \in C[a, b]$, so A is bounded. This verifies that A is indeed an operator.

7.2 Norm of an operator

If A is an operator from a normed space X into some other normed space, we know there is some constant K such that $\|Ax\| \leqslant K\|x\|$ for all $x \in X$. The 'smallest possible' value of K such that this inequality holds provides the measure of the size of A that we mentioned above. Anticipating a little, that value is called the *norm* of A and is denoted by $\|A\|$. We will show soon that this name and the notation are quite consistent with the idea of a norm for a vector space. The following theorems take us logically to that point.

Theorem 7.2.1 *Let A be an operator on a normed vector space X. Set*

$$a = \inf\{K : \|Ax\| \leqslant K\|x\|, \ x \in X\},$$

$$b = \sup\left\{\frac{\|Ax\|}{\|x\|} : x \in X, \ x \neq \theta\right\},$$

$$c = \sup\{\|Ax\| : x \in X, \ \|x\| = 1\},$$

$$d = \sup\{\|Ax\| : x \in X, \ \|x\| \leqslant 1\}.$$

Then

(a) $\|Ax\| \leqslant a\|x\|$ *for all $x \in X$,*

(b) $a = b = c = d$.

The number a here is the number we will later explicitly define to be the norm of A. The theorem shows that any one of the expressions for b, c or d could equally well be chosen as the definition.

To prove (a), we only need to note that, by definition of greatest lower bound (inf), we have $\|Ax\| < (a + \epsilon)\|x\|$, for any $\epsilon > 0$ and all $x \in X$. Then the result follows because ϵ is arbitrary.

We will prove (b) by showing that $a \leqslant b \leqslant c \leqslant d \leqslant a$.

For any nonzero $x \in X$, we have $b \geqslant \|Ax\|/\|x\|$ so $\|Ax\| \leqslant b\|x\|$, and this is true also when $x = \theta$. Thus b belongs to the set

$$\{K : \|Ax\| \leqslant K\|x\|, \ x \in X\}$$

and since a is the greatest lower bound of this set, we have $a \leqslant b$. (This is a common form of argument, used often below.) Next, for $x \in X$, $x \neq \theta$,

$$\frac{\|Ax\|}{\|x\|} = \left\|\frac{1}{\|x\|} Ax\right\| = \left\|A\left(\frac{x}{\|x\|}\right)\right\| \leqslant c,$$

since A is linear and $x/\|x\|$ has norm 1; so $b \leqslant c$. Then we observe that the set $\{x : x \in X, \ \|x\| = 1\}$ is a subset of $\{x : x \in X, \ \|x\| \leqslant 1\}$, so

$c \leqslant d$. Finally, suppose $\|x\| \leqslant 1$ ($x \in X$). Then, by (a), $\|Ax\| \leqslant a$ and so we have $d \leqslant a$. This completes the proof. $\qquad\square$

Let X and Y be normed spaces. It is reasonable to suppose that in general there are many different operators from X into Y. We want to consider in the following paragraphs the set of all such operators, and will denote this set by $B(X, Y)$. This is not a totally new idea: $B(X, Y)$ has some likeness to the set $C[a, b]$, all of whose elements are functions from the interval $[a, b]$ into $\mathbf{R}$.

We will prove that $B(X, Y)$ is a vector space, and that it may be normed. We will use the following natural definitions of addition and scalar multiplication of operators by scalars: if A, A_1 and A_2 are any operators in $B(X, Y)$ and α is any scalar, we define mappings $A_1 + A_2$ and αA by

$$(A_1 + A_2)x = A_1x + A_2x, \quad (\alpha A)x = \alpha Ax,$$

where $x \in X$. We need to show that $A_1 + A_2$ and αA are in fact operators in $B(X, Y)$ and that the axioms of a vector space are satisfied with these definitions.

Since Y is a vector space, it is immediate that $A_1 + A_2$ and αA indeed map X into Y. It is left as an exercise to show that they are linear maps. Since A_1 and A_2 are bounded, there exist constants K_1 and K_2 such that $\|A_1x\| \leqslant K_1\|x\|$ and $\|A_2x\| \leqslant K_2\|x\|$ for all $x \in X$. Then

$$\begin{aligned}
\|(A_1 + A_2)x\| &= \|A_1x + A_2x\| \\
&\leqslant \|A_1x\| + \|A_2x\| \\
&\leqslant K_1\|x\| + K_2\|x\| = (K_1 + K_2)\|x\|
\end{aligned}$$

for all $x \in X$, so $A_1 + A_2$ is also bounded. Similarly,

$$\|(\alpha A)x\| = \|\alpha Ax\| = |\alpha|\,\|Ax\| \leqslant (|\alpha|K)\|x\|$$

for all $x \in X$ and some constant K, since A is bounded, so αA is also bounded. This proves that $A_1 + A_2 \in B(X, Y)$ and $\alpha A \in B(X, Y)$.

The verification of the vector space axioms for $B(X, Y)$ is easy. (The axioms are listed in Definition 1.11.1.) The negative $-A$ of an operator $A \in B(X, Y)$ is the operator $(-1)A$ and the zero vector in $B(X, Y)$ is the operator mapping each point in X into the zero vector in Y. Of the remaining axioms, we will prove here that $A_1 + A_2 = A_2 + A_1$, for any $A_1, A_2 \in B(X, Y)$, and $(\alpha\beta)A = \alpha(\beta A)$, for any $A \in B(X, Y)$ and any scalars α, β. Take any $x \in X$. Then these follow since

$$(A_1 + A_2)x = A_1x + A_2x = A_2x + A_1x = (A_2 + A_1)x$$

and

$$((\alpha\beta)A)x = (\alpha\beta)Ax = \alpha(\beta Ax) = \alpha((\beta A)x) = (\alpha(\beta A))x.$$

In these we have used the vector space properties of Y.

Hence we have proved the following result.

Theorem 7.2.2 *The set $B(X,Y)$ of all operators from X into Y is a vector space.*

The vector space is itself denoted by $B(X,Y)$. We will show soon how $B(X,Y)$ may be normed.

Definition 7.2.3 For any operator $A \in B(X,Y)$, the *norm* of A, denoted by $\|A\|$, is the number

$$\|A\| = \inf\{K : \|Ax\| \leqslant K\|x\|, \ x \in X\}.$$

We have anticipated this. Theorem 7.2.1 gives alternative expressions for $\|A\|$ and proves the important inequality

$$\|Ax\| \leqslant \|A\|\,\|x\|, \quad x \in X.$$

There are many occasions below where we use this inequality.

To find the norm of a given operator, we may use whichever of the expressions in Theorem 7.2.1 is the more convenient. For the operator $A \colon X \to X$ where $Ax = \beta x$, considered above, we have immediately

$$\|A\| = \sup\{\|Ax\| : x \in X, \ \|x\| = 1\}$$
$$= \sup\{|\beta|\,\|x\| : x \in X, \ \|x\| = 1\} = |\beta|.$$

In particular, for the identity operator I, we have

$$\|I\| = 1.$$

Now we are able to complete the development of the normed space $B(X,Y)$.

Theorem 7.2.4 *The vector space $B(X,Y)$ is normed by virtue of the definition of the norm of an operator.*

We must verify (N1), (N2) and (N3) (Definition 6.1.1). We leave the verification of (N1) as an exercise, with the remark that 'obvious' will not do as an answer. To prove (N2), take any operator $A \in B(X,Y)$ and any scalar α. Then, for any $x \in X$,

$$\|(\alpha A)x\| = \|\alpha Ax\| = |\alpha|\,\|Ax\| \leqslant (|\alpha|\,\|A\|)\|x\|,$$

so $\|\alpha A\| \leqslant |\alpha|\,\|A\|$. We will prove also that $\|\alpha A\| \geqslant |\alpha|\,\|A\|$. This is clear when $\alpha = 0$, so we may suppose that $\alpha \neq 0$. Then

$$\|Ax\| = \|(\alpha^{-1}\alpha)Ax\| = \|\alpha^{-1}(\alpha A)x\|$$
$$= |\alpha|^{-1}\|(\alpha A)x\| \leqslant |\alpha|^{-1}\|\alpha A\|\,\|x\|$$

so $\|A\| \leqslant |\alpha|^{-1}\|\alpha A\|$, or $\|\alpha A\| \geqslant |\alpha|\,\|A\|$, as required. Thus, (N2) is verified. For (N3), we have, for any operators $A_1, A_2 \in B(X,Y)$ and all $x \in X$,

$$\|(A_1 + A_2)x\| = \|A_1 x + A_2 x\| \leqslant \|A_1 x\| + \|A_2 x\|$$
$$\leqslant \|A_1\|\,\|x\| + \|A_2\|\,\|x\| = (\|A_1\| + \|A_2\|)\|x\|,$$

so $\|A_1 + A_2\| \leqslant \|A_1\| + \|A_2\|$. This is (N3), completing the proof of the theorem. $\qquad\qquad\qquad\square$

Notice that in proving that $B(X,Y)$ is a normed vector space, we rely very little on the vector space properties of X, but heavily on those of Y. In this light, the following result, which is of great importance in functional analysis, is not as surprising as it first appears.

Theorem 7.2.5 *If Y is a Banach space, then so is $B(X,Y)$.*

This is true regardless of whether X is a Banach space or not! There is a quite standard proof, in which we take any Cauchy sequence in $B(X,Y)$ and show that it converges. However, we will prove the theorem as an application of Theorem 6.2.2, by showing that every absolutely convergent series in $B(X,Y)$ is convergent.

Let $\sum_{k=1}^{\infty} A_k$ be an absolutely convergent series of elements (which are operators) in $B(X,Y)$. Then the real-valued series $\sum_{k=1}^{\infty} \|A_k\|$ converges. Write $y_n = \sum_{k=1}^{n} A_k x$, where x is some fixed element of X. Then $y_n \in Y$ for each $n \in \mathbf{N}$. With $n > m$ for definiteness, we have

$$\|y_n - y_m\| = \left\|\sum_{k=m+1}^{n} A_k x\right\| \leqslant \sum_{k=m+1}^{n} \|A_k x\|$$
$$\leqslant \sum_{k=m+1}^{n} \|A_k\|\,\|x\| \leqslant \epsilon\|x\|$$

for any $\epsilon > 0$ provided m is large enough. This shows that $\{y_n\}$ is a Cauchy sequence in Y and, since Y is a Banach space, the sequence converges. Define a mapping $A\colon X \to Y$ by

$$Ax = \lim y_n = \sum_{k=1}^{\infty} A_k x, \quad x \in X.$$

It is easy to show that A is linear. Further, for any $x \in X$ and any $n \in \mathbf{N}$,

$$\sum_{k=1}^{n} \|A_k x\| \leqslant \sum_{k=1}^{n} \|A_k\| \, \|x\| \leqslant \sum_{k=1}^{\infty} \|A_k\| \, \|x\|,$$

so $\sum_{k=1}^{\infty} \|A_k x\|$ is convergent. Then, using the continuity of $\|\ \|$ (from Exercise 6.4(3)(c)),

$$\|Ax\| = \left\| \lim \sum_{k=1}^{n} A_k x \right\| = \lim \left\| \sum_{k=1}^{n} A_k x \right\|$$

$$\leqslant \lim \sum_{k=1}^{n} \|A_k x\| \leqslant \sum_{k=1}^{\infty} \|A_k\| \, \|x\|,$$

so A is bounded. Hence $A \in B(X, Y)$. Finally, we have, for any $x \in X$ and any $\eta > 0$,

$$\left\| \left(A - \sum_{k=1}^{n} A_k \right) x \right\| = \left\| \sum_{k=n+1}^{\infty} A_k x \right\| \leqslant \sum_{k=n+1}^{\infty} \|A_k x\|$$

$$\leqslant \sum_{k=n+1}^{\infty} \|A_k\| \, \|x\| < \eta \|x\|,$$

when n is large enough, since $\sum A_k$ is absolutely convergent. Hence,

$$\left\| A - \sum_{k=1}^{n} A_k \right\| < \eta,$$

for such n, or $\sum_{k=1}^{n} A_k \to A$. That is, the series $\sum_{k=1}^{\infty} A_k$ is convergent (with sum A), and this completes the proof on applying Theorem 6.2.2. $\square$

We have proved in passing here that if $\{A_n\}$ is a sequence in $B(X, Y)$ and $\sum A_k$ is convergent, then $\|\sum A_k\| \leqslant \sum \|A_k\|$, generalising the triangle inequality in $B(X, Y)$ to infinite series.

7.3 Functionals

The term 'functional' is given to a certain type of mapping.

Definition 7.3.1 Let X be a (real or complex) vector space. A *functional* on X is a mapping $f\colon X \to K$, where K is the set of scalars (either $\mathbf{R}$ or $\mathbf{C}$) for X. The image under f of a point $x \in X$ is denoted by $f(x)$.

Notice that for functionals we revert to the older notation used for real-valued functions, which are of course themselves examples of functionals when their domain is $\mathbf{R}$. The following are further examples:

(a) $f\colon \mathbf{R}^n \to \mathbf{R}$, where $f(x) = \sum_{k=1}^{n} a_k x_k$, $x = (x_1, \ldots, x_n) \in \mathbf{R}^n$ and $(a_1, \ldots, a_n) \in \mathbf{R}^n$ is fixed;

(b) $f\colon C[a,b] \to \mathbf{R}$, where $f(x) = \int_a^b x(t)\, dt$, $x \in C[a,b]$;

(c) $f\colon l_2 \to \mathbf{C}$, where $f(x) = x_j$, $x = (x_1, x_2, \ldots) \in l_2$ and $j \in \mathbf{N}$ is fixed;

(d) $f\colon X \to \mathbf{R}$, where $f(x) = \|x\|$, $x \in X$, if X is a normed space.

As for mappings between vector spaces generally, the functional f is *linear* if

$$f(\alpha_1 x_1 + \alpha_2 x_2) = \alpha_1 f(x_1) + \alpha_2 f(x_2),$$

for any $x_1, x_2 \in X$ and any scalars α_1, α_2. It is left as an exercise to verify that (a), (b) and (c) above give examples of linear functionals, but (d) does not.

The definitions and properties given earlier for mappings between normed spaces carry over to a functional f on X, when X is normed. We quickly repeat these.

The functional f is *continuous* at a point $x \in X$ if whenever $\{x_n\}$ is a sequence in X converging to x then $\{f(x_n)\}$ is a sequence of scalars converging to $f(x)$. If f is linear and continuous at any particular point of X, then it is continuous at all points of X. The functional f is *bounded* in X if there is some constant $M > 0$ such that $|f(x)| \leqslant M\|x\|$ for all $x \in X$. The least such constant M (strictly, the infimum of such constants) is called the *norm* of f, denoted by $\|f\|$. Theorem 7.2.1 implies alternative expressions for $\|f\|$ when f is linear. For all $x \in X$, we have $|f(x)| \leqslant \|f\|\,\|x\|$. For linear functionals on a normed space, the conditions of boundedness and continuity are equivalent.

As above, let K be either $\mathbf{R}$ or $\mathbf{C}$, depending on whether X is a real or complex vector space. Then $B(X, K)$ is the space of all bounded linear functionals on X. As K is complete, Theorem 7.2.5 implies that this space $B(X, K)$ is a Banach space, whether or not X is. The space of functionals $B(X, K)$ is called the *dual space* of the space X, and is usually denoted simply by X'. Rephrasing the above, the dual of a normed vector space is always a Banach space. This is a result with many far-reaching consequences, but they are beyond the scope of this book.

We have stated that a linear functional is continuous if and only if it is bounded. There is another useful necessary and sufficient condition for a linear functional to be continuous. It applies specifically to functionals and not to more general mappings.

Theorem 7.3.2 *A linear functional f on a normed vector space X is continuous on X if and only if the set*

$$N(f) = \{x : x \in X, \ f(x) = 0\}$$

is closed.

The set $N(f)$ is a subset of X, easily shown in fact to be a subspace of X, called the *null space* or *kernel* of f. It is the set of all points of X whose images are 0 under f. (More generally, the null space of a mapping $A \colon X \to Y$, where X and Y are vector spaces, is the set $N(A) = \{x : x \in X, \ Ax = \theta\}$.)

To prove the theorem, we suppose first that f is continuous on X and let $\{x_n\}$ be a sequence in the null space $N(f)$ of f, which, as a sequence in X, converges with limit x, say. To show that $N(f)$ is closed, we must prove that $x \in N(f)$. Now, $f(x_n) = 0$ for all n, so $\lim f(x_n) = 0$. Since f is continuous on X, we must also have $f(x) = \lim f(x_n) = 0$. Thus $x \in N(f)$, as required.

The converse is more difficult to prove. We suppose now that $N(f)$ is closed and must prove that f is continuous on X. By Theorem 7.1.4, it is sufficient to prove that f is continuous at the zero vector θ of X. Then let $\{x_n\}$ be a sequence in X with limit θ. We must show that $f(x_n) \to 0$, since $f(\theta) = 0$.

Possibly, there is a positive integer M such that $x_n \in N(f)$ for all $n > M$. Then $f(x_n) = 0$ when $n > M$, so $f(x_n) \to 0$ as required.

If this is not the case, then for infinitely many terms of $\{x_n\}$ we have $x_n \notin N(f)$. Let $\{y_n\}$ be the subsequence of $\{x_n\}$ resulting from the removal of all terms for which $x_n \in N(f)$. Then $f(y_n) \neq 0$ for any n, and still $y_n \to \theta$. Put

$$t_n = \frac{1}{f(y_n)} y_n$$

for $n \in \mathbf{N}$, so that $f(t_n) = 1$ for all n.

Now $|f(y_n)| \geqslant 0$ for all n, so if we can show that $\overline{\lim} |f(y_n)| = 0$ then it will follow that $\lim |f(y_n)| = 0$. (A review of the notion of limit superior, in Section 1.7, may be required.) The proof will be by contradiction. Suppose $\overline{\lim} |f(y_n)| \neq 0$. Then there must be some number $\delta > 0$ such that $|f(y_n)| > \delta$ for infinitely many n. We may therefore choose a subsequence $\{y_{n_k}\}$ of $\{y_n\}$ with the property that $|f(y_{n_k})| > \delta$ for all $k \in \mathbf{N}$. Then

$$\|t_{n_k}\| = \frac{1}{|f(y_{n_k})|} \|y_{n_k}\| < \frac{1}{\delta} \|y_{n_k}\|$$

for all k, so $t_{n_k} \to \theta$ since $y_{n_k} \to \theta$. We notice that, for any k,

$$f(t_{n_1} - t_{n_k}) = f(t_{n_1}) - f(t_{n_k}) = 1 - 1 = 0,$$

since f is linear, so $t_{n_1} - t_{n_k} \in N(f)$. But $\{t_{n_1} - t_{n_k}\}_{k=1}^{\infty}$ is a convergent sequence in X, all of whose terms belong to $N(f)$, and $N(f)$ is closed. Hence $\lim_{k \to \infty} (t_{n_1} - t_{n_k}) = t_{n_1} - \theta = t_{n_1} \in N(f)$. This contradicts the fact that $f(t_{n_1}) = 1$. Hence $\overline{\lim} |f(y_n)| = 0$. Thus $f(y_n) \to 0$ and so $f(x_n) \to 0$ since $f(x_n) = 0$ when $x_n \neq y_m$ for any m. This completes the proof. □

7.4 Solved problems

(1) For the linear functional $f \colon C[a, b] \to \mathbf{R}$, where

$$f(x) = \int_a^b x(t)\, dt, \quad x \in C[a, b],$$

show that $\|f\| = b - a$.

Solution. The norm for $C[a, b]$ is as usual understood to be the uniform norm. Then, for any $x \in C[a, b]$,

$$|f(x)| = \left| \int_a^b x(t)\, dt \right| \leqslant \int_a^b |x(t)|\, dt$$

$$\leqslant \max_{a \leqslant t \leqslant b} |x(t)| \cdot \int_a^b dt = (b - a)\|x\|,$$

so f is bounded and $\|f\| \leqslant b - a$. Consider the function $x_0 \in C[a, b]$ given by $x_0(t) = 1$ for $a \leqslant t \leqslant b$. We see immediately that $f(x_0) = b - a > 0$ and $\|x_0\| = 1$. If $\|f\| < b - a$, then

$$b - a = |f(x_0)| \leqslant \|f\|\, \|x_0\| < (b - a)\|x_0\| = b - a.$$

This is impossible, so $\|f\| = b - a$, as required. □

For the second of these solved problems, we will need the following definition.

Definition 7.4.1   Let X and Y be normed vector spaces (both real or both complex) and let $A\colon S \to Y$ be a linear mapping from a subspace S of X into Y.

(a) The subset $\{(x, Ax) : x \in S\}$ of $X \times Y$ is called the *graph* of A, denoted by G_A.

(b) Let $\{x_n\}$ be any sequence in S with the following properties: as a sequence in X, $\{x_n\}$ is convergent to x and the sequence $\{Ax_n\}$ in Y is convergent to y. If $x \in S$ and $Ax = y$, then the mapping A is said to be *closed*.

(2) Let X and Y be normed vector spaces and let $A\colon S \to Y$ be a linear mapping from a subspace S of X into Y. Prove the following.

(a) With the definitions

$$(x_1, y_1) + (x_2, y_2) = (x_1 + x_2, y_1 + y_2),$$
$$\alpha(x, y) = (\alpha x, \alpha y)$$

($x_1, x_2, x \in X$, $y_1, y_2, y \in Y$, α scalar), $X \times Y$ is a vector space and the graph G_A of A is a subspace of $X \times Y$.

(b) With the further definition

$$\|(x, y)\| = \|x\| + \|y\|, \quad x \in X, \ y \in Y,$$

$X \times Y$ is a normed vector space. (The norms for X and Y may be different, but we use $\|\ \|$ here for both, and for the norm for $X \times Y$.)

(c) The linear mapping A is closed if and only if its graph G_A is closed.

Solution. (a) It is straightforward to verify that $X \times Y$ is a vector space. (The zero of the space is (θ, θ) where the θ's are the zeros of X and Y, respectively.) To show that G_A is a subspace of $X \times Y$, let $x_1, x_2 \in S$ so $(x_1, Ax_1), (x_2, Ax_2) \in G_A$. Then $x_1 + x_2 \in S$ and

$$(x_1, Ax_1) + (x_2, Ax_2) = (x_1 + x_2, Ax_1 + Ax_2)$$
$$= (x_1 + x_2, A(x_1 + x_2)) \in G_A.$$

Also, for any $x \in S$ and any scalar α, $\alpha x \in S$ and

$$\alpha(x, Ax) = (\alpha x, \alpha Ax) = (\alpha x, A(\alpha x)) \in G_A.$$

We have used the given definitions of addition and multiplication by scalars in $X \times Y$, and the fact that A is a linear mapping.

(b) is left as an exercise.

(c) Suppose first that the mapping A is closed and let $\{(x_n, Ax_n)\}$ be a sequence of points of G_A (so $x_n \in S$ for all n) which converges as a sequence in $X \times Y$. Put $(x, y) = \lim(x_n, Ax_n)$. To show that G_A is closed, we must show that $(x, y) \in G_A$. Given any $\epsilon > 0$, we can find a positive integer N such that

$$\|(x_n, Ax_n) - (x, y)\| < \epsilon$$

when $n > N$. Thus, for such n,

$$\|x_n - x\| + \|Ax_n - y\| = \|(x_n - x, Ax_n - y)\| < \epsilon,$$

by definition of the norm for $X \times Y$. Then both

$$\|x_n - x\| < \epsilon \quad \text{and} \quad \|Ax_n - y\| < \epsilon$$

when $n > N$. Hence $x_n \to x$ and $Ax_n \to y$. But we are given that the mapping A is closed, so we have $x \in S$ and $Ax = y$. Therefore $\lim(x_n, Ax_n) = (x, Ax) \in G_A$, so G_A is closed, as required.

Conversely, suppose G_A is closed. Let $\{x_n\}$ be a sequence of points of S which converges to x as a sequence in X and is such that the sequence $\{Ax_n\}$ in Y converges to y. We must show that $x \in S$ and $Ax = y$. Each term of the sequence $\{(x_n, Ax_n)\}$ is in G_A, and since

$$\|(x_n, Ax_n) - (x, y)\| = \|(x_n - x, Ax_n - y)\|$$
$$= \|x_n - x\| + \|Ax_n - y\|,$$

we must have $(x_n, Ax_n) \to (x, y)$. It follows that $(x, y) \in G_A$, since G_A is closed, and hence that $x \in S$ and $y = Ax$. Thus the mapping A is closed, and the proof is finished. $\qquad\square$

7.5 Exercises

(1) If X, Y are vector spaces and $A\colon X \to Y$ is a linear mapping, show that

    (a) $A(x_1 + x_2) = Ax_1 + Ax_2$, for any $x_1, x_2 \in X$,

    (b) $A(\alpha x) = \alpha Ax$, for any $x \in X$ and any scalar α,

    (c) $A\theta = \theta$.

Show that any mapping A satisfying (a) and (b) is linear.

(2) Define a mapping $A: C[a,b] \to C[a,b]$ by $Ax = y$ where

$$y(s) = \lambda \int_a^b k(s,t)x(t)\,dt, \quad x \in C[a,b], \ a \leqslant s \leqslant b.$$

Here, $C[a,b]$ is considered to be a vector space, with its usual uniform norm. Some analysis in Section 7.1 showed in effect that

$$\|A\| \leqslant |\lambda| M (b-a),$$

where M is the maximum value of $|k(s,t)|$ for $a \leqslant s \leqslant b$ and $a \leqslant t \leqslant b$.

Show that the mapping A is still bounded when considered as a mapping from the normed space $C_1[a,b]$ into itself, and from the normed space $C_2[a,b]$ into itself. That is, consider the effects of the different norms. In each case, show also that the same estimate for $\|A\|$ as that above may be obtained.

(3) Let g be a fixed continuous function on $[a,b]$ and let A be the mapping of $C[a,b]$ into itself defined by $Ax = y$, where $y(t) = g(t)x(t)$, $a \leqslant t \leqslant b$. Show that A is an operator. Do the same when A is considered as a mapping from $C_1[a,b]$ into itself.

(4) Let A_1 and A_2 be linear mappings between vector spaces X and Y. Show that $A_1 + A_2$ and αA, for any scalar α, are also linear mappings from X into Y.

(5) Complete the proof of Theorem 7.2.4 by verifying (N1).

(6) Verify that the functionals of examples (a), (b) and (c) in Section 7.3 are linear, while that of (d) is not.

(7) For the linear functional f of example (c) in Section 7.3, show that $\|f\| = 1$.

(8) Prove (b) in Solved Problem 7.4(2).

(9) If X and Y are normed vector spaces, show that $\| \ \|'$ is a norm for $X \times Y$ where

$$\|(x,y)\|' = \max\{\|x\|, \|y\|\}, \quad x \in X, \ y \in Y,$$

and that $\| \ \|'$ is equivalent to the norm $\| \ \|$ for $X \times Y$ defined in Solved Problem 7.4(2).

(10) If X and Y are Banach spaces, show that $X \times Y$ is also a Banach space, under either of the norms for $X \times Y$ mentioned in the preceding exercise.

(11) Prove that any operator between normed spaces is closed. (In Section 7.10, we will show that the converse is not true: closed linear mappings need not be continuous.)

(12) Show that all operators are uniformly continuous.

(13) Let $A\colon X \to X$ be an operator on a normed space X. Suppose there is a point $x \neq \theta$ and a scalar λ such that $Ax = \lambda x$. Prove that $|\lambda| \leqslant \|A\|$. (If such x and λ exist, then x is called an *eigenvector* of A corresponding to the *eigenvalue* λ.)

. . . .

(14) Let A be a linear mapping from a normed space X into a normed space Y. Prove that A is bounded if and only if A maps bounded sets in X into bounded sets in Y.

(15) Suppose A is a closed operator from a subspace S of a normed space X into a normed space Y. Show that if Y is a Banach space then S is a closed subspace of X.

(16) Let $A\colon X \to Y$ be a closed mapping between normed spaces X and Y, and let S be a compact subset of X. Show that $A(S)$ is a closed subset of Y.

7.6 Inverse mappings

When X and Y are any sets and A is a one-to-one mapping from X onto Y, we know (Definition 1.3.2) that there exists the inverse mapping $A^{-1}\colon Y \to X$, such that $A^{-1}y = x$ when $Ax = y$ ($x \in X$, $y \in Y$). In a formal way at least, this allows us to write down the solution of the equation $Ax = y$ when y is a given point in Y: the solution is just $x = A^{-1}y$, and this solution is unique. In specific applications, although the problem may be easily presented as 'solve $Ax = y$, given y', it is often not easy to determine whether the mapping A is onto and one-to-one, and even if the mapping is such, so that the inverse exists, it may be difficult to exhibit the inverse within the terms of the application. We will be deducing some further conditions which ensure the existence of the inverse of a mapping.

Our first theorem is not in that direction. It simply gives us a useful property of the inverse of a linear mapping, when it exists.

Theorem 7.6.1 *If X and Y are vector spaces, and $A\colon X \to Y$ is a linear mapping for which the inverse A^{-1} exists, then A^{-1} is also a linear mapping.*

To prove this, we must show that

$$A^{-1}(\alpha_1 y_1 + \alpha_2 y_2) = \alpha_1 A^{-1} y_1 + \alpha_2 A^{-1} y_2$$

for any $y_1, y_2 \in Y$ and any scalars α_1, α_2. Let $A^{-1}y_1 = x_1$ and $A^{-1}y_2 = x_2$ (so $x_1, x_2 \in X$). Then $Ax_1 = y_1$ and $Ax_2 = y_2$ and, since A is linear,

$$A(\alpha_1 x_1 + \alpha_2 x_2) = \alpha_1 A x_1 + \alpha_2 A x_2 = \alpha_1 y_1 + \alpha_2 y_2.$$

But this says that

$$A^{-1}(\alpha_1 y_1 + \alpha_2 y_2) = \alpha_1 x_1 + \alpha_2 x_2 = \alpha_1 A^{-1}y_1 + \alpha_2 A^{-1}y_2,$$

so the theorem is proved.   □

Suppose $A\colon X \to Y$ is a linear mapping between vector spaces X and Y with the property that the only solution of the equation $Ax = \theta$ ($x \in X$) is $x = \theta$. In that case, if x_1 and x_2 are points of X such that $Ax_1 = Ax_2$, then $A(x_1 - x_2) = \theta$ and so we must have $x_1 - x_2 = \theta$, or $x_1 = x_2$. This means that the mapping A is one-to-one. If it is also onto, then this property of A is thus sufficient to ensure the existence of the inverse A^{-1}. We can also prove the converse of this result. Suppose $A\colon X \to Y$ is a linear mapping between vector spaces whose inverse A^{-1} exists, and let $x \in X$ be a point for which $Ax = \theta$. Then, uniquely, $x = A^{-1}\theta = \theta$, since A^{-1} is a linear mapping, so $x = \theta$ is the only solution of the equation $Ax = \theta$. We have proved the following.

Theorem 7.6.2 *The inverse A^{-1} of an onto linear mapping $A\colon X \to Y$ between vector spaces X, Y exists if and only if the only solution of the equation $Ax = \theta$, $x \in X$, is $x = \theta$.*

Another way of putting the condition of this theorem is to require that $N(A) = \{\theta\}$, where $N(A)$ is the null space of the mapping A. It then follows by Theorem 7.3.2 that if a linear functional f on a normed space X has an inverse, then f is continuous on X. This is because the subset $\{\theta\}$ of X is certainly closed.

We next give another necessary and sufficient condition for an onto linear mapping between normed spaces to have an inverse.

Theorem 7.6.3 *Let $A\colon X \to Y$ be an onto linear mapping between normed spaces X and Y. The inverse A^{-1} exists, and is bounded, if and only if there is a constant $m > 0$ such that $\|Ax\| \geqslant m\|x\|$ for all $x \in X$.*

Proving this, suppose the inequality holds for all $x \in X$ and some $m > 0$. Then if $Ax = \theta$, we must have $\|x\| = 0$, so $x = \theta$. Hence A^{-1}

exists, by Theorem 7.6.2. Take any $y \in Y$ and put $A^{-1}y = x$. The inequality $\|Ax\| \geqslant m\|x\|$ is, equivalently,

$$\|A^{-1}y\| \leqslant \frac{1}{m}\|y\|.$$

This shows that A^{-1} is bounded (and moreover that $\|A^{-1}\| \leqslant 1/m$).

For the converse, if A^{-1} exists and is bounded, then, for any $y \in Y$, $\|A^{-1}y\| \leqslant \|A^{-1}\|\,\|y\|$. That is, $\|x\| \leqslant \|A^{-1}\|\,\|Ax\|$, where $x = A^{-1}y$. If $y = \theta$, the zero vector in Y, then $x = A^{-1}y = \theta$, the zero vector in X, and trivially in this case $\|Ax\| \geqslant m\|x\|$ for any $m > 0$. Otherwise, $\|y\| > 0$ so, by Theorem 7.6.2, $A^{-1}y = x \neq \theta$ and we have $0 < \|x\| \leqslant \|A^{-1}\|\,\|Ax\|$. Thus $\|A^{-1}\| > 0$ and again we have $\|Ax\| \geqslant m\|x\|$ for all nonzero $x \in X$, if we choose $m = 1/\|A^{-1}\|$, for example. $\qquad\square$

The next theorem is basic to the applications that follow. We recall that I is the identity operator on a normed space X; that is, $Ix = x$ for all $x \in X$.

Theorem 7.6.4 *Let A be an operator from a normed space X into itself and suppose $\|A\| < 1$. Suppose also that the operator $I - A$ is onto. Then the inverse $(I - A)^{-1}$ exists, and*

$$\|(I - A)^{-1}\| \leqslant \frac{1}{1 - \|A\|}.$$

This is a straightforward consequence of the preceding theorem. Using the triangle inequality, for any $x \in X$,

$$\|x\| \leqslant \|x - Ax\| + \|Ax\| \leqslant \|x - Ax\| + \|A\|\,\|x\|.$$

Hence

$$\|(I - A)x\| = \|Ix - Ax\| = \|x - Ax\| \geqslant (1 - \|A\|)\|x\|.$$

By Theorem 7.6.3 and its proof, applied to the operator $I - A$ with $m = 1 - \|A\| > 0$, the result follows. $\qquad\square$

We will prove next that we may drop the assumption above that $I - A$ is onto if we assume instead that X is a Banach space. More specifically, we will prove that $I - A$ must be onto when X is a Banach space. To do this, we need to show that if y is any point in X, then there is some point $x \in X$ such that $(I - A)x = y$. So let $y \in X$ be arbitrary. We introduce the mapping $B \colon X \to X$ by

$$Bx = Ax + y, \qquad x \in X.$$

(If $y \neq \theta$, then B is not linear. Such a mapping as B here, where A is linear, is called *affine*.) For any points $x', x'' \in X$, we have

$$\|Bx' - Bx''\| = \|Ax' - Ax''\| = \|A(x' - x'')\| \leqslant \|A\| \, \|x' - x''\|.$$

When $0 < \|A\| < 1$, this implies that B is a contraction mapping on X. As X is now assumed to be a Banach space, the fixed point theorem (Theorem 3.2.2) tells us that the mapping B has a unique fixed point. That is, there exists a unique point $x \in X$ such that $Bx = x$. But then $Ax + y = x$, or $y = (I - A)x$, as we wished to show.

The fixed point theorem implies further that the solution of the equation $y = (I - A)x$ may be found by successive approximations. Let the successive iterates be $x_0, x_1, x_2, \ldots$ and take $x_0 = y$. Then

$$x_1 = Bx_0 = Ax_0 + y = Ay + y,$$
$$x_2 = Bx_1 = Ax_1 + y = A(Ay + y) + y = A^2y + Ay + y,$$
$$x_3 = Bx_2 = Ax_2 + y = A(A^2y + Ay + y) + y = A^3y + A^2y + Ay + y,$$

and so on; in general,

$$x_n = A^ny + A^{n-1}y + \cdots + A^2y + Ay + y.$$

The sequence $\{x_n\}$ is therefore the sequence of partial sums of the series $\sum_{k=0}^{\infty} A^k y$ (in which by A^0 we mean the identity operator I). Since $\{x_n\}$ converges to the fixed point x of B, the series is convergent with sum x. But on the other hand, $x = (I - A)^{-1}y$.

We summarise all this as follows.

Theorem 7.6.5 *Let A be an operator from a Banach space X into itself, and suppose that $\|A\| < 1$. Then the operator $I - A$ is onto, the inverse $(I - A)^{-1}$ exists, and, for any $y \in X$,*

$$(I - A)^{-1}y = \sum_{k=0}^{\infty} A^k y.$$

Notice that we may look on the final conclusion as a result about the operator A alone:

$$(I - A)^{-1} = \sum_{k=0}^{\infty} A^k \quad \text{if } \|A\| < 1.$$

A full justification of this statement is called for in Exercise 7.9(7). This then appears to be a very satisfying generalisation of sorts of the familiar result on geometric series:

$$(1-a)^{-1} = \sum_{k=0}^{\infty} a^k \quad \text{if } |a| < 1.$$

7.7 Application to integral equations

We have considered the Volterra equation

$$x(s) = \lambda \int_a^s k(s,t)x(t)\,dt + f(s)$$

before, in Chapter 3. Again, λ is an arbitrary nonzero constant, k is a function of two variables which is continuous in the triangle

$$T = \{(s,t) : a \leqslant s \leqslant b, \ a \leqslant t \leqslant s\},$$

and $f \in C[a,b]$. In this section, we will give an alternative approach to the problem of solving the Volterra equation. The corresponding work for the Fredholm equation is easier and the development is left as an exercise.

In the Volterra equation, we suppose $x \in C[a,b]$ and define an operator K from the Banach space $C[a,b]$ into itself by $Kx = y$, where

$$y(s) = \lambda \int_a^s k(s,t)x(t)\,dt.$$

The fact that K is an operator follows as at the end of Section 7.1. The Volterra equation may be written

$$f(s) = x(s) - \lambda \int_a^s k(s,t)x(t)\,dt, \quad a \leqslant s \leqslant b,$$

and so we see that this may be expressed very succinctly as

$$f = (I - K)x.$$

Our aim is then immediately clear: if we can show that the inverse of the operator $I - K$ exists, then the solution of the Volterra equation is

$$x = (I - K)^{-1}f.$$

We will then need a special argument (not required in the analogous treatment of the Fredholm equation) to show that

$$x = \sum_{j=0}^{\infty} K^j f.$$

(The reason for the special treatment is that we can prove that $\|K^n\| < 1$ for n large enough, but cannot prove that $\|K\| < 1$.)

We show first of all that the mappings K^2, K^3, ... may be given similar definitions to that of K. Define a sequence $\{k_n\}$ of functions of two variables by

$$k_1(s,t) = k(s,t),$$

$$k_n(s,t) = \int_t^s k(s,u)k_{n-1}(u,t)\,du, \qquad n = 2,\ 3,\ \ldots,$$

for $a \leqslant t \leqslant s \leqslant b$. Then for the mapping K^n, we have $y = K^n x$, where $x \in C[a,b]$ and

$$y(s) = \lambda^n \int_a^s k_n(s,t)x(t)\,dt.$$

This is proved by induction as follows. When $n = 1$, the result is simply the definition of K. Assume the result is true when $n = m$ and suppose $y = K^{m+1}x$. Then

$$\begin{aligned}
y(s) &= (K^{m+1}x)(s) \\
&= (K(K^m x))(s) \\
&= \lambda \int_a^s k(s,u)\left(\lambda^m \int_a^u k_m(u,t)x(t)\,dt\right)du \\
&= \lambda^{m+1} \int_a^s \int_a^u k(s,u)k_m(u,t)x(t)\,dt\,du \\
&= \lambda^{m+1} \int_a^s \int_t^s k(s,u)k_m(u,t)x(t)\,du\,dt \\
&= \lambda^{m+1} \int_a^s k_{m+1}(s,t)x(t)\,dt,
\end{aligned}$$

This shows the result holds when $n = m + 1$, so our expression for K^n is established.

We now set $M = \max_{(s,t)\in T} |k(s,t)|$ and will prove that

$$|k_n(s,t)| \leqslant \frac{M^n(s-t)^{n-1}}{(n-1)!}, \qquad n \in \mathbf{N},$$

for all (s,t) in the triangle T. Again, we will use induction. When $n = 1$, the result is clear by definition of M. Assume the result is true when

$n = m$. Then, for $n = m + 1$ we have

$$
\begin{aligned}
|k_{m+1}(s, t)| &= \left| \int_t^s k(s, u) k_m(u, t) \, du \right| \\
&\leqslant \int_t^s |k(s, u)| \, |k_m(u, t)| \, du \\
&\leqslant M \frac{M^m}{(m-1)!} \int_t^s (u - t)^{m-1} \, du \\
&= \frac{M^{m+1}}{(m-1)!} \left[\frac{1}{m} (u - t)^m \right]_t^s \\
&= \frac{M^{m+1}(s - t)^m}{m!},
\end{aligned}
$$

and the result for $n = m + 1$ is seen to hold. This induction is now complete.

Next we will use the two preceding results to prove that, for each $n \in \mathbf{N}$, $\|K^n\|$ is bounded and

$$
\|K^n\| \leqslant \frac{|\lambda|^n M^n (b - a)^n}{n!}.
$$

It will then follow that $\|K^n\| < 1$ for all sufficiently large n. Choose any $x \in C[a, b]$. Then

$$
\begin{aligned}
\|K^n x\| &= \max_{a \leqslant s \leqslant b} \left| \lambda^n \int_a^s k_n(s, t) x(t) \, dt \right| \\
&\leqslant \max_{a \leqslant s \leqslant b} |\lambda|^n \int_a^s |k_n(s, t)| \, |x(t)| \, dt \\
&\leqslant |\lambda|^n \int_a^s \frac{M^n (s - t)^{n-1}}{(n-1)!} \, dt \cdot \max_{a \leqslant t \leqslant s} |x(t)| \\
&\leqslant \frac{|\lambda|^n M^n}{(n-1)!} \left[\frac{-1}{n} (s - t)^n \right]_a^s \cdot \max_{a \leqslant t \leqslant b} |x(t)| \\
&= \frac{|\lambda|^n M^n}{n!} (s - a)^n \|x\| \leqslant \frac{|\lambda|^n M^n}{n!} (b - a)^n \|x\|.
\end{aligned}
$$

This implies that the mapping K^n is bounded and furthermore that $\|K^n\| \leqslant (|\lambda| M (b - a))^n / n!$, as required.

It is easy to see that K^n is a linear mapping for each $n \in \mathbf{N}$, so the boundedness of each K^n could be quickly deduced from the following result. If X, Y and Z are normed spaces and $A \colon X \to Y$, $B \colon Y \to Z$ are operators, then the product BA is also an operator, from X into Z, and

$$
\|BA\| \leqslant \|B\| \, \|A\|.
$$

The proof of this is left as an exercise. It follows that if $Y = X$, so that the mappings A^n (for $n = 2, 3, \dots$) exist, then they are in fact operators on X, and $\|A^n\| \leqslant \|A\|^n$. For the operator K above, we could use the fact that $\|K\| \leqslant |\lambda| M(b - a)$ (obtained as in Exercise 7.5(2)) to deduce that $\|K^n\| \leqslant \|K\|^n \leqslant (|\lambda| M(b - a))^n$. This is certainly not as good as the estimate in the preceding paragraph.

Take any $n \in \mathbf{N}$ and any $x \in C[a, b]$. By repeated use of the rules for combining operators, given after the proof of Theorem 7.2.1, and using the result $\|I\| = 1$, we have

$$
\begin{aligned}
\|(I - K^n)x\| &= \|(I + K + K^2 + \cdots + K^{n-1})((I - K)x)\| \\
&\leqslant \|I + K + K^2 + \cdots + K^{n-1}\| \, \|(I - K)x\| \\
&\leqslant (\|I\| + \|K\| + \|K^2\| + \cdots + \|K^{n-1}\|) \, \|(I - K)x\| \\
&\leqslant \left(1 + \sum_{j=1}^{n-1} \frac{|\lambda|^j M^j (b - a)^j}{j!}\right) \|(I - K)x\| \\
&< \sum_{j=0}^{\infty} \frac{(|\lambda| M(b - a))^j}{j!} \|(I - K)x\| \\
&= e^{|\lambda| M(b-a)} \|(I - K)x\|.
\end{aligned}
$$

Hence

$$
\|(I - K)x\| > e^{-|\lambda| M(b-a)} \|(I - K^n)x\|.
$$

In particular, choose n so that $\|K^n\| < 1$ and put $q = \|K^n\|$. Then $\|K^n x\| \leqslant q\|x\|$ and

$$
\begin{aligned}
\|(I - K^n)x\| &= \|x - K^n x\| \geqslant \|x\| - \|K^n x\| \\
&\geqslant \|x\| - q\|x\| = (1 - q)\|x\|.
\end{aligned}
$$

Thus

$$
\|(I - K)x\| > e^{-|\lambda| M(b-a)} (1 - q)\|x\|
$$

for all $x \in C[a, b]$.

We are going to apply Theorem 7.6.3, with $m = e^{-|\lambda| M(b-a)}(1 - q)$, to show that $(I - K)^{-1}$ exists. This can be done as soon as we show that $I - K$ is an onto operator. Since n has been chosen so that $\|K^n\| < 1$, Theorem 7.6.5 assures us that the operator $I - K^n$ is onto. Thus, for any $y \in C[a, b]$ we know there exists $x \in C[a, b]$ such that $(I - K^n)x = y$. But then

$$
(I - K^n)x = (I - K)((I + K + K^2 + \cdots + K^{n-1})x) = y,
$$

implying the existence of a function $z \in C[a, b]$ such that $(I - K)z = y$. This means $I - K$ is onto, and Theorem 7.6.3 may be applied.

As we indicated at the beginning of this section, the solution of the Volterra equation, written as $f = (I - K)x$, is thus $x = (I - K)^{-1}f$. To show now that this solution is given by

$$x = \sum_{j=0}^{\infty} K^j f$$

it is sufficient to return to the inequalities

$$\|K^j\| \leqslant \frac{(|\lambda| M(b - a))^j}{j!}, \qquad j \in \mathbf{N}.$$

By a simple comparison test (Theorem 1.8.6), it then follows that the series $\sum \|K^j f\|$ is convergent. Thus the series $\sum K^j f$ is absolutely convergent and so, by Theorem 6.2.2 since $C[a, b]$ is a Banach space, it is also convergent. To show that the sum of the series $\sum_{j=0}^{\infty} K^j f$ is x, we may use the continuity of the operator $I - K$ as follows:

$$(I - K)\left(\sum_{j=0}^{\infty} K^j f\right) = (I - K)\left(\lim_{n \to \infty} \sum_{j=0}^{n} K^j f\right)$$

$$= \lim(I - K)\left(\sum_{j=0}^{n} K^j f\right)$$

$$= \lim\left(\sum_{j=0}^{n} K^j f - \sum_{j=0}^{n} K^{j+1} f\right)$$

$$= \lim(If - K^{n+1} f)$$

$$= f - \lim K^{n+1} f = f,$$

since $\|K^n f\| \leqslant \|K^n\| \|f\| \to 0$. Hence $\sum_{j=0}^{\infty} K^j f = (I - K)^{-1}f = x$.

The solution of the Volterra equation

$$x(s) = \lambda \int_a^s k(s, t)x(t)\, dt + f(s)$$

thus always exists uniquely and is given by

$$x(s) = f(s) + \sum_{j=1}^{\infty} \lambda^j \int_a^s k_j(s, t)f(t)\, dt.$$

In practice, it is convenient to invert the order of summation and integration and so write

$$x(s) = f(s) + \int_a^s f(t) \sum_{j=1}^{\infty} \lambda^j k_j(s,t)\, dt.$$

Using Theorem 1.10.7, term-by-term integration is indeed permissible here because, by an earlier result,

$$|\lambda^j k_j(s,t)| \leqslant \frac{|\lambda|^j M^j (s-t)^{j-1}}{(j-1)!} \leqslant |\lambda| M \frac{(|\lambda| M (b-a))^{j-1}}{(j-1)!}$$

for $j \in \mathbf{N}$ and all $(s,t) \in T$, and, by the Weierstrass M-test (Theorem 1.10.8), the series $\sum_{j=1}^{\infty} \lambda^j k_j(s,t)$ is uniformly convergent in t.

As an example, we will solve the equation

$$x(s) = \int_0^s (t-s)x(t)\, dt + e^s.$$

Here we have $\lambda = 1$, $k(s,t) = t - s$ and $f(s) = e^s$. We obtain

$$k_1(s,t) = t - s,$$

$$k_2(s,t) = \int_t^s (u-s)(t-u)\, du$$

$$= -\int_0^{t-s} (t-s-v)v\, dv \qquad [t - u = v]$$

$$= -\left[\frac{1}{2}(t-s)v^2 - \frac{1}{3}v^3\right]_0^{t-s}$$

$$= -\frac{1}{6}(t-s)^3,$$

$$k_3(s,t) = -\int_t^s (u-s) \cdot \frac{1}{6}(t-u)^3\, du$$

$$= \frac{1}{6}\int_0^{t-s} (t-s-v)v^3\, dv$$

$$= \frac{1}{6}\left[\frac{1}{4}(t-s)v^4 - \frac{1}{5}v^5\right]_0^{t-s}$$

$$= \frac{1}{120}(t-s)^5,$$

and in general, as should be verified by induction,

$$k_j(s,t) = \frac{(-1)^{j+1}}{(2j-1)!}(t-s)^{2j-1}.$$

Hence

$$x(s) = e^s + \int_0^s e^t \sum_{j=1}^{\infty} (-1)^{j+1} \frac{(t-s)^{2j-1}}{(2j-1)!} \, dt$$

$$= e^s + \int_0^s e^t \sin(t-s) \, dt$$

$$= \frac{1}{2}(e^s + \sin s + \cos s).$$

7.8 Application to numerical analysis

Before going into this further application of Theorem 7.6.5, we require a little more information about products of mappings.

We know that the associative law is satisfied: if X, Y, Z, W are any sets and $A\colon X \to Y$, $B\colon Y \to Z$ and $C\colon Z \to W$ are mappings, then $C(BA) = (CB)A$.

When the sets are vector spaces, we may ask whether the distributive laws are satisfied. The answer is interesting. We can easily prove that if X, Y, Z are vector spaces and $A\colon Y \to Z, B\colon Y \to Z, C\colon X \to Y$ are any mappings, then

$$(A + B)C = AC + BC.$$

We simply note that, for any $x \in X$,

$$((A + B)C)x = (A + B)(Cx) = A(Cx) + B(Cx)$$
$$= (AC)x + (BC)x = (AC + BC)x.$$

There is however another distributive law, and this second law is not generally satisfied. We can show this much: if X, Y, Z are vector spaces and $A\colon Y \to Z, B\colon X \to Y, C\colon X \to Y$ are mappings, then, provided A is linear,

$$A(B + C) = AB + AC.$$

To do this, take any $x \in X$. Then

$$(A(B + C))x = A((B + C)x) = A(Bx + Cx).$$

Now, because A is linear,

$$A(Bx + Cx) = A(Bx) + A(Cx) = (AB)x + (AC)x = (AB + AC)x.$$

Of course, both distributive laws are satisfied if we are concerned throughout only with linear mappings, or operators in particular. (The second distributive law only requires that $A(y_1 + y_2) = Ay_1 + Ay_2$ for all $y_1, y_2 \in Y$. Such mappings are called *additive*. If $A(\alpha y) = \alpha Ay$ for all $y \in Y$ and all scalars α, then A is called *homogeneous*. A mapping is linear if and only if it is both additive and linear. See Exercise 7.5(1).)

For the next few preliminary results, we suppose that A maps a set X onto itself.

It is clear that if I is the identity mapping on X, then $IA = A$ and $AI = A$.

If A^{-1} exists, then for any $x \in X$ we have

$$A^{-1}(Ax) = x \quad \text{and} \quad A(A^{-1}x) = x$$

so that we may write

$$A^{-1}A = I \quad \text{and} \quad AA^{-1} = I.$$

We now prove the following converse of this result. Two cases need to be identified. If B maps X into itself and $BA = I$, then the inverse of A exists and $A^{-1} = B$; if C maps X onto itself and $AC = I$, then the inverse of A exists and $A^{-1} = C$.

To prove this, note first that the second statement follows from the first since it implies that the inverse of C exists and $C^{-1} = A$; but then $C = (C^{-1})^{-1} = A^{-1}$. Now suppose $BA = I$. Since A is an onto map, for any given $y \in X$ there must be at least one $x \in X$ such that $Ax = y$. For any such x,

$$x = Ix = (BA)x = B(Ax) = By.$$

This implies that there is in fact just one such x, since it is the image of y under B. That is, the equation $Ax = y$ has a unique solution for x. Hence A^{-1} exists, and

$$A^{-1} = IA^{-1} = (BA)A^{-1} = B(AA^{-1}) = BI = B.$$

Finally, we prove that if A and B both map X onto itself and both have inverses, then the product BA has an inverse, and

$$(BA)^{-1} = A^{-1}B^{-1}.$$

This follows from the preceding result, since $A^{-1}B^{-1}$ certainly maps X into itself and, using the associative law twice,

$$(A^{-1}B^{-1})(BA) = ((A^{-1}B^{-1})B)A = (A^{-1}(B^{-1}B))A = A^{-1}A = I.$$

Our interest in this section is in finding bounds for the relative errors that occur in the kinds of approximations which we must often make in practice. This question has been considered previously, in Section 3.4, for the particular type of approximating mapping known as a perturbation, and for a particular type of problem. In the context now of a normed space X, we were at that time concerned with solving an equation of the form $Ax = x$ $(x \in X)$ for some mapping A on X, and we considered the effect of using an approximating mapping $\widetilde{A}$ for which $\|\widetilde{A}w - Aw\| < \epsilon$ for all $w \in X$ and some number $\epsilon > 0$.

We begin here with a different problem: that of solving for $x \in X$ the equation $Ax = v$, where v is a given nonzero point in X. We suppose now that X is a Banach space and that A is an operator on X whose inverse A^{-1} exists and is bounded (so A^{-1} is also an operator). Of course, we have simply $x = A^{-1}v$. But knowing that A^{-1} exists does not imply that we can actually find it in a given practical situation. Furthermore, and this is the particular aspect we will consider, the operator A itself may not be known with any certainty. This is so, for example, when measured quantities are involved. If A is approximated by a mapping $\widetilde{A}$, which we also assume to be bounded and linear and having an inverse, then we must investigate the difference $A^{-1}v - \widetilde{A}^{-1}v$. In general, we can do no more than obtain an estimate for the *absolute normed error* $\|A^{-1}v - \widetilde{A}^{-1}v\|$, or, preferably, the *relative normed error* $\|A^{-1}v - \widetilde{A}^{-1}v\|/\|A^{-1}v\|$.

Our assumptions on $\widetilde{A}$ imply that there is an operator E on X such that $\widetilde{A} = A + E$. We prove that, provided

$$\|E\| < \frac{1}{\|A^{-1}\|},$$

then automatically the inverse $(A + E)^{-1}$ exists and

$$\|(A + E)^{-1}\| \leqslant \frac{\|A^{-1}\|}{1 - \|A^{-1}\|\,\|E\|}.$$

To do this, we define a mapping B on X by $B = -A^{-1}E$. Then B is easily seen to be linear and bounded, and

$$\|B\| = \|A^{-1}E\| \leqslant \|A^{-1}\|\,\|E\| < 1.$$

Hence Theorem 7.6.5 applies: the operator $I - B$ has an inverse and, from Theorem 7.6.4,

$$\|(I - B)^{-1}\| \leqslant \frac{1}{1 - \|B\|} \leqslant \frac{1}{1 - \|A^{-1}\|\,\|E\|}.$$

We may write, using the associative law and the second distributive law,

$$A + E = AI + (AA^{-1})E = A(I + A^{-1}E) = A(I - B),$$

expressing $A+E$ as a product of operators, each of which has an inverse. Using our preliminary work, we then know that $(A + E)^{-1}$ exists, and $(A + E)^{-1} = (I - B)^{-1}A^{-1}$. Hence

$$\|(A + E)^{-1}\| = \|(I - B)^{-1}A^{-1}\|$$

$$\leqslant \|(I - B)^{-1}\|\,\|A^{-1}\| \leqslant \frac{\|A^{-1}\|}{1 - \|A^{-1}\|\,\|E\|},$$

which is the desired result.

Suppose $x = A^{-1}v$ is to be approximated by $(A + E)^{-1}v$, which we will call y, say. Then, as we mentioned, we want an estimate of the relative error $\|x - y\|/\|x\|$. (Since $v \neq \theta$, of course $x \neq \theta$.) To obtain such an estimate, we write

$$\begin{aligned}
x - y &= A^{-1}v - (A + E)^{-1}v \\
&= ((A + E)^{-1}(A + E))A^{-1}v - (A + E)^{-1}v \\
&= (A + E)^{-1}(((A + E)A^{-1})v - Iv) \\
&= (A + E)^{-1}(Iv + (EA^{-1})v - v) \\
&= (A + E)^{-1}(Ex),
\end{aligned}$$

so that

$$\|x - y\| \leqslant \|(A + E)^{-1}\|\,\|E\|\,\|x\|.$$

Then, using the preceding result,

$$\frac{\|x - y\|}{\|x\|} \leqslant \frac{\|A^{-1}\|\,\|E\|}{1 - \|A^{-1}\|\,\|E\|}$$

and this is a result of considerable practical significance.

The quotient $\|E\|/\|A\|$ is a measure of the relative error in replacing the operator A by the operator $A + E$. The estimate of the relative error in x may be expressed in terms of this:

$$\frac{\|x - y\|}{\|x\|} \leqslant \frac{\|A\|\,\|A^{-1}\|}{1 - \|A\|\,\|A^{-1}\|(\|E\|/\|A\|)} \cdot \frac{\|E\|}{\|A\|}.$$

Writing

$$k(A) = \|A\|\,\|A^{-1}\|,$$

we have

$$\frac{\|x - y\|}{\|x\|} \leqslant \frac{k(A)}{1 - k(A)(\|E\|/\|A\|)} \cdot \frac{\|E\|}{\|A\|}.$$

The number $k(A)$ is called the *condition number* of the operator A. It arises in a number of numerical applications of the above type. To see its significance, write $\gamma = k(A)\|E\|/\|A\|$. Then $\gamma = \|A^{-1}\|\,\|E\| < 1$, so $\gamma/(1-\gamma)$ may be expanded in a geometric series:

$$\frac{\gamma}{1-\gamma} = \gamma + \gamma^2 + \gamma^3 + \cdots.$$

Thus, to a first-order approximation in which we ignore γ^2 and higher powers of γ, the relative error in x is $k(A)$ times the relative error in A.

Notice that

$$1 = \|I\| = \|AA^{-1}\| \leqslant \|A\|\,\|A^{-1}\| = k(A),$$

so that the condition number $k(A)$, which may be defined for any operator A having a bounded inverse, always satisfies $k(A) \geqslant 1$. If A is such that $k(A) = 1$, then A is said to be *perfectly conditioned*, while operators with large condition numbers are called *ill-conditioned*.

The most common numerical application of the condition number occurs when solving systems of linear equations. To illustrate this, we will consider the equations

$$
\begin{aligned}
x_1 + 2x_2 &= 4,\\
1.0001x_1 + 2.001x_2 &= 4.001.
\end{aligned}
$$

It may be checked that their solution is $x_1 = 2.5$, $x_2 = 0.75$. Superficially, it would appear that a good approximation to the solution would be obtained by considering instead the equations

$$
\begin{aligned}
y_1 + 2y_2 &= 4,\\
y_1 + 2.001y_2 &= 4.001,
\end{aligned}
$$

in which there is only a very slight change in one of the coefficients. We find that $y_1 = 2$, $y_2 = 1$, which is a considerable change in the solution. This is an example of an ill-conditioned system: a slight change in the data gives rise to a large change in the solution. To make the situation even more drastic, we could argue that the solution of the original system should be roughly like that of

$$
\begin{aligned}
z_1 + 2z_2 &= 4,\\
z_1 + 2z_2 &= 4.001;
\end{aligned}
$$

but this of course has no solution at all. Or we could say that both equations are roughly just

$$u_1 + 2u_2 = 4,$$

and this, as a system in its own right, has both pairs $(2.5, 0.75)$ and $(2, 1)$ as solutions (u_1, u_2), among infinitely many others.

Now we will relate this example to the preceding theory. Consider the mapping $A \colon \mathbf{R}^n \to \mathbf{R}^n$ given by $Ax = y$, where A is determined by the $n \times n$ matrix (a_{jk}) of real numbers a_{jk} and $x = (x_1, x_2, \ldots, x_n)^T \in \mathbf{R}^n$. As usual, we denote the matrix also by A. Then $y = (y_1, y_2, \ldots, y_n)^T$, where

$$y_j = \sum_{k=1}^{n} a_{jk} x_k, \qquad j = 1, 2, \ldots, n.$$

Considering $\mathbf{R}^n$ as a real vector space, it is easy to see that A is a linear mapping. For simplicity in what follows, we will assume that $\mathbf{R}^n$ is normed by

$$\|x\| = \max_{1 \leqslant j \leqslant n} |x_j|.$$

The mapping A is bounded, since

$$\|Ax\| = \|y\| = \max_{1 \leqslant j \leqslant n} \left| \sum_{k=1}^{n} a_{jk} x_k \right|$$
$$\leqslant \max_{1 \leqslant j \leqslant n} \sum_{k=1}^{n} |a_{jk}| \, |x_k| \leqslant \left(\max_{1 \leqslant j \leqslant n} \sum_{k=1}^{n} |a_{jk}| \right) \|x\|.$$

Hence A is an operator and

$$\|A\| \leqslant \max_{1 \leqslant j \leqslant n} \sum_{k=1}^{n} |a_{jk}|.$$

To see that in fact we have equality here, suppose

$$\max_{1 \leqslant j \leqslant n} \sum_{k=1}^{n} |a_{jk}| = \sum_{k=1}^{n} |a_{mk}|.$$

That is, suppose the maximum occurs when $j = m$, and consider the point $x' = (x'_1, x'_2, \ldots, x'_n) \in \mathbf{R}^n$, where

$$x'_k = \begin{cases} \dfrac{a_{mk}}{|a_{mk}|}, & a_{mk} \neq 0, \\ 1, & a_{mk} = 0, \end{cases}$$

for $k = 1, 2, \ldots, n$. We see that $\|x'\| = 1$, and then

$$\|A\| = \|A\| \, \|x'\| \geqslant \|Ax'\|$$

$$= \max_{1 \leqslant j \leqslant n} \left| \sum_{k=1}^{n} a_{jk} x'_k \right| \geqslant \left| \sum_{k=1}^{n} a_{mk} x'_k \right| = \sum_{k=1}^{n} |a_{mk}|.$$

Hence

$$\|A\| = \max_{1 \leqslant j \leqslant n} \sum_{k=1}^{n} |a_{jk}|,$$

the greatest of the row-sums of the absolute values of elements of the matrix A.

Note finally that if the inverse of the operator A exists, then it is determined by the inverse matrix A^{-1}.

Our example concerned the operator $A : \mathbf{R}^2 \to \mathbf{R}^2$ with matrix

$$A = \begin{pmatrix} 1 & 2 \\ 1.0001 & 2.001 \end{pmatrix}.$$

In general, the matrix $\begin{pmatrix} a & b \\ c & d \end{pmatrix}$ has inverse

$$\frac{1}{ad - bc} \begin{pmatrix} d & -b \\ -c & a \end{pmatrix}$$

when $ad \neq bc$, so

$$A^{-1} = \frac{1}{0.0008} \begin{pmatrix} 2.001 & -2 \\ -1.0001 & 1 \end{pmatrix}.$$

We deduce that

$$\|A\| = 3.0011, \qquad \|A^{-1}\| = \frac{4.001}{0.0008} = 5001.25,$$

so the condition number $k(A)$ exceeds 15,000. This is large.

In the example, we approximated the equation

$$A \begin{pmatrix} x_1 \\ x_2 \end{pmatrix} = \begin{pmatrix} 4 \\ 4.001 \end{pmatrix}$$

by

$$(A + E) \begin{pmatrix} y_1 \\ y_2 \end{pmatrix} = \begin{pmatrix} 4 \\ 4.001 \end{pmatrix},$$

where $E = \begin{pmatrix} 0 & 0 \\ -0.0001 & 0 \end{pmatrix}$. Then

$$\|E\| = 0.0001, \qquad \|E\|\,\|A^{-1}\| = 0.500125 < 1,$$

so the estimate of relative error in the solution that we obtained above may be applied. If $x = (x_1, x_2)^T$ and $y = (y_1, y_2)^T$, then

$$\frac{\|x - y\|}{\|x\|} \leqslant \frac{\|A^{-1}\|\,\|E\|}{1 - \|A^{-1}\|\,\|E\|} = \frac{0.500125}{0.499875},$$

which is just greater than 1. In fact,

$$\|x\| = 2.5, \qquad \|x - y\| = \|(2.5 - 2, 0.75 - 1)\| = 0.5,$$

so $\|x - y\|/\|x\| = 0.2$.

It should be realised that the condition number for an operator depends on the norm adopted. Both the actual and the estimated relative errors in the above problem likewise depend on the norm. We chose in the example a norm for $\mathbf{R}^n$ that is simple to evaluate in terms of the matrix defining an operator. The result,

$$\|A\| = \max_{1 \leqslant j \leqslant n} \sum_{k=1}^{n} |a_{jk}|,$$

is one example of a *matrix norm*, and others may be obtained by taking different norms for $\mathbf{R}^n$. In particular, if we choose the Euclidean norm for $\mathbf{R}^n$, then the corresponding matrix norm turns out to be given by $\|A\| = \sqrt{|\lambda_M|}$, where λ_M is the eigenvalue of the matrix $A^T A$, greatest in absolute value. (See Exercise 7.5(13). If, there, $X = \mathbf{R}^n$ and A is defined by a matrix as here, then the notions of eigenvalue and eigenvector, of an operator and of a matrix, coincide.) Another example of a matrix norm is given in Exercise 7.9(10).

We end this section with another approximation problem in which the condition number arises. Again suppose that X is a normed space (not necessarily Banach) and that A is an operator on X having a bounded inverse. As before, let v be a given nonzero point in X and again suppose we wish to solve the equation $Ax = v$ for $x \in X$. This time, suppose $y \in X$ is tried as an approximation to x. We will obtain bounds on the relative error $\|x - y\|/\|x\|$.

We note first that

$$\|v\| = \|Ax\| \leqslant \|A\|\,\|x\| \quad \text{and} \quad \|x\| = \|A^{-1}v\| \leqslant \|A^{-1}\|\,\|v\|.$$

Putting $Ay = w$, we also have

$$\|v - w\| = \|Ax - Ay\| = \|A(x - y)\| \leqslant \|A\| \, \|x - y\|,$$
$$\|x - y\| = \|A^{-1}v - A^{-1}w\| = \|A^{-1}(v - w)\| \leqslant \|A^{-1}\| \, \|v - w\|.$$

Then

$$\frac{\|v - w\|}{\|A\|} \cdot \frac{1}{\|A^{-1}\| \, \|v\|} \leqslant \frac{\|x - y\|}{\|x\|} \leqslant \|A^{-1}\| \, \|v - w\| \cdot \frac{\|A\|}{\|v\|}$$

and this may be written

$$\frac{1}{k(A)} \frac{\|v - w\|}{\|v\|} \leqslant \frac{\|x - y\|}{\|x\|} \leqslant k(A) \frac{\|v - w\|}{\|v\|}.$$

In particular, when $k(A) = 1$ we see that

$$\frac{\|x - y\|}{\|x\|} = \frac{\|v - w\|}{\|v\|}.$$

7.9 Exercises

(1) Let $K \colon C[a, b] \to C[a, b]$ be the operator A of Exercise 7.5(2).

  (a) Show that the Fredholm equation

$$x(s) = \lambda \int_a^b k(s, t)x(t)\, dt + f(s)$$

  may be written simply as $f = (I - K)x$.

  (b) Prove that $(I - K)^{-1}$ exists provided $|\lambda| < 1/M(b - a)$, and in that case the solution of the integral equation is

$$x = \sum_{j=0}^{\infty} K^j f.$$

(2) Continuing, define a sequence $\{k_n\}$ of functions of two variables by

$$k_1(s, t) = k(s, t),$$

$$k_n(s, t) = \int_a^b k(s, u)k_{n-1}(u, t)\, du, \qquad n = 2,\ 3,\ \dots,$$

where $a \leqslant s \leqslant b$, $a \leqslant t \leqslant b$. Prove that

  (a) if $y = K^n x$, for $x \in C[a, b]$, $n \in \mathbf{N}$, then

$$y(s) = \lambda^n \int_a^b k_n(s, t)x(t)\, dt,$$

  (b) $|k_n(s, t)| \leqslant M^n (b - a)^{n-1}$.

244   *7 Mappings on Normed Spaces*

(3) Continuing, show that the Fredholm equation in (1)(a) has solution

$$x(s) = f(s) + \int_a^b f(t) \sum_{j=1}^{\infty} \lambda^j k_j(s,t) \, dt,$$

provided $|\lambda| < 1/M(b-a)$.

(4) Solve the following Fredholm integral equations by the above
method:

(a) $x(s) = \dfrac{1}{2} \int_0^1 st x(t) \, dt + \dfrac{5s}{6}$,

(b) $x(s) = \dfrac{1}{4} \int_0^{\pi/2} st x(t) \, dt + \sin s$,

(c) $x(s) = \dfrac{1}{4} \int_1^2 \dfrac{s}{t} x(t) \, dt + f(s)$, for any function f that is
continuous on $[1,2]$.

(5) Solve the following Volterra integral equations:

(a) $x(s) = \displaystyle\int_0^s (t-s)x(t) \, dt + s$,

(b) $x(s) = \displaystyle\int_1^s \dfrac{s}{t} x(t) \, dt + se^s$,

(c) $x(s) = \displaystyle\int_1^s \dfrac{s}{t} x(t) \, dt + s$.

(6) Let X, Y, Z be normed spaces and let $A\colon X \to Y$, $B\colon Y \to Z$
be operators. Prove that

(a) the product BA is an operator that maps X into Z, and

$$\|BA\| \leqslant \|B\| \, \|A\|,$$

(b) if $Y = X$, then $\|A^k\| \leqslant \|A\|^k$, for $k = 2, 3, \ldots$,

(c) if A has an inverse, then A^k ($k = 2, 3, \ldots$) has an inverse, $(A^k)^{-1} = (A^{-1})^k$ (which we write as A^{-k}), and
$\|A^{-k}\| \geqslant \|A\|^{-k}$.

(7) Let A be an operator from a Banach space X into itself. Show
that $\sum_{k=0}^{\infty} A^k$ is convergent if $\|A\| < 1$, and that then

$$\left(\sum_{k=0}^{\infty} A^k \right) x = \sum_{k=0}^{\infty} A^k x$$

for any $x \in X$.

(8) Let X be a normed space and let A be a mapping on X for which A^{-1} exists.

    (a) Prove that $(\gamma A)^{-1} = \gamma^{-1} A^{-1}$ for any scalar $\gamma \neq 0$.

    (b) Let $E = \alpha A$ for a scalar $\alpha \neq -1$. Prove that $(A + E)^{-1}$ exists.

    (c) Let $v \in X$ be given, $v \neq \theta$. If $Ax = v$ and $(A + E)y = v$, prove that

$$\frac{\|x - y\|}{\|x\|} = \frac{|\alpha|}{|1 + \alpha|}.$$

(9) Define an operator $A: \mathbf{R}^2 \to \mathbf{R}^2$ by

$$A(x_1, x_2) = \begin{pmatrix} 1 & 1 \\ 1 & 0.991 \end{pmatrix} \begin{pmatrix} x_1 \\ x_2 \end{pmatrix}.$$

Find the condition number of A. Compare the solutions of the systems

$$\begin{array}{ll} x_1 + \quad x_2 = 3, & \\ x_1 + 0.991 x_2 = 2.98, & \end{array} \quad \text{and} \quad \begin{array}{l} y_1 + \quad y_2 = 3, \\ y_1 + 0.99 y_2 = 2.98, \end{array}$$

both exactly and using the estimate of Section 7.8. (Assume $\mathbf{R}^2$ is normed by $\|(x_1, x_2)\| = \max\{|x_1|, |x_2|\}$.)

(10) Let $A: \mathbf{R}^n \to \mathbf{R}^n$ be a mapping defined by an $n \times n$ matrix (a_{jk}), and suppose $\mathbf{R}^n$ is normed by $\|x\| = \sum_{k=1}^{n} |x_k|$, where $x = (x_1, x_2, \ldots, x_n)$. Show that A is bounded and deduce the matrix norm:

$$\|A\| = \max_{1 \leqslant k \leqslant n} \sum_{j=1}^{n} |a_{jk}|.$$

(Hint: Show that $\|A\| \leqslant \max_{1 \leqslant k \leqslant n} \sum_{j=1}^{n} |a_{jk}| = \sum_{j=1}^{n} |a_{jm}|$, say, and deduce that equality must hold by considering the point $(0, \ldots, 0, 1, 0, \ldots, 0) \in \mathbf{R}^n$, where the mth component is 1 and all others are 0.)

(11) Let A be an operator on a normed space X and suppose A has a bounded inverse. Let B and C be operators on X such that $AB = C$. Suppose A and C are known and B is to be approximated by an operator $\widetilde{B}$. Prove that

$$\frac{\|\widetilde{B} - B\|}{\|B\|} \leqslant k(A) \frac{\|A\widetilde{B} - C\|}{\|C\|}.$$

. . . .

(12) Let A be an operator on a normed space X for which A^{-1} exists and is bounded.

    (a) Prove that if λ is an eigenvalue of A, then λ^{-1} is an eigenvalue of A^{-1}. (See Exercise 7.5(13).)

    (b) Write

$$L = \sup\{|\lambda| : Ax = \lambda x, \ x \in X, \ x \neq \theta\},$$
$$l = \inf\{|\lambda| : Ax = \lambda x, \ x \in X, \ x \neq \theta\}.$$

    Prove that $k(A) \geqslant L/l$.

(13) Let $A \colon X \to Y$ be a mapping between normed spaces X and Y.

    (a) If A is additive, show that $A(\rho x) = \rho A x$ for all $x \in X$ and all $\rho \in \mathbf{Q}$.

    (b) If, further, A is continuous, prove that A is homogeneous, and hence linear.

7.10 Unbounded mappings

In this chapter, we have been almost solely concerned with operators, that is, bounded linear mappings. There is a much fuller general theory for operators than for unbounded linear mappings. Perhaps this is to be expected, since the latter are not continuous (Theorem 7.1.5).

Fortunately, many problems involving unbounded mappings can be rearranged to involve only bounded ones. We will shortly see that the mapping which takes a differentiable function into its derivative is unbounded. However, problems involving such mappings can often be reorganised to involve mappings defined by integrals, like those in Section 7.7, and these are bounded. This happened in effect in Section 3.3 where the existence theorem for second-order linear differential equations was established by transforming the differential equation into a Volterra integral equation.

We will not give much theory here for unbounded mappings, but will be content to indicate through an example that the appearance of unbounded mappings is sometimes unavoidable.

Let us denote by $C'[a, b]$ the real vector space of functions that have a continuous derivative on $[a, b]$. Previously, we have used the space $C^{(1)}[a, b]$ of differentiable functions defined on $[a, b]$. These spaces are not the same: we have $C'[a, b] \subseteq C^{(1)}[a, b]$, but not the reverse. The following function f illustrates this. Take

$$f(x) = \begin{cases} x^2 \sin \dfrac{1}{x}, & -1 \leqslant x \leqslant 1, \ x \neq 0, \\ 0, & x = 0. \end{cases}$$

We have

$$f'(0) = \lim_{h \to 0} \frac{h^2 \sin(1/h) - 0}{h} = 0,$$

while, when $x \neq 0$,

$$f'(x) = 2x \sin \frac{1}{x} - \cos \frac{1}{x}.$$

Hence $f'(x) \not\to f'(0)$ as $x \to 0$, so f belongs to $C^{(1)}[-1, 1]$ but not to $C'[-1, 1]$.

We define $C'[a, b]$ to have the uniform norm that we use for $C[a, b]$. Let $D \colon C'[a, b] \to C[a, b]$ be the mapping defined by

$$Df = f'.$$

It is easy to see that D is linear. We will show that it is unbounded. For this purpose, consider the function $g \in C'[a, b]$ where $g(x) = \sin \omega x$, for some positive real number ω. We have

$$\|g\| = \max_{a \leqslant x \leqslant b} |\sin \omega x| = 1$$

(assuming $b > a + 2\pi/\omega$), and

$$\|Dg\| = \|g'\| = \max_{a \leqslant x \leqslant b} |\omega \cos \omega x| = |\omega| = \omega \|g\|.$$

Hence we cannot have $\|Dg\| \leqslant K\|g\|$ for some fixed number K and all $g \in C'[a, b]$, since ω may be arbitrarily large. This shows that the mapping D is indeed unbounded.

As as alternative demonstration of this, consider the sequence $\{f_n\}$ of functions in $C'[a, b]$ given by

$$f_n(x) = \frac{\sin n^2 x}{n}.$$

We have $\lim f_n = \theta$ (the zero vector of $C'[a, b]$, the function identically zero in $[a, b]$), but $f_n'(x) = n \cos nx$ so $Df_n \not\to D\theta$. That is, D is not continuous at θ so D is unbounded, by Theorem 7.1.5.

In Definition 7.4.1(b), we introduced the idea of a closed mapping. Any operator between normed spaces is closed (Exercise 7.5(11)), but we will show now that the converse of this is not true. Specifically, we will show that the mapping D above is closed, although we saw it to be unbounded.

To do this, let $\{f_n\}$ be a sequence of functions in $C'[a,b]$ for which $f_n \to f$ and $Df_n \to g$. For D to be closed, we must show that $f \in C'[a,b]$ and that $g = Df$. Refer to Theorem 1.10.5. For each n, the derivatives f_n' belong to $C[a,b]$ by definition of the space $C'[a,b]$, and the sequence $\{f_n'\}$ is uniformly convergent on $[a,b]$ since this is what convergence means with the uniform norm. Hence $\lim f_n' = f'$, so $f' = g$. Further, f' is continuous on $[a,b]$ since it is the limit of a uniformly convergent sequence of continuous functions (Theorem 1.10.3). That is, $f \in C'[a,b]$, and so D is closed.

Our main example in this section is from quantum mechanics. It makes use of the following theorem.

Theorem 7.10.1 *Let A and B be linear mappings from a normed vector space into itself and suppose that $AB - BA = \alpha I$ for some scalar $\alpha \neq 0$. Then A and B cannot both be bounded.*

To prove this, we will suppose that A and B are both bounded. Note first that if $\|B\| = 0$ then

$$|\alpha| = \|\alpha I\| = \|AB - BA\| \leqslant \|AB\| + \|BA\| \leqslant 2\|A\|\,\|B\| = 0.$$

Since $\alpha \neq 0$, this is impossible, so $\|B\| \neq 0$. We use induction to prove that

$$AB^n - B^n A = \alpha n B^{n-1}, \qquad n \in \mathbf{N}.$$

When $n = 1$, this is clear. (As usual, B^0 is I.) Assume the result when $n = m$. Then, when $n = m + 1$, using the distributive laws for linear mappings,

$$\begin{aligned}
AB^{m+1} - B^{m+1}A &= (AB^m - B^m A)B + B^m(AB - BA) \\
&= \alpha m B^{m-1}B + B^m(\alpha I) = \alpha(m+1)B^m,
\end{aligned}$$

as required. Then, for $n \in \mathbf{N}$,

$$\begin{aligned}
\|\alpha n B^{n-1}\| &= \|AB^n - B^n A\| \\
&\leqslant \|AB^n\| + \|B^n A\| \\
&\leqslant \|A\|\,\|B^n\| + \|B^n\|\,\|A\| \\
&= 2\|A\|\,\|B^{n-1}B\| \\
&\leqslant 2\|A\|\,\|B^{n-1}\|\,\|B\|,
\end{aligned}$$

so

$$\|A\|\,\|B\|\,\|B^{n-1}\| \geqslant \frac{1}{2}\|\alpha n B^{n-1}\| = \frac{1}{2}|\alpha|\,n\,\|B^{n-1}\|.$$

There are now two cases. First, perhaps $\|B^n\| \neq 0$ for any integer $n \geqslant 2$. Then $\|A\|\,\|B\| \geqslant \frac{1}{2}|\alpha|n$ and this is impossible if A and B are both bounded, as n may be arbitrarily large. Alternatively, $\|B^m\| = 0$ for some integer $m \geqslant 2$. In that case, from the above,

$$|\alpha|\,m\,\|B^{m-1}\| \leqslant 2\|A\|\,\|B^m\|,$$

so that also $\|B^{m-1}\| = 0$, and in the same way $\|B^{m-2}\| = 0, \ldots,$ $\|B^2\| = 0$, $\|B\| = 0$, and again we arrive at a contradiction. Hence, as required, A and B are not both bounded. $\qquad\square$

In quantum mechanics, there are natural (and philosophical) difficulties involved in measuring quantities associated with the the motion of atomic particles. These quantities, called *observables*, are, by the axioms of quantum mechanics, represented by certain mappings. The mappings allow us to speak, for example, of the statistical distribution of the possible velocities of a particle, rather than of its actual velocity. *Heisenberg's uncertainty principle* claims that it is fundamentally impossible to describe precisely all aspects of the motion of any particle, essentially because the act of measuring one aspect of the motion necessarily changes other aspects. It is therefore important to know if there are quantities that can be measured simultaneously.

It turns out that if A and B are mappings associated with certain observables, then simultaneous measurement of those observables is possible if and only if $AB = BA$. We can show here only that the basic mappings of quantum mechanics cannot all be bounded.

Let ψ be a one-dimensional wave function. For our purposes, this is any function of position x and of time t, which is such that both ψ and $\partial\psi/\partial x$, for fixed t, belong to $C'[a,b]$. The momentum and position mappings P and X, respectively, are defined on $C'[a,b]$ into itself by

$$P\psi = -i\hbar\frac{\partial}{\partial x}\psi, \qquad X\psi = x\psi,$$

where $\hbar = h/2\pi$ (h is Planck's constant) and $i = \sqrt{-1}$. (For the purely illustrative purpose of this discussion, we treat i as an ordinary constant.) We see that

$$(PX)\psi = P(X\psi)$$
$$= P(x\psi)$$
$$= -i\hbar\frac{\partial}{\partial x}(x\psi)$$
$$= -i\hbar\left(x\frac{\partial}{\partial x}\psi + \psi\right)$$
$$= x\left(-i\hbar\frac{\partial}{\partial x}\psi\right) - i\hbar\psi$$
$$= X(P\psi) - i\hbar I\psi$$
$$= ((XP) + (-i\hbar)I)\psi,$$

and hence

$$PX - XP = -i\hbar I.$$

Theorem 7.10.1 then implies that at least one of the mappings P and X must be unbounded. (In fact, P is a straightforward differential mapping and, like D above, can be shown directly to be unbounded.)

The earlier remarks imply that the position and momentum of an atomic particle cannot be measured simultaneously.

8

Inner Product Spaces

8.1 Definitions; simple consequences

We introduced normed spaces with a discussion on the desirability of being able to add together the elements of a metric space. For that reason we began working with vector spaces rather than arbitrary sets. The same argument as in the earlier discussion could apply to the desirability of being able to multiply together the elements of a metric space, and it is not necessary to repeat it here.

There are various ways of defining a product, each way serving its own end and yet each generalising the notion of the product of real numbers. One way is to suppose that the underlying set of all we have developed so far is no longer a vector space only but also has the properties of a ring or a field in which multiplication of elements is already defined. This line can be developed into the theory of *Banach algebras*.

What we do here requires no such basic structural alteration: we will continue to work in a vector space and will say that a product is defined whenever we have a function of pairs of elements of the space that satisfies four axioms or requirements to be listed below. Specifically, this is called an *inner product*, and may be viewed more easily as a generalisation of the familiar scalar product of ordinary three-dimensional vectors. The common definition of the scalar product of two vectors uses the angle between the vectors. Working in reverse, we can use the inner product to define the angle between elements of quite arbitrary vector spaces. Except in one important instance, this does not generally give rise to any useful interpretations.

As usual, we will assume that the scalars in a general vector space are complex numbers unless we specify otherwise. We will denote the inner product of two points x, y in a vector space X by $\langle x, y \rangle$, and this may be looked upon as the image of a peculiar-looking mapping $\langle \ , \ \rangle$ from $X \times X$ into $\mathbf{C}$. Thus: $\langle \ , \ \rangle (x, y) = \langle x, y \rangle$. Note that the inner product of two vectors is a complex number.

Definition 8.1.1   An *inner product space* is a vector space X together with a mapping $\langle \ , \ \rangle : X \times X \to \mathbf{C}$ with the properties

(IP1) $\langle x, x \rangle > 0$ for all $x \in X$, $x \neq \theta$,

(IP2) $\langle x, y \rangle = \overline{\langle y, x \rangle}$ for all $x, y \in X$,

(IP3) $\langle \alpha x, y \rangle = \alpha \langle x, y \rangle$ for all $x, y \in X$ and every scalar α,

(IP4) $\langle x + y, z \rangle = \langle x, z \rangle + \langle y, z \rangle$ for all $x, y, z \in X$.

This inner product space is denoted by $(X, \langle \ , \ \rangle)$ and the mapping $\langle \ , \ \rangle$ is called the *inner product* for the space.

If $\langle \ , \ \rangle_1, \langle \ , \ \rangle_2, \ldots$ denote different inner products for the same vector space X, then $(X, \langle \ , \ \rangle_1)$, $(X, \langle \ , \ \rangle_2)$, $\ldots$ are different inner product spaces. Only rarely will we consider more than one inner product for any vector space, so we will always write X, say, by itself to denote the inner product space, with inner product assumed to be $\langle \ , \ \rangle$.

An inner product is sometimes called a *scalar product*, and alternative names for an inner product space are *Euclidean space, unitary space* and *pre-Hilbert space*. Other notations in common use for the inner product are $\langle \ | \ \rangle$, $(\ , \ )$ and $(\ | \ )$. Certain authors replace the main parts of (IP3) and (IP4) by $\langle x, \alpha y \rangle = \alpha \langle x, y \rangle$ and $\langle x, y + z \rangle = \langle x, y \rangle + \langle x, z \rangle$, respectively. This has some significance in the case of (IP3), but not for (IP4), as will become apparent from (a) and (b), immediately below.

The bar over $\langle y, x \rangle$ in (IP2) denotes the complex conjugate. It follows from (IP2), with $x = y$, that $\langle x, x \rangle$ is always a real number, and in (IP1) we specify that this number must be positive when $x \neq \theta$. Definition 8.1.1 applies equally well when X is a real vector space, the only difference being that the inner product of two vectors is then always a real number. In that case (IP2) becomes in essence $\langle x, y \rangle = \langle y, x \rangle$, and we speak of a real inner product space.

There are a number of immediate consequences of Definition 8.1.1. These include:

(a) $\langle x, \beta y \rangle = \overline{\beta} \langle x, y \rangle$ for all $x, y \in X$ and every scalar β,

(b) $\langle x, y + z \rangle = \langle x, y \rangle + \langle x, z \rangle$ for all $x, y, z \in X$,

(c) $\langle x, \theta \rangle = \langle \theta, y \rangle = 0$ for all $x, y \in X$.

We prove (a) as follows:

$$\langle x, \beta y \rangle = \overline{\langle \beta y, x \rangle} = \overline{\beta \langle y, x \rangle} = \overline{\beta} \, \overline{\langle y, x \rangle} = \overline{\beta} \langle x, y \rangle .$$

To show that $\langle \theta, y \rangle = 0$, we use (IP4) in writing

$$\langle \theta, y \rangle = \langle \theta + \theta, y \rangle = \langle \theta, y \rangle + \langle \theta, y \rangle,$$

and the result is clear. The proofs of (b) and the other half of (c) are left as exercises.

Another consequence, encompassing both (IP3) and (IP4), is

(d) $\left\langle \sum_{k=1}^{n} \alpha_k x_k, y \right\rangle = \sum_{k=1}^{n} \alpha_k \langle x_k, y \rangle$ for all $x_1, x_2, \ldots, x_n, y \in X$ and all scalars $\alpha_1, \alpha_2, \ldots, \alpha_n$.

This is proved by mathematical induction. When $n = 1$, the result is simply (IP3). Suppose the result is true when $n = m$. Then, when $n = m + 1$,

$$\left\langle \sum_{k=1}^{m+1} \alpha_k x_k, y \right\rangle = \left\langle \sum_{k=1}^{m} \alpha_k x_k + \alpha_{m+1} x_{m+1}, y \right\rangle$$

$$= \left\langle \sum_{k=1}^{m} \alpha_k x_k, y \right\rangle + \langle \alpha_{m+1} x_{m+1}, y \rangle$$

$$= \sum_{k=1}^{m} \alpha_k \langle x_k, y \rangle + \alpha_{m+1} \langle x_{m+1}, y \rangle$$

$$= \sum_{k=1}^{m+1} \alpha_k \langle x_k, y \rangle,$$

as required.

In the same vein, we also have

(e) $\left\langle x, \sum_{j=1}^{m} \beta_j y_j \right\rangle = \sum_{j=1}^{m} \overline{\beta}_j \langle x, y_j \rangle$ for all $x, y_1, y_2, \ldots, y_m \in X$ and all scalars $\beta_1, \beta_2, \ldots, \beta_m$;

and, most generally,

(f) $\left\langle \sum_{k=1}^{n} \alpha_k x_k, \sum_{j=1}^{m} \beta_j y_j \right\rangle = \sum_{k=1}^{n} \sum_{j=1}^{m} \alpha_k \overline{\beta}_j \langle x_k, y_j \rangle.$

The proof of (f) is left as an exercise. We will subsequently use (a) to (f) without specific reference to this list.

As our first example of an inner product space, we take the vector space $\mathbf{C}^n$ of n-tuples of complex numbers and define for it an inner product by

$$\langle x, y \rangle = \sum_{k=1}^{n} x_k \overline{y}_k,$$

where $x = (x_1, x_2, \ldots, x_n)$, $y = (y_1, y_2, \ldots, y_n)$ are any elements of $\mathbf{C}^n$.

It is necessary to verify that this does indeed define an inner product. For (IP1), we have

$$\langle x, x \rangle = \sum_{k=1}^{n} x_k \overline{x}_k = \sum_{k=1}^{n} |x_k|^2 > 0$$

when $x \neq \theta$. For (IP2),

$$\overline{\langle y, x \rangle} = \overline{\sum_{k=1}^{n} y_k \overline{x}_k} = \sum_{k=1}^{n} \overline{y_k \overline{x}_k} = \sum_{k=1}^{n} \overline{y}_k x_k = \langle x, y \rangle.$$

For (IP3),

$$\langle \alpha x, y \rangle = \sum_{k=1}^{n} \alpha x_k \overline{y}_k = \alpha \sum_{k=1}^{n} x_k \overline{y}_k = \alpha \langle x, y \rangle,$$

where $\alpha \in \mathbf{C}$. Finally, for (IP4), if $z = (z_1, z_2, \ldots, z_n) \in \mathbf{C}^n$,

$$\langle x + y, z \rangle = \sum_{k=1}^{n} (x_k + y_k) \overline{z}_k = \sum_{k=1}^{n} x_k \overline{z}_k + \sum_{k=1}^{n} y_k \overline{z}_k = \langle x, z \rangle + \langle y, z \rangle.$$

As a final extension of our symbolism, this inner product space is itself denoted by $\mathbf{C}^n$. This is a natural notation, for a reason that will appear.

Notice that it would not be sufficient to define the inner product for the vector space $\mathbf{C}^n$ by the equation

$$\langle x, y \rangle = \sum_{k=1}^{n} x_k y_k,$$

since then neither (IP1) nor (IP2) would be satisfied. But this equation does define an inner product for the real vector space $\mathbf{R}^n$ and the resulting real inner product space is itself denoted by $\mathbf{R}^n$. Here we have the expected generalisation of the equation

$$\mathbf{x} \cdot \mathbf{y} = x_1 y_1 + x_2 y_2 + x_3 y_3,$$

where $\mathbf{x} = x_1 \mathbf{i} + x_2 \mathbf{j} + x_3 \mathbf{k}$ and $\mathbf{y} = y_1 \mathbf{i} + y_2 \mathbf{j} + y_3 \mathbf{k}$ are familiar three-dimensional vectors. Thinking in reverse, it is the need to have a similar definition of an inner product for $\mathbf{C}^n$ to this one for $\mathbf{R}^n$, and the need to maintain the condition in (IP1), that led to the axiom (IP3), in which taking the complex conjugate at first sight seems odd.

For the vector space l_2 of complex-valued sequences $(x_1, x_2, \dots)$ for which the series $\sum_{k=1}^{\infty} |x_k|^2$ converges, we define an inner product by

$$\langle x, y \rangle = \sum_{k=1}^{\infty} x_k \overline{y}_k$$

where $x = (x_1, x_2, \dots)$ and $y = (y_1, y_2, \dots)$ belong to l_2. To verify that this series of complex numbers always converges, we make use of the Cauchy–Schwarz inequality (Theorem 2.2.1). If $m \leqslant n$, we have

$$\left| \sum_{k=m}^{n} x_k \overline{y}_k \right| \leqslant \sum_{k=m}^{n} |x_k \overline{y}_k| = \sum_{k=m}^{n} |x_k| \, |y_k| \leqslant \sqrt{\sum_{k=m}^{n} |x_k|^2} \sqrt{\sum_{k=m}^{n} |y_k|^2},$$

and the result follows using Theorem 1.8.2, since $x, y \in l_2$. The verification of (IP1) to (IP4) for this inner product is similar to that for the inner product for $\mathbf{C}^n$. This definition will be the only one to be defined on l_2 and, as before, the resulting inner product space will also be denoted by l_2.

For our final examples at this time, we turn to function spaces. It may have been noticed that in $\mathbf{C}^n$, $\mathbf{R}^n$ and l_2, we have in each case that $\langle x, x \rangle = \|x\|^2$, an equation relating the inner product to the norm for these spaces. Indeed, this is why it is natural to maintain the same notation for the inner product spaces as for the normed spaces. We will indicate later that there is no way to define an inner product for the normed space $C[a, b]$ of continuous functions on $[a, b]$ so that the same equation holds, the norm for $C[a, b]$ of course being the uniform norm. As we make explicit below, it is desirable for this equation always to hold, so we will have little further use for the space $C[a, b]$.

However, we can define, for continuous functions x, y on $[a, b]$,

$$\langle x, y \rangle = \int_a^b x(t) y(t) \, dt,$$

and then

$$\langle x, x \rangle = \int_a^b (x(t))^2 \, dt,$$

which is $\|x\|^2$ for the normed space $C_2[a, b]$. That this does define an inner product is easily verified, and so we speak of the real inner product space $C_2[a, b]$.

The above discussion suggests that perhaps any inner product space X can be considered as a normed space if we define

$$\|x\| = \sqrt{\langle x, x \rangle}, \qquad x \in X.$$

To show that this is in fact true, we must verify the axioms for a norm (Definition 6.1.1) for the mapping whose value at x is $\sqrt{\langle x, x \rangle}$. For (N1), we certainly have $\langle \theta, \theta \rangle = 0$ and, if $x \neq \theta$, $\langle x, x \rangle > 0$ by (IP1). For (N2), we have

$$\sqrt{\langle \alpha x, \alpha x \rangle} = \sqrt{\alpha \bar{\alpha} \langle x, x \rangle} = \sqrt{|\alpha|^2 \langle x, x \rangle} = |\alpha| \sqrt{\langle x, x \rangle}.$$

Only (N3) remains to be verified.

The verification of (N3) follows easily once we have established the general Cauchy–Schwarz inequality, which we state tentatively as follows: *For any vectors x, y in an inner product space,*

$$|\langle x, y \rangle|^2 \leqslant \langle x, x \rangle \langle y, y \rangle.$$

This generalises the earlier forms of the Cauchy–Schwarz inequality in Theorem 2.2.1 and Exercise 2.4(6). Then (N3) is derived as follows. If $x, y \in X$,

$$
\begin{aligned}
\langle x + y, x + y \rangle &= \langle x, x \rangle + \langle x, y \rangle + \langle y, x \rangle + \langle y, y \rangle \\
&= \langle x, x \rangle + \langle x, y \rangle + \overline{\langle x, y \rangle} + \langle y, y \rangle \\
&= \langle x, x \rangle + 2 \operatorname{Re} \langle x, y \rangle + \langle y, y \rangle \\
&\leqslant \langle x, x \rangle + 2 |\langle x, y \rangle| + \langle y, y \rangle \\
&\leqslant \langle x, x \rangle + 2 \sqrt{\langle x, x \rangle \langle y, y \rangle} + \langle y, y \rangle \\
&= (\sqrt{\langle x, x \rangle} + \sqrt{\langle y, y \rangle})^2;
\end{aligned}
$$

that is,

$$\sqrt{\langle x + y, x + y \rangle} \leqslant \sqrt{\langle x, x \rangle} + \sqrt{\langle y, y \rangle}.$$

This indeed is (N3).

We now specify that the only norm ever to be used in conjunction with a given inner product space X will be that defined by

$$\|x\| = \sqrt{\langle x, x \rangle}, \qquad x \in X.$$

This is similar in intent to the statement that a normed space is only considered as a metric space when the metric is given by

$$d(x, y) = \|x - y\|, \qquad x, y \in X.$$

The reasoning behind our maintaining the names $\mathbf{C}^n$, $\mathbf{R}^n$, l_2 and $C_2[a,b]$ for certain inner product spaces as well as normed spaces and metric spaces should now be very clear.

Now we must prove the Cauchy–Schwarz inequality that we have just used. By virtue of the specification made for norms on inner product spaces, we may state the inequality differently as follows.

Theorem 8.1.2 (General Cauchy–Schwarz Inequality) *For any points x, y in an inner product space,*

$$|\langle x,y\rangle| \leqslant \|x\|\,\|y\|.$$

Note that the proof we now give is quite different in approach from that given for Theorem 2.2.1. If $y = \theta$, the inequality is certainly true, so we may suppose that $y \neq \theta$. Then $\|y\| > 0$. Let α be any scalar. Then

$$\begin{aligned}
0 \leqslant \|x+\alpha y\|^2 &= \langle x+\alpha y, x+\alpha y\rangle \\
&= \langle x,x\rangle + \langle x,\alpha y\rangle + \langle \alpha y,x\rangle + \langle \alpha y,\alpha y\rangle \\
&= \langle x,x\rangle + \overline{\alpha}\langle x,y\rangle + \alpha\langle y,x\rangle + \alpha\overline{\alpha}\langle y,y\rangle \\
&= \|x\|^2 + \alpha\overline{\langle x,y\rangle} + \overline{\alpha}(\langle x,y\rangle + \alpha\|y\|^2).
\end{aligned}$$

Now set $\alpha = -\langle x,y\rangle/\|y\|^2$. We see that $\langle x,y\rangle + \alpha\|y\|^2 = 0$ and so

$$0 \leqslant \|x\|^2 - \frac{\langle x,y\rangle\,\overline{\langle x,y\rangle}}{\|y\|^2} = \|x\|^2 - \frac{|\langle x,y\rangle|^2}{\|y\|^2}.$$

Thus the inequality is proved. □

We said before that there is no way the normed space $C[a,b]$ can be considered as an inner product space in a consistent fashion. This is a consequence of the next theorem, in which we establish what is known as the *parallelogram law* for inner product spaces. It generalises the statement that the sum of the squares of the diagonals of a parallelogram equals the sum of the squares of its sides.

Theorem 8.1.3 *For any points x, y in an inner product space,*

$$\|x+y\|^2 + \|x-y\|^2 = 2\|x\|^2 + 2\|y\|^2.$$

The proof is a direct calculation. We have

$$\begin{aligned}
\|x+y\|^2 &= \langle x+y, x+y \rangle \\
&= \langle x,x \rangle + \langle x,y \rangle + \langle y,x \rangle + \langle y,y \rangle \\
&= \|x\|^2 + \|y\|^2 + \langle x,y \rangle + \langle y,x \rangle.
\end{aligned}$$

Expand $\|x-y\|^2$ in a similar way and add the expressions to give the theorem. □

It is easy to show by an example that the parallelogram law does not hold for $C[a,b]$, so $C[a,b]$ is not an inner product space. We give an example in $C[0,1]$. Define functions x and y by

$$x(t) = t, \quad y(t) = 1-t, \qquad 0 \leqslant t \leqslant 1.$$

Then $(x+y)(t) = 1$ and $(x-y)(t) = 2t-1$. We see that

$$\|x\| = \max_{0 \leqslant t \leqslant 1} |t| = 1,$$

and similarly $\|y\| = \|x+y\| = \|x-y\| = 1$.

Since every inner product space is also a normed space, we can speak of convergent sequences in an inner product space, and of Cauchy sequences and so on, by introducing into the space the norm that is defined by the inner product. To illustrate this idea, we will prove:

Theorem 8.1.4 *If $\{x_n\}$ and $\{y_n\}$ are sequences in an inner product space, which converge to x and y, respectively, then $\{\langle x_n, y_n \rangle\}$ is a convergent sequence in $\mathbf{C}$, with limit $\langle x, y \rangle$.*

To say that the sequence $\{x_n\}$ converges to x means, as usual, that $\|x_n - x\| \to 0$, where now we understand that the inner product space has been normed by taking $\|w\| = \sqrt{\langle w,w \rangle}$ for each w in the space. Similarly for the sequence $\{y_n\}$. To prove that $\langle x_n, y_n \rangle \to \langle x, y \rangle$, we write

$$\begin{aligned}
\langle x_n, y_n \rangle - \langle x, y \rangle &= \langle x_n, y_n \rangle - \langle x_n, y \rangle + \langle x_n, y \rangle - \langle x, y \rangle \\
&= \langle x_n, y_n - y \rangle + \langle x_n - x, y \rangle,
\end{aligned}$$

so that

$$\begin{aligned}
|\langle x_n, y_n \rangle - \langle x, y \rangle| &\leqslant |\langle x_n, y_n - y \rangle| + |\langle x_n - x, y \rangle| \\
&\leqslant \|x_n\|\,\|y_n - y\| + \|x_n - x\|\,\|y\|,
\end{aligned}$$

using the Cauchy–Schwarz inequality. Every convergent sequence is bounded, so $\|x_n\| < C$ for some constant C and all n, and $\|y_n - y\| \to 0$, $\|x_n - x\| \to 0$. Hence, as required, $\langle x_n, y_n \rangle \to \langle x, y \rangle$. $\qquad\qquad\square$

8.2 Orthonormal vectors

We have mentioned that introducing an inner product into a vector space allows us to generalise the notion of angle between two vectors. The definition is suggested by recalling that if $\mathbf{x} = x_1\,\mathbf{i} + x_2\,\mathbf{j} + x_3\,\mathbf{k}$ and $\mathbf{y} = y_1\,\mathbf{i} + y_2\,\mathbf{j} + y_3\,\mathbf{k}$ are ordinary nonzero three-dimensional vectors, then the angle between them is ω, where $0 \leqslant \omega \leqslant \pi$ and

$$\cos \omega = \frac{\mathbf{x} \cdot \mathbf{y}}{|\mathbf{x}|\,|\mathbf{y}|} = \frac{x_1 y_1 + x_2 y_2 + x_3 y_3}{\sqrt{x_1^2 + x_2^2 + x_3^2}\,\sqrt{y_1^2 + y_2^2 + y_3^2}}.$$

With the standard definition of norm and inner product for $\mathbf{R}^3$, the right-hand side here is precisely $\langle x, y \rangle /(\|x\|\,\|y\|)$ (writing x for $\mathbf{x}$ and y for $\mathbf{y}$). Thus we say that for any nonzero vectors x, y in a real inner product space, the angle between them is the number

$$\cos^{-1} \frac{\langle x, y \rangle}{\|x\|\,\|y\|}.$$

(Since in any case we make little use of this notion, we are restricting ourselves to real spaces here. Certain difficulties arise with the analogous idea for complex inner product spaces.) It is a consequence of the Cauchy–Schwarz inequality that this angle always exists, because that inequality states that $-1 \leqslant \langle x, y \rangle /(\|x\|\,\|y\|) \leqslant 1$. By definition of the inverse cosine function, the angle is in the interval $[0, \pi]$.

However, this concept has little application in general. The major exception is in the notion of orthogonality. If the ordinary vectors $\mathbf{x}$, $\mathbf{y}$ above are perpendicular (or *orthogonal*), then the angle between them is $\pi/2$ and $\mathbf{x} \cdot \mathbf{y} = 0$. The first statement in the following definition is a natural generalisation of this.

Definition 8.2.1 Two vectors x, y in an inner product space X are called *orthogonal* if $\langle x, y \rangle = 0$. We then write $x \perp y$. A subset S of X is called an *orthogonal set* in X if $x \perp y$ for all $x, y \in S$ $(x \neq y)$. If, moreover, $\|x\| = 1$ for all $x \in S$, then S is called an *orthonormal set* in X.

Notice that $\theta \perp x$ for any x in an inner product space. Clearly, $y \perp x$ if and only if $x \perp y$.

The familiar unit vectors **i**, **j**, **k** of ordinary vector analysis provide an example of an orthonormal set in $\mathbf{R}^3$. These vectors may of course also be written as $(1, 0, 0)$, $(0, 1, 0)$, $(0, 0, 1)$, respectively. Another example of an orthonormal set in $\mathbf{R}^3$ is

$$\left\{ \left(\frac{1}{\sqrt{3}}, \frac{1}{\sqrt{3}}, \frac{1}{\sqrt{3}} \right), \left(\frac{-1}{\sqrt{2}}, \frac{1}{\sqrt{2}}, 0 \right), \left(\frac{-1}{\sqrt{6}}, \frac{-1}{\sqrt{6}}, \frac{2}{\sqrt{6}} \right) \right\}.$$

In the inner product space l_2, an example of an orthonormal set is $\{(1, 0, 0, \dots), (0, 1, 0, \dots), (0, 0, 1, 0, \dots), \dots\}$. Any subset of this set is also an orthonormal set in l_2.

The set $\{\cos t, \cos 2t, \cos 3t, \dots\}$ $(-\pi \leqslant t \leqslant \pi)$ of functions in the real inner product space $C_2[-\pi, \pi]$ is an orthogonal set, since

$$\langle \cos mt, \cos nt \rangle = \int_{-\pi}^{\pi} \cos mt \cos nt \, dt = \begin{cases} \pi, & m = n, \\ 0, & m \neq n, \end{cases}$$

for $m, n \in \mathbf{N}$. Clearly the set is orthonormal once each member is divided by $\sqrt{\pi}$. A 'bigger' orthogonal set in the same space is the set $\{1, \sin t, \cos t, \sin 2t, \cos 2t, \sin 3t, \cos 3t, \dots\}$ $(-\pi \leqslant t \leqslant \pi)$ and of course any subset of this set will again be an orthogonal set in $C_2[-\pi, \pi]$.

Before moving on now to some general results, we need to extend parts of Definition 1.11.3, dealing with linear independence and the span of a set of vectors, so that those notions may be applied to infinite sets of vectors.

Definition 8.2.2 Let S be a nonempty set of vectors in a vector space V.

(a) The set S is called *linearly independent* if every finite subset of it is linearly independent in the original sense.

(b) The *span* of S, denoted by $\mathrm{Sp}\, S$, is the subspace of V consisting of all linear combinations of finite numbers of vectors in S.

In (b), it is easy to verify that $\mathrm{Sp}\, S$ is indeed a subspace of V, in satisfaction of Definition 1.11.2.

As an example, in the real vector space $C[a, b]$, consider the infinite set $S = \{1, t, t^2, t^3, \dots\}$ $(a \leqslant t \leqslant b)$. This is linearly independent, since a linear combination of finitely many vectors in S is a polynomial function on $[a, b]$, and this is the zero function on $[a, b]$ only when all coefficients are 0. The span of S is the subspace of $C[a, b]$ consisting of all polynomial functions defined on $[a, b]$.

From the next two results, we will be assured that a finite-dimensional inner product space always has a basis which is an orthonormal set.

Theorem 8.2.3 *An orthogonal set of nonzero vectors in an inner product space is linearly independent.*

To prove this, let S be the orthogonal set and let $\{x_1, x_2, \ldots, x_n\}$ be an arbitrary finite subset of S. The result will follow when we show that this subset is linearly independent. Suppose that

$$\alpha_1 x_1 + \alpha_2 x_2 + \cdots + \alpha_n x_n = \theta$$

for some scalars $\alpha_1, \alpha_2, \ldots, \alpha_n$. We take the inner product of both sides with x_k for $k = 1, 2, \ldots, n$ in turn, obtaining

$$\langle \alpha_1 x_1 + \alpha_2 x_2 + \cdots + \alpha_n x_n, x_k \rangle = \langle \theta, x_k \rangle = 0.$$

Expanding the left-hand side,

$$\alpha_1 \langle x_1, x_k \rangle + \alpha_2 \langle x_2, x_k \rangle + \cdots + \alpha_n \langle x_n, x_k \rangle = 0.$$

But $\{x_1, x_2, \ldots, x_n\}$ is an orthogonal set and $x_k \neq \theta$ for any k, so only one of the inner products on the left is nonzero, namely $\langle x_k, x_k \rangle$. Thus we have $\alpha_k \langle x_k, x_k \rangle = 0$, which implies that $\alpha_k = 0$. Since this is true for all $k = 1, 2, \ldots, n$, the set $\{x_1, x_2, \ldots, x_n\}$ is linearly independent, as required. □

Theorem 8.2.4 *If S is a linearly independent countable set of vectors in an inner product space X, then there exists an orthogonal set T in X such that $\operatorname{Sp} T = \operatorname{Sp} S$.*

To relate this to the introductory comment, we consider the special case in which S is a basis for X (implying that S is a finite set and that X is finite-dimensional). Then $\operatorname{Sp} S = X$. Conceivably, the set T of the theorem includes the zero vector θ of X, but if that is the case then it is clear that $\operatorname{Sp}(T \backslash \{\theta\}) = \operatorname{Sp} T$. In either case, we have an orthogonal set of nonzero vectors which, by the theorem, spans X. But that set is linearly independent by Theorem 8.2.3, and hence is a basis for X.

The theorem says somewhat more than this, in that X need not be finite-dimensional and S need not be a finite set. In the proof, we actually construct the set T from the given set S. The method is known as the *Gram–Schmidt orthonormalisation process*.

We suppose that S is an infinite set, so we may write $S = \{x_1, x_2, \ldots\}$. If S is a finite set, the procedure described below is clearly equally valid.

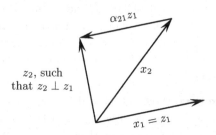

Figure 13

We will construct first an orthogonal set $\{z_1, z_2, \dots\}$ in X that spans
the same subspace as S does and will then set $y_k = z_k/\|z_k\|$, for each
$k \in \mathbf{N}$. (The construction will ensure that $\|z_k\| \neq 0$.) Then $\|y_k\| = 1$
for each k and $\{y_1, y_2, \dots\}$ will be our orthonormal set T such that
$\operatorname{Sp} T = \operatorname{Sp} S$. We proceed step by step to indicate the general method.

First we set $z_1 = x_1$, and then $z_2 = x_2 + \alpha_{21} z_1$, where α_{21} is a scalar
to be chosen such that $\langle z_2, z_1 \rangle = 0$. This requires $\langle x_2 + \alpha_{21} z_1, z_1 \rangle = 0$,
or $\langle x_2, z_1 \rangle + \alpha_{21} \langle z_1, z_1 \rangle = 0$, so that we take

$$\alpha_{21} = -\frac{\langle x_2, z_1 \rangle}{\|z_1\|^2}.$$

We cannot have $\|z_1\| = 0$, for then $x_1 = z_1 = \theta$ and a linearly indepen-
dent set of vectors, such as S, cannot include the zero vector. Thus z_2
is a certain linear combination of x_2 and z_1, that is, of x_2 and x_1. The
significance of α_{21} is indicated in Figure 13.

We next set $z_3 = x_3 + \alpha_{32} z_2 + \alpha_{31} z_1$, and will choose α_{32} and α_{31}
so that $\langle z_3, z_1 \rangle = 0$ and $\langle z_3, z_2 \rangle = 0$. To have $\langle z_3, z_1 \rangle = 0$, we require
$\langle x_3 + \alpha_{32} z_2 + \alpha_{31} z_1, z_1 \rangle = 0$, or

$$\langle x_3, z_1 \rangle + \alpha_{32} \langle z_2, z_1 \rangle + \alpha_{31} \langle z_1, z_1 \rangle = 0;$$

but $\langle z_2, z_1 \rangle = 0$ by construction, so

$$\alpha_{31} = -\frac{\langle x_3, z_1 \rangle}{\|z_1\|^2}.$$

To have $\langle z_3, z_2 \rangle = 0$, we require $\langle x_3 + \alpha_{32} z_2 + \alpha_{31} z_1, z_2 \rangle = 0$, or

$$\langle x_3, z_2 \rangle + \alpha_{32} \langle z_2, z_2 \rangle + \alpha_{31} \langle z_1, z_2 \rangle = 0$$

and so

$$\alpha_{32} = -\frac{\langle x_3, z_2 \rangle}{\|z_2\|^2}.$$

We cannot have $\|z_2\| = 0$, for then $z_2 = x_2 + \alpha_{21} z_1 = x_2 + \alpha_{21} x_1 = \theta$ so that $x_2 = -\alpha_{21} x_1$. This is not possible since x_1 and x_2 are vectors in a linearly independent set.

We now have enough to suggest the general approach. We write $z_1 = x_1$ and

$$z_n = x_n + \sum_{k=1}^{n-1} \alpha_{nk} z_k, \qquad n = 2, \ 3, \ \ldots,$$

and verify by induction that if

$$\alpha_{nk} = -\frac{\langle x_n, z_k \rangle}{\|z_k\|^2},$$

then $\langle z_n, z_m \rangle = 0$ for $m, n \in \mathbf{N}$ ($m \neq n$). As above, we cannot have $\|z_k\| = 0$ for any k since this would imply that $\{x_1, x_2, \ldots, x_k\}$ is a linearly dependent set of vectors. The induction argument is as follows. We have already settled the first few cases. Now suppose that $\langle z_k, z_m \rangle = 0$ for all integer values of k and m from 1 to $n-1$ ($k \neq m$). Then we have, for any $m = 1, 2, \ldots, n-1$,

$$\langle z_n, z_m \rangle = \left\langle x_n + \sum_{k=1}^{n-1} \alpha_{nk} z_k, z_m \right\rangle$$

$$= \langle x_n, z_m \rangle + \sum_{k=1}^{n-1} \alpha_{nk} \langle z_k, z_m \rangle$$

$$= \langle x_n, z_m \rangle + \alpha_{nm} \langle z_m, z_m \rangle$$

$$= \langle x_n, z_m \rangle - \frac{\langle x_n, z_m \rangle}{\|z_m\|^2} \|z_m\|^2$$

$$= 0.$$

Hence $\{z_1, z_2, \ldots\}$ is an orthogonal set in X. It is clear that each vector z_n is a linear combination of $x_1, x_2, \ldots, x_n$ and that each vector x_n is a linear combination of $z_1, z_2, \ldots, z_n$. It follows that $\mathrm{Sp}\,T = \mathrm{Sp}\,S$, completing the proof of Theorem 8.2.4. $\qquad \square$

The construction used in this proof is perhaps as important as the theorem itself. So much so, that we will state it more explicitly.

Theorem 8.2.5 *Let* $\{x_1, x_2, \ldots\}$ *be a linearly independent set of vectors in an inner product space. Put*

$$z_1 = x_1, \qquad z_n = x_n - \sum_{k=1}^{n-1} \frac{\langle x_n, z_k \rangle}{\|z_k\|^2} z_k, \quad \text{for } n = 2, \ 3, \ \ldots,$$

and $y_n = z_n / \|z_n\|$ *for each* n. *Then* $\{y_1, y_2, \ldots\}$ *is an orthonormal set in the space.*

The Gram–Schmidt orthonormalisation process may be applied to the basic power functions $\{t^k : k = 0, 1, \ldots, \ a \leqslant t \leqslant b\}$ in the real inner product space $C_2[a, b]$. When $a = -1$, $b = 1$, the polynomial functions that result are known as the *Legendre polynomials*. Denoting these by P_0, P_1, $\ldots$, they are therefore such that

$$\int_{-1}^{1} P_j(t) P_k(t) \, dt = 0, \ j \neq k; \qquad \int_{-1}^{1} (P_k(t))^2 \, dt = 1.$$

The calculations for the first few Legendre polynomials are left as an exercise, being a little simpler than those in the example we will shortly do. The first five are

$$P_0(t) = \frac{\sqrt{2}}{2},$$

$$P_1(t) = \frac{\sqrt{6}}{2} t,$$

$$P_2(t) = \frac{\sqrt{10}}{4} (3t^2 - 1),$$

$$P_3(t) = \frac{\sqrt{14}}{4} (5t^3 - 3t),$$

$$P_4(t) = \frac{3\sqrt{2}}{16} (35t^4 - 30t^2 + 3),$$

all on $[-1, 1]$.

The Legendre polynomials are one instance of a number of sets of polynomial functions that have received much attention. All of these arise as particular cases of a different definition for the inner product on the set of continuous functions on $[a, b]$. Let w be a given integrable function defined on (a, b), and such that $w(t) > 0$ for all $t \in (a, b)$. It is easily verified that

$$\langle x, y \rangle = \int_{a}^{b} w(t) x(t) y(t) \, dt$$

defines an inner product for continuous functions x, y on $[a, b]$. The resulting real inner product space is said to have *weight function* w. We will denote it by $C^w[a, b]$. Thus $C_2[a, b] = C^w[a, b]$ with $w(t) = 1$.

The various sets of polynomial functions just referred to are the results of taking different weight functions with special values for a and b. We will take

$$a = -1,\ b = 1,\quad w(t) = \frac{1}{\sqrt{1 - t^2}},\ -1 < t < 1,$$

and will now apply the Gram–Schmidt process to the functions 1, t, $t^2, \ldots$, $-1 \leqslant t \leqslant 1$.

Use the notation of Theorem 8.2.5. For $k \in \mathbf{N}$, define functions x_k by $x_k(t) = t^{k-1}$, $-1 \leqslant t \leqslant 1$. Then $z_1 = x_1$ and

$$z_2 = x_2 - \frac{\langle x_2, z_1 \rangle}{\|z_1\|^2}\, z_1.$$

Now,

$$\|z_1\|^2 = \int_{-1}^{1} \frac{dt}{\sqrt{1 - t^2}} = 2\left[\sin^{-1} t\right]_0^1 = \pi,$$

while

$$\langle x_2, z_1 \rangle = \int_{-1}^{1} w(t)x_2(t)z_1(t)\, dt = \int_{-1}^{1} \frac{t\, dt}{\sqrt{1 - t^2}} = 0,$$

since the integrand is an odd function (a common argument, used often below). Thus $z_2 = x_2$, or $z_2(t) = t$, $-1 \leqslant t \leqslant 1$. Next,

$$z_3 = x_3 - \frac{\langle x_3, z_1 \rangle}{\|z_1\|^2}\, z_1 - \frac{\langle x_3, z_2 \rangle}{\|z_2\|^2}\, z_2,$$

and

$$\langle x_3, z_1 \rangle = \int_{-1}^{1} \frac{t^2\, dt}{\sqrt{1 - t^2}} = \int_{-\pi/2}^{\pi/2} \sin^2 \phi\, d\phi = \frac{\pi}{2}\quad [t = \sin \phi],$$

$$\|z_2\|^2 = \int_{-1}^{1} \frac{t^2\, dt}{\sqrt{1 - t^2}} = \frac{\pi}{2},$$

$$\langle x_3, z_2 \rangle = \int_{-1}^{1} \frac{t^3\, dt}{\sqrt{1 - t^2}} = 0.$$

Thus

$$z_3(t) = x_3(t) - \frac{\pi/2}{\pi} z_1(t) - 0 = t^2 - \frac{1}{2},\quad -1 \leqslant t \leqslant 1.$$

Next,

$$z_4 = x_4 - \frac{\langle x_4, z_1 \rangle}{\|z_1\|^2} z_1 - \frac{\langle x_4, z_2 \rangle}{\|z_2\|^2} z_2 - \frac{\langle x_4, z_3 \rangle}{\|z_3\|^2} z_3,$$

and

$$\langle x_4, z_1 \rangle = \int_{-1}^{1} \frac{t^3 \, dt}{\sqrt{1 - t^2}} = 0,$$

$$\langle x_4, z_2 \rangle = \int_{-1}^{1} \frac{t^4 \, dt}{\sqrt{1 - t^2}} = \int_{-\pi/2}^{\pi/2} \sin^4 \phi \, d\phi = \frac{3\pi}{8},$$

$$\|z_3\|^2 = \int_{-1}^{1} \frac{(t^2 - \frac{1}{2})^2}{\sqrt{1 - t^2}} \, dt = \int_{-1}^{1} \frac{t^4 - t^2 + \frac{1}{4}}{\sqrt{1 - t^2}} \, dt = \frac{3\pi}{8} - \frac{\pi}{2} + \frac{\pi}{4} = \frac{\pi}{8},$$

$$\langle x_4, z_3 \rangle = \int_{-1}^{1} \frac{t^3(t^2 - \frac{1}{2})}{\sqrt{1 - t^2}} \, dt = 0.$$

Thus

$$z_4(t) = x_4(t) - 0 - \frac{3\pi/8}{\pi/2} z_2(t) - 0 = t^3 - \frac{3}{4}t, \quad -1 \leqslant t \leqslant 1.$$

Also,

$$\|z_4\|^2 = \int_{-1}^{1} \frac{(t^3 - \frac{3}{4}t)^2}{\sqrt{1 - t^2}} \, dt = \int_{-1}^{1} \frac{t^6 - \frac{3}{2}t^4 + \frac{9}{16}t^2}{\sqrt{1 - t^2}} \, dt$$

$$= \int_{-\pi/2}^{\pi/2} \sin^6 \phi \, d\phi - \frac{9\pi}{16} + \frac{9\pi}{32} = \frac{5\pi}{16} - \frac{9\pi}{32} = \frac{\pi}{32}.$$

The first four required polynomial functions, orthonormal in this space $C^w[-1, 1]$, are $y_n = z_n/\|z_n\|$, for $n = 1, 2, 3, 4$. That is,

$$y_1(t) = \sqrt{\frac{1}{\pi}},$$

$$y_2(t) = \sqrt{\frac{2}{\pi}} \, t,$$

$$y_3(t) = \sqrt{\frac{8}{\pi}} \left(t^2 - \frac{1}{2} \right) = \sqrt{\frac{2}{\pi}} \, (2t^2 - 1),$$

$$y_4(t) = \sqrt{\frac{32}{\pi}} \left(t^3 - \frac{3}{4}t \right) = \sqrt{\frac{2}{\pi}} (4t^3 - 3t),$$

all on $[-1, 1]$.

It will be observed that $y_1, \ldots, y_4$ here are multiples of the Chebyshev polynomials $T_0, \ldots, T_3$ of Section 6.7. Those polynomials were defined by

$$T_n(t) = \cos n\theta \quad \text{where } t = \cos\theta, \ n = 0, \ 1, \ \ldots.$$

We can show in general that the polynomials T_n are orthogonal in the space $C^w[-1, 1]$ (with $w(t) = 1/\sqrt{1 - t^2}$) by noting that

$$\int_0^\pi \cos m\theta \cos n\theta \, d\theta = \begin{cases} \pi, & m = n = 0, \\ \dfrac{\pi}{2}, & m = n \neq 0, \\ 0, & m \neq n. \end{cases}$$

Substituting $t = \cos\theta$, we have

$$\langle T_m, T_n \rangle = \int_{-1}^1 \frac{T_m(t)T_n(t)}{\sqrt{1 - t^2}} \, dt = \begin{cases} \pi, & m = n = 0, \\ \dfrac{\pi}{2}, & m = n \neq 0, \\ 0, & m \neq n. \end{cases}$$

It is clear that with the factors $\sqrt{1/\pi}$, $\sqrt{2/\pi}$, as in y_1, ..., y_4, the Chebyshev polynomials are orthonormal.

Thus the Gram–Schmidt orthonormalisation process applied to the powers 1, t, t^2, ... leads to the Legendre polynomials in $C_2[-1, 1]$ and the Chebyshev polynomials in $C^w[-1, 1]$ (with $w(t) = 1/\sqrt{1 - t^2}$). These are included in the following table detailing various classes of orthonormal polynomials that have been studied. (Where the table implies $a = -\infty$, say, the inner product is to be defined in a natural way by a certain improper integral. There will be no problems concerning the convergence of the integrals or the verification of (IP1) to (IP4).)

a	b	$w(t)$	*Name*
-1	1	1	Legendre polynomials
-1	1	$1/\sqrt{1 - t^2}$	Chebyshev polynomials
-1	1	$\sqrt{1 - t^2}$	Chebyshev polynomials (of the second kind)
-1	1	$(1 - t)^\lambda (1 + t)^\mu; \ \lambda, \mu > -1$	Jacobi polynomials
0	∞	$t^\lambda e^{-t}, \ \lambda > -1$	Laguerre polynomials
$-\infty$	∞	e^{-t^2}	Hermite polynomials

8.3 Least squares approximation

In Section 6.6, we considered the problem of best approximation in a normed space. In Theorem 6.6.3, we stated that a unique solution exists to this problem when the space is strictly convex. (Recall that a normed space X is strictly convex if, whenever $\|x+y\| = \|x\| + \|y\|$, $x, y \in X$, $x \neq \theta$, $y \neq \theta$, we must have $x = \beta y$ for some number $\beta > 0$.) It is therefore very pleasing that we can show that any inner product space is strictly convex. This is the content of our first theorem below. We then deduce a simple formula that gives the best approximation and apply this in a further discussion of least squares polynomial approximations. The term *least squares approximation* is used generally for approximation problems in inner product spaces, for a reason that will become apparent.

Theorem 8.3.1 *Inner product spaces are strictly convex.*

Let X be an inner product space and suppose $\|x+y\| = \|x\| + \|y\|$ for some nonzero vectors $x, y \in X$. To prove the theorem, we must show that $x - \beta y = \theta$ for some number $\beta > 0$. Since

$$\|x+y\|^2 = \langle x+y, x+y \rangle = \langle x, x \rangle + \langle x, y \rangle + \langle y, x \rangle + \langle y, y \rangle$$
$$= \|x\|^2 + 2 \operatorname{Re} \langle x, y \rangle + \|y\|^2$$

and

$$(\|x\| + \|y\|)^2 = \|x\|^2 + 2\|x\| \, \|y\| + \|y\|^2,$$

the condition $\|x+y\| = \|x\| + \|y\|$ implies that $\operatorname{Re} \langle x, y \rangle = \|x\| \, \|y\|$. But then, using the Cauchy–Schwarz inequality,

$$\|x\| \, \|y\| = \operatorname{Re} \langle x, y \rangle \leqslant |\langle x, y \rangle| \leqslant \|x\| \, \|y\|,$$

so we must have $|\langle x, y \rangle| = \|x\| \, \|y\|$. It follows (see the proof of Theorem 8.1.2) that

$$\left\| x - \frac{\langle x, y \rangle}{\|y\|^2} y \right\|^2 = \|x\|^2 - \frac{|\langle x, y \rangle|^2}{\|y\|^2} = 0,$$

and hence we take $\beta = \langle x, y \rangle / \|y\|^2$, completing the proof. (Note that β is real and positive, since $\operatorname{Re} \langle x, y \rangle = |\langle x, y \rangle| = \|x\| \, \|y\| > 0$.) $\qquad \square$

As indicated, we can now invoke Theorem 6.6.3. We do that in the next theorem, in which we show also how to obtain the best approximation.

Theorem 8.3.2 *A finite-dimensional subspace S of an inner product space X contains a unique best approximation of any point $x \in X$. If $\{y_1, y_2, \ldots, y_n\}$ is a basis for S and is orthonormal, then the best approximation of x is $\sum_{k=1}^{n} \langle x, y_k \rangle y_k$.*

We are assuming of course that the subspace S has dimension n. The existence of a basis $\{y_1, \ldots, y_n\}$ that is an orthonormal set in X is a consequence of Theorem 8.2.4, because, given any basis, such a basis can be obtained by the Gram–Schmidt process, Theorem 8.2.5.

For the second statement of the theorem, take any point $\sum_{k=1}^{n} \alpha_k y_k$ $(\alpha_1, \ldots, \alpha_n \in \mathbf{C})$ in S. Then

$$
\left\| x - \sum_{k=1}^{n} \alpha_k y_k \right\|^2 = \left\langle x - \sum_{k=1}^{n} \alpha_k y_k, x - \sum_{k=1}^{n} \alpha_k y_k \right\rangle
$$

$$
= \langle x, x \rangle - \left\langle x, \sum_{k=1}^{n} \alpha_k y_k \right\rangle - \left\langle \sum_{k=1}^{n} \alpha_k y_k, x \right\rangle
$$

$$
+ \left\langle \sum_{k=1}^{n} \alpha_k y_k, \sum_{j=1}^{n} \alpha_j y_j \right\rangle
$$

$$
= \|x\|^2 - \sum_{k=1}^{n} \overline{\alpha}_k \langle x, y_k \rangle - \sum_{k=1}^{n} \alpha_k \overline{\langle x, y_k \rangle} + \sum_{k=1}^{n} \alpha_k \overline{\alpha}_k,
$$

since $\langle y_k, y_j \rangle = 0$ when $j \neq k$ and $\langle y_k, y_k \rangle = 1$. For any complex numbers z_1 and z_2, we have

$$
|z_1 - z_2|^2 = (z_1 - z_2)(\overline{z}_1 - \overline{z}_2) = |z_1|^2 - z_1 \overline{z}_2 - \overline{z}_1 z_2 + |z_2|^2,
$$

so

$$
\left\| x - \sum_{k=1}^{n} \alpha_k y_k \right\|^2 = \sum_{k=1}^{n} \left(|\alpha_k|^2 - \alpha_k \overline{\langle x, y_k \rangle} - \overline{\alpha}_k \langle x, y_k \rangle + |\langle x, y_k \rangle|^2 \right)
$$

$$
+ \|x\|^2 - \sum_{k=1}^{n} |\langle x, y_k \rangle|^2
$$

$$
= \sum_{k=1}^{n} |\alpha_k - \langle x, y_k \rangle|^2 + \|x\|^2 - \sum_{k=1}^{n} |\langle x, y_k \rangle|^2.
$$

Clearly, the final expression is least when we choose $\alpha_k = \langle x, y_k \rangle$ for each k. Since our problem is to find $p \in S$ such that $\|x - p\|$ is a minimum, we conclude that

$$
p = \sum_{k=1}^{n} \langle x, y_k \rangle y_k,
$$

as required. $\qquad \square$

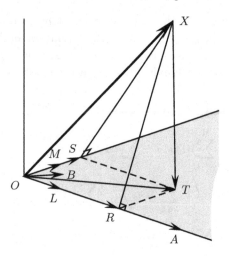

Figure 14

Because the solution p here is unique, the same point would have been obtained whatever orthonormal basis we began with. Thus

$$\sum_{k=1}^{n} \langle x, y_k \rangle \, y_k = \sum_{k=1}^{n} \langle x, y'_k \rangle \, y'_k$$

for any orthonormal sets $\{y_1, \ldots, y_n\}$, $\{y'_1, \ldots, y'_n\}$ spanning the same subspace of an inner product space X, when $x \in X$. This is not an easy result to prove if we do not take into account that either expression gives the best approximation of x in the subspace.

The next theorem is a direct consequence of these calculations.

Theorem 8.3.3 *If $\{y_1, y_2, \ldots, y_n\}$ is an orthonormal set in an inner product space X, and $x \in X$, then*

$$\sum_{k=1}^{n} |\langle x, y_k \rangle|^2 \leqslant \|x\|^2.$$

This is known as *Bessel's inequality*. To prove it, we simply note that

$$0 \leqslant \left\| x - \sum_{k=1}^{n} \langle x, y_k \rangle \, y_k \right\|^2 = \|x\|^2 - \sum_{k=1}^{n} |\langle x, y_k \rangle|^2. \qquad \square$$

Theorem 8.3.2 has a simple geometric interpretation. Consider Figure 14, in which we want the best approximation of $\overrightarrow{OX}$ (in $\mathbf{R}^3$) by a vector in the horizontal plane (the subspace of $\mathbf{R}^3$ in which all vectors have third component 0, shown shaded). The vectors $\overrightarrow{OA}$, $\overrightarrow{OB}$ span this plane, and from them are constructed orthonormal vectors $\overrightarrow{OL}$, $\overrightarrow{OM}$. The required best approximation of $\overrightarrow{OX}$ is $\overrightarrow{OT}$ since any other vector from X to the plane has length exceeding $|\overrightarrow{XT}|$. We obtain $\overrightarrow{OT}$ as $\overrightarrow{OR} + \overrightarrow{RT}$ ($= \overrightarrow{OR} + \overrightarrow{OS}$), where $\overrightarrow{OR}$ is the projection of $\overrightarrow{OX}$ on $\overrightarrow{OL}$ and $\overrightarrow{OS}$ is the projection of $\overrightarrow{OX}$ on $\overrightarrow{OM}$. From ordinary vector algebra, $\overrightarrow{OR} = (\overrightarrow{OX} \cdot \overrightarrow{OL})\overrightarrow{OL}$ and $\overrightarrow{OS} = (\overrightarrow{OX} \cdot \overrightarrow{OM})\overrightarrow{OM}$, so

$$\overrightarrow{OT} = (\overrightarrow{OX} \cdot \overrightarrow{OL})\overrightarrow{OL} + (\overrightarrow{OX} \cdot \overrightarrow{OM})\overrightarrow{OM},$$

which is precisely the answer given by Theorem 8.3.2. Bessel's inequality in this situation is also clear:

$$|\overrightarrow{OR}|^2 + |\overrightarrow{OS}|^2 = |\overrightarrow{OR}|^2 + |\overrightarrow{RT}|^2 = |\overrightarrow{OT}|^2$$
$$= |\overrightarrow{OX}|^2 - |\overrightarrow{XT}|^2 \leqslant |\overrightarrow{OX}|^2.$$

In practice, a common method for determining the best least squares approximation is indicated in the following theorem. The need to construct an orthonormal basis is avoided.

Theorem 8.3.4 *Let* $\{x_1, x_2, \ldots, x_n\}$ *be a basis for a subspace S of an inner product space X. Let $x \in X$ be given. If $\sum_{k=1}^n \beta_k x_k$ is the best least squares approximation in S of x, then*

$$\sum_{k=1}^n \beta_k \langle x_k, x_i \rangle = \langle x, x_i \rangle, \qquad i = 1,\ 2,\ \ldots,\ n.$$

These equations, called the *normal equations*, are a system of n linear equations in n unknowns, from which the coefficients $\beta_1, \ldots, \beta_n$ may be obtained.

To prove the theorem, let $\{y_1, \ldots, y_n\}$ be an orthonormal basis for S, so that $\sum_{k=1}^n \langle x, y_k \rangle y_k$ is another expression for the best least squares approximation in S of x. Note that, for any $j = 1, 2, \ldots, n$,

$$\left\langle x - \sum_{k=1}^n \langle x, y_k \rangle y_k, y_j \right\rangle = \langle x, y_j \rangle - \sum_{k=1}^n \langle x, y_k \rangle \langle y_k, y_j \rangle$$
$$= \langle x, y_j \rangle - \langle x, y_j \rangle = 0.$$

Then, if $x_i = \sum_{j=1}^{n} \gamma_{ij} y_j$ gives x_i $(i = 1, \ldots, n)$ as a linear combination of $y_1, \ldots, y_n$,

$$\left\langle x - \sum_{k=1}^{n} \langle x, y_k \rangle y_k, x_i \right\rangle = \left\langle x - \sum_{k=1}^{n} \langle x, y_k \rangle y_k, \sum_{j=1}^{n} \gamma_{ij} y_j \right\rangle$$

$$= \sum_{j=1}^{n} \overline{\gamma}_{ij} \left\langle x - \sum_{k=1}^{n} \langle x, y_k \rangle y_k, y_j \right\rangle$$

$$= 0,$$

so that $\langle x - \sum_{k=1}^{n} \beta_k x_k, x_i \rangle = 0$ for any $i = 1, \ldots, n$. Hence

$$\langle x, x_i \rangle - \sum_{k=1}^{n} \beta_k \langle x_k, x_i \rangle = 0, \qquad i = 1, \ldots, n,$$

as we wished to show. $\qquad\qquad\qquad\qquad\qquad\qquad\qquad\qquad\qquad\qquad \square$

As an illustration of this theorem, we will obtain the best least squares linear approximation to the function sin on $[0, \pi/2]$. This will be a function whose graph is the line $y = \beta_1 + \beta_2 t$, shown in Figure 12, at the end of Chapter 6. Relating this problem to Theorem 8.3.4, we are considering the function $\sin \in C_2[0, \pi/2]$ and approximating it in the subspace spanned by $\{x_1, x_2\}$, where $x_1(t) = 1$, $x_2(t) = t$, $0 \leqslant t \leqslant \pi/2$. By that theorem, the best least squares approximation in this subspace is $\beta_1 x_1 + \beta_2 x_2$, where

$$\beta_1 \langle x_1, x_1 \rangle + \beta_2 \langle x_2, x_1 \rangle = \langle \sin, x_1 \rangle,$$
$$\beta_1 \langle x_1, x_2 \rangle + \beta_2 \langle x_2, x_2 \rangle = \langle \sin, x_2 \rangle.$$

Now,

$$\langle x_1, x_1 \rangle = \int_0^{\pi/2} (x_1(t))^2 \, dt = \int_0^{\pi/2} dt = \frac{\pi}{2},$$

$$\langle x_2, x_1 \rangle = \int_0^{\pi/2} x_2(t) x_1(t) \, dt = \int_0^{\pi/2} t \, dt = \frac{1}{2} \left(\frac{\pi}{2} \right)^2,$$

$$\langle x_2, x_2 \rangle = \int_0^{\pi/2} (x_2(t))^2 \, dt = \int_0^{\pi/2} t^2 \, dt = \frac{1}{3} \left(\frac{\pi}{2} \right)^3,$$

$$\langle \sin, x_1 \rangle = \int_0^{\pi/2} \sin(t) x_1(t) \, dt = \int_0^{\pi/2} \sin t \, dt = 1,$$

$$\langle \sin, x_2 \rangle = \int_0^{\pi/2} \sin(t) x_2(t) \, dt = \int_0^{\pi/2} t \sin t \, dt = 1.$$

The system of equations becomes

$$\frac{\pi}{2}\beta_1 + \frac{\pi^2}{8}\beta_2 = 1,$$

$$\frac{\pi^2}{8}\beta_1 + \frac{\pi^3}{24}\beta_2 = 1,$$

from which

$$\beta_1 = \frac{8\pi - 24}{\pi^2}, \qquad \beta_2 = \frac{96 - 24\pi}{\pi^3}.$$

The line $y = \beta_1 x_1(t) + \beta_2 x_2(t)$ is thus $y = 8(\pi - 3)/\pi^2 + 24(4 - \pi)t/\pi^3$, as stated in Solved Problem 6.9(2).

8.4 The Riesz representation theorem

As another application of the Gram–Schmidt orthonormalisation process, we will prove an important result known as the Riesz representation theorem. This gives us a characterisation, or representation, of the set of all linear functionals on a finite-dimensional inner product space. It will be recalled (Definition 7.3.1) that a linear functional on such a space X is a linear mapping from X into $\mathbf{C}$, the set of scalars for X. (If X is a real inner product space, only minor changes need to be made to what follows.)

Let $v \in X$ be some fixed vector. An example of a linear functional on X is the mapping f given by

$$f(x) = \langle x, v \rangle, \qquad x \in X.$$

Since inner products of points in X are complex numbers, this is indeed a functional. It is linear, because

$$f(\alpha_1 x_1 + \alpha_2 x_2) = \langle \alpha_1 x_1 + \alpha_2 x_2, v \rangle$$
$$= \alpha_1 \langle x_1, v \rangle + \alpha_2 \langle x_2, v \rangle = \alpha_1 f(x_1) + \alpha_2 f(x_2),$$

for any $x_1, x_2 \in X$ and $\alpha_1, \alpha_2 \in \mathbf{C}$. The Riesz theorem says that there are in fact no other types of linear functionals on a finite-dimensional inner product space: any linear functional on the space X above must have the form $\langle x, w \rangle$ for some unique point $w \in X$. Specifically:

Theorem 8.4.1 (Riesz Representation Theorem) *Let X be a finite-dimensional inner product space and let f be a linear functional on X. Then there exists a unique point $v \in X$ such that $f(x) = \langle x, v \rangle$ for all $x \in X$.*

The proof follows. Suppose the dimension of X is n and that the set $\{x_1, x_2, \ldots, x_n\}$ is a basis for X. By virtue of the Gram–Schmidt orthonormalisation process, we may assume that this is an orthonormal set, for if it were not then the process would allow us to construct another basis which was an orthonormal set. Consider the vector

$$v = \sum_{k=1}^{n} \overline{f(x_k)} x_k$$

and define a functional f_v on X by $f_v(x) = \langle x, v \rangle$. As above, f_v is linear. We will show that the functionals f and f_v coincide: $f(x) = f_v(x)$ for all $x \in X$. We note first that for any basis vector x_j we have

$$f_v(x_j) = \langle x_j, v \rangle = \left\langle x_j, \sum_{k=1}^{n} \overline{f(x_k)} x_k \right\rangle = \sum_{k=1}^{n} f(x_k) \langle x_j, x_k \rangle = f(x_j),$$

since $\langle x_j, x_k \rangle = 0$ for $k \neq j$ and $\langle x_j, x_j \rangle = \|x_j\|^2 = 1$. Thus f_v and f agree for any basis vector. Then, for any $x = \sum_{k=1}^{n} \alpha_k x_k \in X$,

$$f_v(x) = f_v \left(\sum_{k=1}^{n} \alpha_k x_k \right) = \sum_{k=1}^{n} \alpha_k f_v(x_k)$$

$$= \sum_{k=1}^{n} \alpha_k f(x_k) = f \left(\sum_{k=1}^{n} \alpha_k x_k \right) = f(x).$$

Thus, f_v and f indeed coincide on X. It remains to show that no vector other than v has the same effect. To do this, suppose $u \in X$ is such that $f(x) = \langle x, u \rangle = \langle x, v \rangle$ for all $x \in X$. Then we have $\langle x, u - v \rangle = 0$ for all $x \in X$. In particular, then $\langle u - v, u - v \rangle = 0$, so $u - v = \theta$. That is, $u = v$ so v is unique and this completes the proof. $\square$

The following is a simple consequence of this theorem.

Theorem 8.4.2 *All linear functionals on finite-dimensional inner product spaces are bounded.*

We now know that if f is a linear functional on a finite-dimensional space X, then there is a point $v \in X$ such that $f(x) = \langle x, v \rangle$ for all $x \in X$. But then, by the Cauchy–Schwarz inequality (Theorem 8.1.2),

$$|f(x)| = |\langle x, v \rangle| \leqslant \|x\| \, \|v\|$$

so f is bounded. $\square$

We can say more. The inequality $|f(x)| \leqslant \|v\| \, \|x\|$ implies that

$\|f\| \leqslant \|v\|$. In fact, we show that $\|f\| = \|v\|$. This is seen by noting that $|f(v)| = |\langle v, v \rangle| = \|v\|^2$ so that we cannot have $|f(x)| < \|v\| \, \|x\|$ for all $x \in X$.

There are many other versions of the Riesz representation theorem, giving corresponding types of results in other spaces. We will deduce another in connection with Hilbert space, later. The benefit in being able to characterise a whole class of entities (in the above, the class of all linear functionals on finite-dimensional inner product spaces) should by now be recognised. The Riesz theorem, and a variation known as the *Riesz–Fréchet theorem*, are the springboard for many important results in advanced analysis and functional analysis.

8.5 Solved problems

(1) Find the best least squares quadratic approximation on $[-1, 1]$ of the function f, where

$$f(x) = \frac{1}{1 + x^2}, \qquad -1 \leqslant x \leqslant 1.$$

(Equivalently, we could say: Find numbers a_0, a_1, a_2 such that

$$\int_{-1}^{1} \left(\frac{1}{1 + x^2} - a_0 - a_1 x - a_2 x^2 \right)^2 dx$$

is a minimum.)

Solution. Since the Legendre polynomials P_0, P_1, ... form an orthonormal set in $C_2[-1, 1]$, Theorem 8.3.2 assures us that the best least squares quadratic approximation of f on $[-1, 1]$ is the function

$$g_1 = \sum_{k=0}^{2} \langle f, P_k \rangle P_k.$$

Then

$$g_1(x) = \sum_{k=0}^{2} \langle f, P_k \rangle P_k(x) = \sum_{k=0}^{2} P_k(x) \int_{-1}^{1} \frac{P_k(t)}{1 + t^2} dt.$$

We need the integrals

$$\int_{-1}^{1} \frac{dt}{1 + t^2} = \frac{\pi}{2}, \qquad \int_{-1}^{1} \frac{t \, dt}{1 + t^2} = 0, \qquad \int_{-1}^{1} \frac{t^2 \, dt}{1 + t^2} = 2 - \frac{\pi}{2}.$$

Substituting the expressions for P_0, P_1, P_2 in Section 8.2, we have

$$g_1(x) = \frac{\sqrt{2}}{2} \cdot \frac{\pi}{2\sqrt{2}} + 0 + \frac{\sqrt{10}}{4}(3x^2 - 1) \cdot \frac{\sqrt{10}}{4}\left(3\left(2 - \frac{\pi}{2}\right) - \frac{\pi}{2}\right)$$

$$= \frac{3\pi}{2} - \frac{15}{4} - \frac{15}{4}(\pi - 3)x^2.$$

To three decimal places,

$$g_1(x) = 0.962 - 0.531x^2. \qquad \square$$

Note that another method is available: find the normal equations as in the example following Theorem 8.3.4.

The function g_1 obtained here may be compared with the best uniform quadratic approximation g_2 of f, given by

$$g_2(x) = \frac{\sqrt{2}}{2} + \frac{1}{4} - \frac{1}{2}x^2$$

$$= 0.957 - 0.5x^2, \qquad -1 \leqslant x \leqslant 1,$$

to three decimal places. This was obtained in Exercise 6.10(16)(a). The best least squares Chebyshev quadratic approximation g_3 of f is given by

$$g_3(x) = \frac{7\sqrt{2}}{2} - 4 - 2(3\sqrt{2} - 4)x^2$$

$$= 0.950 - 0.485x^2, \qquad -1 \leqslant x \leqslant 1,$$

to three decimal places. This is obtained in the same way as the function g_1, using the normalised Chebyshev polynomials y_1, y_2, y_3, given towards the end of Section 8.2, and the weight function $w(t) = 1/\sqrt{1 - t^2}$, $-1 < t < 1$. The integral

$$\int_{-1}^{1} \frac{dx}{(1 + x^2)\sqrt{1 - x^2}} = \frac{\pi}{\sqrt{2}}$$

is required. The error functions $E_k = f - g_k$, for $k = 1, 2, 3$, are plotted in Figure 15.

(2) Let $\{x_1, x_2, \dots\}$ be an (infinite) orthonormal set in an inner product space X and let u be a given point in X. Define a sequence $\{u_n\}$ in X by

$$u_n = \sum_{k=1}^{n} \langle u, x_k \rangle x_k, \qquad n \in \mathbf{N}.$$

Show that $\{u_n\}$ is a Cauchy sequence.

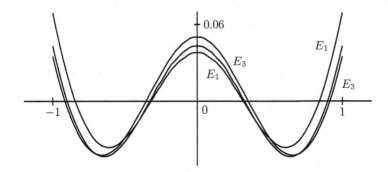

Figure 15

Solution. We need to consider $\|u_n - u_m\|$. Taking $n > m$ for definiteness,

$$
\begin{aligned}
\|u_n - u_m\|^2 &= \langle u_n - u_m, u_n - u_m \rangle \\
&= \left\langle \sum_{k=m+1}^{n} \langle u, x_k \rangle \, x_k, \; \sum_{j=m+1}^{n} \langle u, x_j \rangle \, x_j \right\rangle \\
&= \sum_{k=m+1}^{n} \sum_{j=m+1}^{n} \langle u, x_k \rangle \, \overline{\langle u, x_j \rangle} \, \langle x_k, x_j \rangle \\
&= \sum_{k=m+1}^{n} \langle u, x_k \rangle \, \overline{\langle u, x_k \rangle} \\
&= \sum_{k=m+1}^{n} |\langle u, x_k \rangle|^2,
\end{aligned}
$$

using the fact that the set $\{x_1, x_2, \dots\}$ is orthonormal in X. By Bessel's inequality (Theorem 8.3.3),

$$
\sum_{k=1}^{n} |\langle u, x_k \rangle|^2 \leqslant \|u\|^2,
$$

so the series $\sum |\langle u, x_k \rangle|^2$ converges. The result follows using Theorem 1.8.2. $\qquad\square$

8.6 Exercises

(1) For vectors in an inner product space, prove that

   (a) $\langle x, y + z \rangle = \langle x, y \rangle + \langle x, z \rangle$,   (b) $\langle x, \theta \rangle = 0$.

(2) Prove (f) in Section 8.1.

(3) For vectors in a complex inner product space, prove that

$$\langle x, y \rangle + \langle y, x \rangle = \frac{1}{2} \left(\|x + y\|^2 - \|x - y\|^2 \right)$$

and

$$\langle x, y \rangle - \langle y, x \rangle = \frac{i}{2} \left(\|x + iy\|^2 - \|x - iy\|^2 \right),$$

and hence deduce the *polarisation identity*:

$$\langle x, y \rangle = \frac{1}{4} \sum_{k=1}^{4} i^k \|x + i^k y\|^2.$$

(4) Let X be a finite-dimensional vector space. Show that an inner product may be defined for X.

(5) Let $\{x_n\}$, $\{y_n\}$ be Cauchy sequences in an inner product space. Prove that $\{\langle x_n, y_n \rangle\}$ is a convergent sequence in $\mathbf{C}$.

(6) Let $\{y_1, y_2, \ldots, y_n\}$ be a subset of an inner product space X and suppose $x \perp y_k$ for some $x \in X$ and all $k = 1, \ldots, n$. Prove that $x \perp \sum_{k=1}^{n} \alpha_k y_k$ for any scalars $\alpha_1, \ldots, \alpha_n$.

(7) Let $\{x_n\}$ be a convergent sequence in an inner product space X, with $\lim x_n = x$. If there exists $y \in X$ such that $\langle x_n, y \rangle = 0$ for all $n \in \mathbf{N}$, prove that $\langle x, y \rangle = 0$.

(8) Let $\{x_n\}$, $\{y_n\}$ be sequences in an inner product space such that $x_n \to \theta$ and $\{y_n\}$ is bounded. Prove that $\langle x_n, y_n \rangle \to 0$.

(9) In an inner product space, show that if x and y are orthonormal vectors, then $x + y$ and $x - y$ are orthogonal. Interpret this result geometrically.

(10) Let X be an inner product space. If $x, y \in X$ and $x \perp y$, prove that $\|x + y\|^2 = \|x\|^2 + \|y\|^2$. More generally, if $\{x_1, x_2, \ldots, x_n\}$ is an orthogonal set in X, prove that

$$\left\| \sum_{k=1}^{n} x_k \right\|^2 = \sum_{k=1}^{n} \|x_k\|^2.$$

(These are the *Pythagorean identities*. They are generalisations of Pythagoras' theorem of ordinary geometry.)

(11) Apply the Gram–Schmidt orthonormalisation process

    (a) to the vectors $(1,1,0)$, $(1,0,1)$, $(0,1,1)$ in $\mathbf{R}^3$ to find an orthonormal basis for this space;

    (b) to the vectors $(1,1,i)$, $(1,i,0)$, $(i,0,0)$ in $\mathbf{C}^3$ to find an orthonormal basis for this space;

    (c) to find an orthonormal set of vectors that spans the same subspace of $\mathbf{R}^4$ as the set

$$\{(2,0,1,0),(0,0,1,1),(0,1,1,0)\}.$$

(12)   (a) Verify that the first four Legendre polynomials $P_0, \ldots, P_3$ are as given in Section 8.2.

    (b) Find the first four Hermite polynomials. (Note: the integral $\int_{-\infty}^{\infty} e^{-t^2}\, dt = \sqrt{\pi}$ will be required.)

(13) Find the linear function that is the best least squares approximation of the function $\sqrt{x}$ on the interval (a) $[0,1]$, (b) $[1,4]$.

(14) Use Legendre polynomials to show that the best least squares quadratic polynomial approximation of $|x|$, for $-1 \leqslant x \leqslant 1$, is $(3 + 15x^2)/16$.

. . . .

(15) Let $\{x_n\}$ be a sequence in an inner product space X such that $\|x_n\| \to \|x\|$ (in $\mathbf{R}$) for some $x \in X$ and $\langle x_n, x \rangle \to \langle x, x \rangle$ (in $\mathbf{C}$). Prove that $x_n \to x$.

(16) For points in a real inner product space, show that $\langle x, y \rangle = 0$ if and only if $\|ax + y\| \geqslant \|y\|$ for all real numbers α.

(17) For x, y, z in an inner product space, prove that

$$\|x - z\| = \|x - y\| + \|y - z\|$$

if and only if there is a real number a, $0 \leqslant a \leqslant 1$, such that $y = ax + (1 - a)z$.

(18) Let S be any nonempty subset of an inner product space X. The set $\{x : x \in X,\ \langle x, y \rangle = 0 \text{ for all } y \in S\}$ is called the *annihilator* of S, denoted by $S^{\perp}$. Prove that (a) $\{\theta\}^{\perp} = X$, (b) whatever the set S, the set $S^{\perp}$ is a closed subspace of X.

(19) Carry out the calculations to find the function g_3 in Solved Problem 8.5(1).

(20) Let X be the vector space of all functions continuous on a closed interval $[a, b]$. Define a mapping $(\ \ |\ \ )$ on $X \times X$ by

$$(x \mid y) = \sum_{k=1}^{n} x(t_k)y(t_k), \qquad x, y \in X,$$

for some fixed $n \in \mathbf{N}$ and fixed points $t_1, t_2, \ldots, t_n \in [a, b]$.

    (a) Show that this does not define an inner product for X, but the equation

$$\nu(x) = \sqrt{(x \mid x)}, \qquad x \in X,$$

       defines a seminorm ν for X. (See Exercise 6.4(12).)

    (b) Define orthogonality with respect to $(\ \ |\ \ )$ on X as for inner products. Take $a = 0$, $b = 2\pi$ and prove that the set $\{1, \cos t, \cos 2t, \ldots, \cos(N-1)t\}$ in X is orthogonal if $n = 2N$ and $t_k = k\pi/N$ for $k = 1, 2, \ldots, n$. (Hint: Reduce the sum to a finite geometric series by use of the identities $\cos(A + B) + \cos(A - B) = 2\cos A \cos B$ and $e^{i\theta} = \cos\theta + i\sin\theta$.)

(Such seminorms and 'inner products' as in this exercise have considerable application in approximation theory.)

(21) The $n \times n$ matrix $(\langle x_j, x_k \rangle)$ is called the *Gram matrix* of the vectors $x_1, x_2, \ldots, x_n$ in an inner product space. Show that these vectors are linearly independent if and only if their Gram matrix has nonzero determinant. (Hint: See Theorem 8.3.4.)

9

Hilbert Space

9.1 Definition of Hilbert space

A Hilbert space has the same relationship to an inner product space as a Banach space has to a normed space. Since any inner product space is a normed space, the norm being determined by the inner product as in Section 8.1, nothing new is involved in the notion of completeness for an inner product space.

Definition 9.1.1 A complete inner product space is called a *Hilbert space*.

It follows that every Hilbert space is also a Banach space, though the converse is certainly not true: $C[a, b]$ is a Banach space but not a Hilbert space, since it is not even an inner product space. Any finite-dimensional inner product space is complete (Theorem 6.5.4), so all finite-dimensional inner product spaces are Hilbert spaces.

As a metric space, we have seen that $C_2[a, b]$ is not complete, and so it cannot be complete as an inner product space. That is, $C_2[a, b]$ is not a Hilbert space. As our main example of a Hilbert space, except for finite-dimensional ones like $\mathbf{C}^n$, we are thus left with the space l_2. There is an important analogue of the space $C_2[a, b]$, where the integral for the space is developed in a different manner from the usual (Riemann) integral, so that, as one consequence, the space turns out to be complete. The *Lebesgue integral* is an example of such an integral, but its treatment is beyond the scope of this book.

However, much of what follows will be valid for inner product spaces in general, and, although we have only one example here of an infinite-dimensional Hilbert space, there is plenty to discuss just with regard to l_2. This was in fact the space originally studied by David Hilbert, and it was attempts to generalise it that led to the notion of Hilbert space.

An essential part of any discussion of Hilbert space is the idea of *separability*. In Section 9.3, we will define what we mean by a separable Hilbert space. It turns out that, in a sense to be made clear, l_2 is in fact the only separable Hilbert space. The function space that we alluded to above, the analogue of $C_2[a, b]$, appears then as just a version of l_2. It was largely the realisation of this that showed that the original theories of quantum mechanics, Heisenberg's matrix formulation and Schrödinger's wave formulation, were equivalent.

9.2 The adjoint operator

In Section 8.4, we proved the Riesz representation theorem for finite-dimensional inner product spaces. Our first aim here is to give the corresponding theorem for Hilbert spaces. Then the knowledge that all (bounded) linear functionals on a Hilbert space have a specific form will gives rise to the highly important notion of an operator *adjoint* to a given operator.

Theorem 9.2.1 *If f is a bounded linear functional on a Hilbert space X, then there exists a unique point $v \in X$ such that $f(x) = \langle x, v \rangle$ for all $x \in X$.*

Note that here we require the functional to be bounded, whereas in the former case boundedness was a consequence of the Riesz theorem. In Section 7.3, we wrote X' to denote the space of all bounded linear functionals on the normed space X, and called it the dual space of X. This theorem therefore says that if X is a Hilbert space, then $f \in X'$ only if $f(x) = \langle x, v \rangle$ for some $v \in X$ and all $x \in X$. The converse of Theorem 9.2.1 is also true, its proof being similar to the argument following Theorem 8.4.2.

The proof of Theorem 9.2.1 is considerably more involved than that for Theorem 8.4.1. We give it in a number of steps.

(a) As in the proof of Theorem 8.4.1, we can rest assured that if any such point v can be found, then it will be unique.

(b) Recall (see Theorem 7.3.2) that the null space $N(f)$ of f is the subspace $\{x : x \in X, \ f(x) = 0\}$. Suppose $N(f) = X$. Then we may take $v = \theta$ since $f(x) = 0 = \langle x, \theta \rangle$ for all $x \in X$. Therefore, for the rest of the proof we suppose that $N(f) \neq X$.

(c) We need the following result. *Given a closed subspace S of a Hilbert space X and a point $x \in X$, there exists a point $p \in S$ such that $\|p - x\|$ is a minimum.* (This has separate applications in approximation theory. Compare it with Theorem 4.3.3.)

For the proof, we set $d = \inf_{y \in S} \|y - x\|$. By definition of greatest lower bound, there must be a sequence $\{y_n\}$ in S such that $\|y_n - x\| \to d$. By Theorem 8.1.3 (the parallelogram law), for any $m, n \in \mathbf{N}$,

$$\|(y_n - x) + (y_m - x)\|^2 + \|(y_n - x) - (y_m - x)\|^2$$
$$= 2\|y_n - x\|^2 + 2\|y_m - x\|^2,$$

from which

$$\|y_n - y_m\|^2 = 2\|y_n - x\|^2 + 2\|y_m - x\|^2 - 4\left\|\frac{y_n + y_m}{2} - x\right\|^2.$$

Since S is a subspace of X, we must have $\frac{1}{2}(y_n + y_m) \in S$, and therefore $\|\frac{1}{2}(y_n + y_m) - x\| \geq d$. Then

$$\|y_n - y_m\|^2 \leq 2\|y_n - x\|^2 + 2\|y_m - x\|^2 - 4d^2 < \epsilon$$

for any $\epsilon > 0$, provided m and n are large enough. Hence $\{y_n\}$ is a Cauchy sequence which, since X is complete (being a Hilbert space), must converge. Set $p = \lim y_n$. Since S is closed, we have $p \in S$. Finally, since $\| \ \|$ is a continuous mapping on X (Exercise 6.4(3)(c)),

$$\|p - x\| = \|\lim y_n - x\| = \lim \|y_n - x\| = d,$$

and our result is proved.

(d) Since f is bounded it is continuous, and so, by Theorem 7.3.2, the subspace $N(f)$ of X is closed. We have assumed $N(f) \neq X$. Choose a point $x \in X$ such that $x \notin N(f)$. By (c), there is a point $p \in N(f)$ such that $\|p - x\| = d = \min_{y \in N(f)} \|y - x\|$. Put $w = p - x$. We cannot have $w = \theta$, for then $p = x$; but $p \in N(f)$ and $x \notin N(f)$. We will show that $w \perp z$ for every $z \in N(f)$.

Take any particular point $z \in N(f)$, $z \neq \theta$, and any scalar α. Then $p + \alpha z \in N(f)$, and

$$\|w + \alpha z\| = \|(p + \alpha z) - x\| \geq d = \|w\|.$$

Therefore,

$$0 \leq \|w + \alpha z\|^2 - \|w\|^2 = \langle w + \alpha z, w + \alpha z \rangle - \langle w, w \rangle$$
$$= \overline{\alpha} \langle w, z \rangle + \alpha \langle z, w \rangle + |\alpha|^2 \|z\|^2.$$

In particular, with $\alpha = - \langle w, z \rangle / \|z\|^2$, we have

$$0 \leqslant -\frac{\overline{\langle w, z \rangle}}{\|z\|^2} \langle w, z \rangle - \frac{\langle w, z \rangle}{\|z\|^2} \langle z, w \rangle + \frac{|\langle w, z \rangle|^2}{\|z\|^4} \|z\|^2 = -\frac{|\langle w, z \rangle|^2}{\|z\|^2}.$$

Clearly, this can only be possible if $\langle w, z \rangle = 0$.

(e) To complete the proof of Theorem 9.2.1, we let w be a point in X such that $w \neq \theta$ and $w \perp z$ for every $z \in N(f)$. This is in accord with (d). Put

$$v = \frac{\overline{f(w)}w}{\|w\|^2}.$$

It is a matter of verifying now that $f(x) = \langle x, v \rangle$ for all $x \in X$. We need to observe that, for any $x \in X$, we have $f(x)w - f(w)x \in N(f)$, since

$$f(f(x)w - f(w)x) = f(x)f(w) - f(w)f(x) = 0.$$

Then $w \perp (f(x)w - f(w)x)$, or

$$0 = \langle f(x)w - f(w)x, w \rangle = f(x)\|w\|^2 - f(w)\langle x, w \rangle.$$

It follows that

$$f(x) - \langle x, v \rangle = f(x) - \left\langle x, \frac{\overline{f(w)}w}{\|w\|^2} \right\rangle = f(x) - \frac{f(w)}{\|w\|^2} \langle x, w \rangle = 0,$$

and the proof is finished.   □

The notion of an adjoint operator is arrived at as follows.

We take any operator A mapping a Hilbert space X into a Hilbert space Y. Thus $A \in B(X, Y)$, as defined in Section 7.2. To avoid confusion, we will write the inner products for X and Y as $\langle \ , \ \rangle_X$ and $\langle \ , \ \rangle_Y$, respectively. Let y be an arbitrary fixed point in Y and define a functional f on X by

$$f(x) = \langle Ax, y \rangle_Y, \qquad x \in X.$$

It is easy to check that f is linear. Moreover, f is bounded since

$$|f(x)| = |\langle Ax, y \rangle_Y| \leqslant \|Ax\| \, \|y\| \leqslant (\|A\| \, \|y\|)\|x\|,$$

using the Cauchy–Schwarz inequality (Theorem 8.1.2). Then Theorem 9.2.1, above, may be employed: there must exist a unique point $v \in X$ such that $f(x) = \langle x, v \rangle_X$ for all $x \in X$. Hence we can write

$$\langle Ax, y \rangle_Y = \langle x, v \rangle_X$$

for some $v \in X$ and all $x \in X$. The point $v \in X$ is determined by the choice of the point $y \in Y$. Let A^* be the mapping from Y into X which associates $v \in X$ with $y \in Y$. That is, $A^* : Y \to X$ is defined by $A^*y = v$, where $\langle Ax, y \rangle_Y = \langle x, v \rangle_X$ for all $x \in X$. This mapping A^* is called the *adjoint* of the operator A. We will repeat this below, after showing that A^* is also linear and bounded.

For $y_1, y_2 \in Y$, suppose $A^*y_1 = v_1$, $A^*y_2 = v_2$. Take $\alpha_1, \alpha_2 \in \mathbf{C}$ and set $A^*(\alpha_1 y_1 + \alpha_2 y_2) = w$. Then, for any $x \in X$,

$$
\begin{aligned}
\langle x, w \rangle_X &= \langle Ax, \alpha_1 y_1 + \alpha_2 y_2 \rangle_Y \\
&= \overline{\alpha}_1 \langle Ax, y_1 \rangle_Y + \overline{\alpha}_2 \langle Ax, y_2 \rangle_Y \\
&= \overline{\alpha}_1 \langle x, v_1 \rangle_X + \overline{\alpha}_2 \langle x, v_2 \rangle_X \\
&= \langle x, \alpha_1 v_1 + \alpha_2 v_2 \rangle_X ,
\end{aligned}
$$

so that, as in the uniqueness part of the proof of Theorem 8.4.1, we have $w = \alpha_1 v_1 + \alpha_2 v_2$. That is,

$$
A^*(\alpha_1 y_1 + \alpha_2 y_2) = \alpha_1 A^* y_1 + \alpha_2 A^* y_2
$$

so A^* is linear. To show that A^* is bounded, we reintroduce the functional f above. As we did following the proof of Theorem 8.4.2, we can show that $\|f\| = \|v\|$. Also, from the above, we have $\|f\| \leqslant \|A\| \|y\|$, where $A^*y = v$. Hence, for any $y \in Y$,

$$
\|A^*y\| = \|v\| = \|f\| \leqslant \|A\| \|y\|.
$$

This shows that A^* is bounded and that $\|A^*\| \leqslant \|A\|$.

We are therefore justified in calling A^* an operator in the following definition.

Definition 9.2.2 If X and Y are Hilbert spaces, the *adjoint* of the operator $A \in B(X, Y)$ is the operator $A^* \in B(Y, X)$ determined by the equation

$$
\langle Ax, y \rangle_Y = \langle x, A^*y \rangle_X , \qquad x \in X, \ y \in Y.
$$

When $Y = X$, we call the operator A *self-adjoint* if $A^* = A$.

For a self-adjoint operator A on a Hilbert space X it is clear that $\langle Ax, y \rangle_X = \langle x, Ay \rangle_X$ for all $x, y \in X$.

Just above, we showed that $\|A^*\| \leqslant \|A\|$ for any operator A between Hilbert spaces. We will show now that in fact $\|A^*\| = \|A\|$. By A^{**} below, we mean the adjoint of the operator A^*.

Theorem 9.2.3 *For any operator* A *between Hilbert spaces, we have* $A^{**} = A$ *and* $\|A^*\| = \|A\|$.

Let A map X into Y. Using the definition of an adjoint operator twice, we have

$$\langle y, Ax \rangle_Y = \overline{\langle Ax, y \rangle}_Y = \overline{\langle x, A^*y \rangle}_X = \langle A^*y, x \rangle_X = \langle y, A^{**}x \rangle_Y \,,$$

for all $x \in X$ and $y \in Y$. Hence $Ax = A^{**}x$ for all $x \in X$, so $A^{**} = A$, as required. Furthermore, as well as the inequality $\|A^*\| \leqslant \|A\|$, we now also have $\|A\| = \|A^{**}\| \leqslant \|A^*\|$, so $\|A^*\| = \|A\|$.   □

A vast amount of theory has been developed for adjoint operators, self-adjoint operators and associated concepts. In particular, the ideas have been extended to include unbounded mappings. As an indication of the need for this theory, we mention that in quantum mechanics, for example, all the mappings that are associated with observable quantities are self-adjoint.

We have previously referred briefly to the eigenvalues of a mapping: for any linear mapping A from a vector space X into itself, a scalar λ is an *eigenvalue* of A if there is a nonzero vector $x \in X$ such that $Ax = \lambda x$; and then x is an *eigenvector* of A corresponding to the eigenvalue λ. If the mapping $A - \lambda I$, where as usual I is the identity mapping on X, is onto, then it follows from Theorem 7.6.2 that the mapping $(A - \lambda I)^{-1}$ exists if and only if λ is not an eigenvalue of A.

We can quickly obtain some useful information on the eigenvalues and eigenvectors of a self-adjoint operator on a Hilbert space.

Theorem 9.2.4 *Let* A *be a self-adjoint operator on a Hilbert space. Then*

(a) *the eigenvalues of* A *are real,*

(b) *eigenvectors of* A *corresponding to distinct eigenvalues are orthogonal.*

Let the Hilbert space here be X. To prove (a), suppose λ is an eigenvalue of A, so that $Ax = \lambda x$ for some $x \in X$, $x \neq \theta$. Since A is self-adjoint, we have $\langle Ax, y \rangle = \langle x, Ay \rangle$ for all $y \in X$, and in particular this is true when $y = x$. Then

$$\lambda \langle x, x \rangle = \langle \lambda x, x \rangle = \langle Ax, x \rangle = \langle x, Ax \rangle = \langle x, \lambda x \rangle = \overline{\lambda} \langle x, x \rangle .$$

As $x \neq \theta$, we have $\lambda = \overline{\lambda}$ and so λ must be real.

For (b), we suppose λ and μ are eigenvalues of A, with $\lambda \neq \mu$, and that x and y, respectively, are corresponding eigenvectors. Then

$$\lambda \langle x, y \rangle = \langle \lambda x, y \rangle = \langle Ax, y \rangle = \langle x, Ay \rangle = \langle x, \mu y \rangle = \overline{\mu} \langle x, y \rangle.$$

Now, $\mu = \overline{\mu}$ by (a), and $\lambda \neq \mu$, so we must have $\langle x, y \rangle = 0$. This completes the proof. □

We end this section by deriving the adjoint operator for any matrix operator on $\mathbf{C}^n$.

Consider the elements of $\mathbf{C}^n$ to be column vectors and let the operator $A \colon \mathbf{C}^n \to \mathbf{C}^n$ be given by the $n \times n$ matrix $\mathcal{A} = (a_{jk})$. (It is convenient here to depart from our usual practice and denote the matrix differently from the operator.) We will show that the adjoint A^* is given by the $n \times n$ matrix $\mathcal{A}^* = \overline{\mathcal{A}}^T$. Here, $\overline{\mathcal{A}}$ is the conjugate of $\mathcal{A}$, defined in Section 1.11.

With the above interpretation of the elements of $\mathbf{C}^n$, it is clear that, for $x, y \in \mathbf{C}^n$,

$$\langle x, y \rangle = x^T \overline{y}.$$

Then, since $Ax = \mathcal{A}x$,

$$\langle Ax, y \rangle = \langle \mathcal{A}x, y \rangle = (\mathcal{A}x)^T \overline{y} = x^T \mathcal{A}^T \overline{y}$$

$$= x^T \overline{\overline{\mathcal{A}}^T} y = x^T \overline{\mathcal{A}^* y} = \langle x, \mathcal{A}^* y \rangle.$$

But $\langle Ax, y \rangle = \langle x, A^* y \rangle$ by definition of A^*, so $A^* y = \mathcal{A}^* y$ for all $y \in \mathbf{C}^n$, and hence A^* is determined by the matrix $\mathcal{A}^*$.

The matrix $\mathcal{A}$ above is called *Hermitian* when $\mathcal{A}^* = \mathcal{A}$. In that case, the corresponding operator A is self-adjoint. In the following paragraph, we will illustrate the preceding theorem by taking

$$\mathcal{A} = \begin{pmatrix} 2 & 1-i & i \\ 1+i & 1 & 0 \\ -i & 0 & 1 \end{pmatrix},$$

which is clearly seen to be Hermitian.

The eigenvalues and eigenvectors of an operator determined by a matrix are taken as belonging to the matrix itself, so that their definitions then coincide with those given in linear algebra courses. We want to find those scalars λ such that $\mathcal{A}x = \lambda x$ for some nonzero $x \in \mathbf{C}^3$. Writing this equation as $(\mathcal{A} - \lambda I)x = \theta$, where I here is the 3×3 identity matrix, we therefore employ one of the conditions that a homogeneous system of

9 Hilbert Space

linear equations have a nontrivial solution, namely that the coefficient matrix $\mathcal{A} - \lambda I$ have determinant 0. That is,

$$\begin{vmatrix} 2 - \lambda & 1 - i & i \\ 1 + i & 1 - \lambda & 0 \\ -i & 0 & 1 - \lambda \end{vmatrix} = 0,$$

or, expanding the determinant,

$$\lambda^3 - 4\lambda^2 + 2\lambda + 1 = 0.$$

The roots of this equation are 1 and $(3 \pm \sqrt{13})/2$, so that these are the eigenvalues of $\mathcal{A}$. Note that they are all real, in accord with Theorem 9.2.4(a). For the eigenvectors, we find in turn nontrivial solutions of the equation $(\mathcal{A} - \lambda I)x = \theta$, where λ has each of the three values just given. It may be checked that we may take

$$x_1 = \begin{pmatrix} 0 \\ 1 \\ 1 + i \end{pmatrix}, \quad x_2 = \begin{pmatrix} \frac{1}{2}(\sqrt{13} + 1) \\ 1 + i \\ -i \end{pmatrix}, \quad x_3 = \begin{pmatrix} \frac{1}{2}(\sqrt{13} - 1) \\ -(1 + i) \\ i \end{pmatrix},$$

as eigenvectors corresponding to the eigenvalues 1, $(3 + \sqrt{13})/2$ and $(3 - \sqrt{13})/2$, respectively. Note that, for $j \neq k$, $\langle x_j, x_k \rangle = x_j^T \overline{x}_k = 0$, in accord with Theorem 9.2.4(b).

The preceding discussion applies equally well, with simplifications, to matrix operators on $\mathbf{R}^n$. The matrix for the adjoint of such an operator is then just the transpose of the original matrix. A matrix operator on $\mathbf{R}^n$ is self-adjoint if and only if its matrix is *symmetric*, that is, equal to its transpose.

9.3 Separability

Before we can go further in a discussion of Hilbert space, we need the notion of *separability*. We will define this term for topological spaces. Since any inner product space is a normed space, any normed space is a metric space, and, by virtue of the metric topology, any metric space is a topological space, we will be able to carry the definition through to inner product spaces. We need to recall the definition of the closure of a subset S of a topological space (Definition 5.1.3): the closure of S is the intersection of all closed sets that contain S. Properties of the closure of a set were given in Exercises 5.7(5) and 5.7(6).

We also recall, from Definition 2.7.2, that S is a sequentially closed subset of a metric space X if S is a nonempty subset of X such that all sequences in S that converge as sequences in X have their limits also in S. It is an easy matter conceptually to make S closed if it is not already closed: simply add to S the limits of those sequences in it that converge to points not in it. The result is the closure of S. This process is simply making use of results in the above-mentioned exercises: $\overline{S} = S \cup S'$, and S is closed if and only if $S = \overline{S}$.

The relevant definitions will now be given in topological space, and then quickly related to metric space.

Definition 9.3.1

(a) A subset S of a topological space X is said to be *dense* in X if $\overline{S} = X$.

(b) A topological space X is called *separable* if there is a subspace of X which is countable and dense in X.

Roughly speaking, a subset S of a metric space X is therefore dense in X if it consists of all of X except at most for the limits of sequences in S which converge in X. We say that the rationals are dense in the reals because any irrational number can always be given as the limit of a sequence of rational numbers, or, equivalently, because there always exists a rational number arbitrarily close to any given irrational number. This is made more precise, and more general, in the following theorem.

Theorem 9.3.2 *A subset S of a metric space (X, d) is dense in X if and only if for every $x \in X$ and every number $\epsilon > 0$ there exists a point $y \in S$ such that $d(x, y) < \epsilon$.*

For the proof, suppose first that S is dense in X. Choose any $x \in X$ and any $\epsilon > 0$. If $x \in S$, then simply take $y = x$. Otherwise, since $X = \overline{S} = S \cup S'$, x is a cluster point of S so certainly there exists $y \in S$ such that $d(x, y) < \epsilon$. For the converse, given $x \in X$, take $\epsilon = 1/n$ for each $n = 1, 2, \ldots$ in turn, and so generate a sequence $\{y_n\}$ in S such that $d(x, y_n) < 1/n$. Then $y_n \to x$, so $x \in S \cup S' = \overline{S}$. That is, $X \subseteq \overline{S}$, so S is dense in X. $\qquad\square$

Now we can turn to examples of separable spaces. These all depend on two facts: the set $\mathbf{Q}$ of rational numbers is countable (Theorem 1.4.3(a)), and the cartesian product of any finite number of countable sets is again a countable set (Theorem 1.4.2(a)).

We stated just above that the rationals are dense in the reals; furthermore, the rationals are countable. This is all that is required for our first example: the metric space (normed space, inner product space) $\mathbf{R}$ is separable. It has a countable dense subset, namely $\mathbf{Q}$. The space $\mathbf{R}^n$ is separable, since the set of all points $(x_1, x_2, \ldots, x_n) \in \mathbf{R}^n$, in which each x_k is rational, is a countable dense subset of $\mathbf{R}^n$. (This subset is just $\mathbf{Q}^n$.) Also, the space $\mathbf{C}^n$ is separable, since the set of all points $(x_1, x_2, \ldots, x_n) \in \mathbf{C}^n$, in which each x_k has rational real and imaginary parts, is a countable dense subset of $\mathbf{C}^n$. (The proofs that the indicated subsets are dense are similar to the proof below in the case of l_2.)

By invoking the special form of the Weierstrass approximation theorem given in Theorem 6.8.4, we can prove that the metric space (or the normed space) $C[a, b]$ is separable. A countable dense subset is the set of all polynomial functions on $[a, b]$ with rational coefficients. The details are left as an exercise.

Finally, we show that the metric space l_2 is separable. As above, we must exhibit a countable dense subset of l_2. We show that such a subset is the set S of all sequences $(y_1, y_2, \ldots)$ of complex numbers in which each y_k has rational real and imaginary parts and for which there is some positive integer m (depending on $y \in S$) such that $y_{m+1} = y_{m+2} = \cdots = 0$. This set S is indeed a subset of l_2, since

$$\sum_{k=1}^{\infty} |y_k|^2 = \sum_{k=1}^{m} |y_k|^2,$$

and this is certainly finite. Let $x = (x_1, x_2, \ldots)$ be any point of l_2. Then, by definition of l_2, for any number $\epsilon > 0$ there exists a positive integer n so that

$$\sum_{k=n+1}^{\infty} |x_k|^2 < \frac{\epsilon^2}{2}.$$

Because the rationals are dense in the reals, it follows that we can find a point $y \in S$ such that

$$|x_k - y_k| < \frac{\epsilon}{\sqrt{2n}}$$

for $k = 1, 2, \ldots, n$, with $y_{n+1} = y_{n+2} = \cdots = 0$. We then have, if d is the metric for l_2,

$$d(x, y) = \sqrt{\sum_{k=1}^{\infty} |x_k - y_k|^2} = \sqrt{\sum_{k=1}^{n} |x_k - y_k|^2 + \sum_{k=n+1}^{\infty} |x_k|^2}$$

$$< \sqrt{n \cdot \frac{\epsilon^2}{2n} + \frac{\epsilon^2}{2}} = \epsilon.$$

By Theorem 9.3.2, thus S is dense in l_2. Furthermore, S is countable and so we have proved the following.

Theorem 9.3.3 *The Hilbert space l_2 is separable.*

9.4 Solved problems

(1) Let A be a linear mapping from an inner product space X into itself. Prove that

$$\langle x, Ay\rangle + \langle y, Ax\rangle = \frac{1}{2}(\langle x + y, A(x + y)\rangle - \langle x - y, A(x - y)\rangle)$$

and

$$\langle x, Ay\rangle - \langle y, Ax\rangle = \frac{i}{2}(\langle x + iy, A(x + iy)\rangle - \langle x - iy, A(x - iy)\rangle)$$

for any vectors $x, y \in X$.

Solution. The identities follow readily by expanding the right-hand sides. For the second one, for example, we have

$$\begin{aligned}
\langle x+iy, &A(x + iy)\rangle - \langle x - iy, A(x - iy)\rangle \\
&= \langle x + iy, Ax + iAy\rangle - \langle x - iy, Ax - iAy\rangle \\
&= \langle x, Ax\rangle - i\langle x, Ay\rangle + i\langle y, Ax\rangle + \langle y, Ay\rangle \\
&\quad - (\langle x, Ax\rangle + i\langle x, Ay\rangle - i\langle y, Ax\rangle + \langle y, Ay\rangle) \\
&= -2i(\langle x, Ay\rangle - \langle y, Ax\rangle). \qquad \square
\end{aligned}$$

(2) Show that the following conditions on an operator A on a Hilbert space X are equivalent:

  (a) A is self-adjoint,
  (b) $\langle Ax, x\rangle = \langle x, Ax\rangle$ for all $x \in X$,
  (c) for all $x \in X$, $\langle Ax, x\rangle$ is real.

Solution. We show that the three statements are equivalent by showing that, schematically, (a) $\Rightarrow$ (b) $\Rightarrow$ (c) $\Rightarrow$ (a). (When this is done, either (b) or (c) may be taken as a necessary and sufficient condition for the operator A to be self-adjoint.)

If A is self-adjoint, then $\langle Ax, y\rangle = \langle y, Ax\rangle$ for all $x, y \in X$, so in particular, when $y = x$, we have $\langle Ax, x\rangle = \langle x, Ax\rangle$ for all $x \in X$. Thus (a) implies (b).

By definition of an inner product, for any $x \in X$, $\overline{\langle Ax, x\rangle} = \langle x, Ax\rangle$.

If, further, (b) is true, then we have $\overline{\langle Ax, x \rangle} = \langle Ax, x \rangle$, and so $\langle Ax, x \rangle$ is a real number. Thus (b) implies (c).

The final step is not quite as easy. The right-hand sides of the equations in Solved Problem (1) contain complex numbers of the form $\langle z, Az \rangle$ for some $z \in X$. In each case, when we assume (c) to be true, we have $\langle z, Az \rangle = \overline{\langle Az, z \rangle} = \langle Az, z \rangle$. Then we see that those right-hand sides are unchanged by interchanging x and Ax, and y and Ay. The same must be true of the left-hand sides, so that, for all $x, y \in X$,

$$\langle x, Ay \rangle + \langle y, Ax \rangle = \langle Ax, y \rangle + \langle Ay, x \rangle,$$
$$\langle x, Ay \rangle - \langle y, Ax \rangle = \langle Ax, y \rangle - \langle Ay, x \rangle.$$

Adding these equations gives $\langle x, Ay \rangle = \langle Ax, y \rangle$, so (c) implies (a).   □

(3) Let S be a subspace of a Hilbert space X. Suppose $\langle x, y \rangle = 0$ for all $x \in S$ only when $y = \theta$. Prove that S is dense in X. Conversely, suppose S is dense in X. Prove that if $\langle x, y \rangle = 0$ for all $x \in S$, then $y = \theta$, uniquely.

Solution. We suppose first that $\langle x, y \rangle = 0$ for all $x \in S$ only when $y = \theta$. The proof that S is then dense in X will be by contradiction, so suppose that $\overline{S} \neq X$. Since S is a subspace of X, then $\overline{S}$ is also a subspace of X by Exercise 9.5(5) (to be proved). Part (d) in the proof of Theorem 9.2.1 applies equally well for any closed subspace of X which is a proper subset of X (like $N(f)$ there, and $\overline{S}$ here), so we may conclude here that there exists a nonzero point $w \in X$ such that $\langle x, w \rangle = 0$ for all $x \in S$. This contradicts the hypothesis, so S is dense in X, as required.

For the converse, suppose S is dense in X, so $\overline{S} = X$. Suppose also that $\langle x, y \rangle = 0$ for all $x \in S$ and some $y \in X$. Any point $z \in X$ is the limit of some sequence $\{z_n\}$ in S. Now, $\langle z_n, y \rangle = 0$ for all n, so $\langle z, y \rangle = 0$ by Exercise 8.6(7). That is, $\langle z, y \rangle = 0$ for all $z \in X$. In particular, when $z = y$, we have $\langle y, y \rangle = 0$ so $y = \theta$, and the result is proved.   □

9.5 Exercises

(1) Prove that the identity operator on a Hilbert space is self-adjoint.

(2) If A is any operator from a Hilbert space into itself, and A^* is its adjoint, prove that the operators A^*A, $A + A^*$ and $i(A - A^*)$ are self-adjoint.

(3) Let X be a Hilbert space and A and B operators from X into itself. Prove that

    (a) $(AB)^* = B^*A^*$,

    (b) if A and B are self-adjoint then AB is self-adjoint if and only if $AB = BA$.

(4) Verify that the matrix

$$\begin{pmatrix} 1 & \frac{2}{3}i\sqrt{3} & 0 \\ -\frac{2}{3}i\sqrt{3} & 0 & 1 \\ 0 & 1 & 2 \end{pmatrix}$$

is Hermitian, and find its eigenvalues and corresponding eigenvectors.

(5) Prove that the closure of a subspace of a normed space is also a subspace of that space.

(6) Show that the metric space (X, d) is separable, when $X = \mathbf{R}^n$ and $d(x, y) = \sum_{k=1}^{n} |x_k - y_k|$, with $x = (x_1, x_2, \ldots, x_n) \in \mathbf{R}^n$, $y = (y_1, y_2, \ldots, y_n) \in \mathbf{R}^n$.

(7) Prove that the normed space $C[a, b]$ is separable. (Hint: Use Theorem 6.8.4.)

(8) Define operators A and B from l_2 into itself by

$$A(x_1, x_2, \ldots) = (x_2, x_3, x_4, \ldots),$$
$$B(x_1, x_2, \ldots) = (x_1, \tfrac{1}{2}x_2, \tfrac{1}{3}x_3, \ldots).$$

    (a) Show that any number λ with $|\lambda| < 1$ is an eigenvalue of A. Find corresponding eigenvectors.

    (b) Find the adjoint A^* and show that A^* has no eigenvalues.

    (c) Show that B is self-adjoint. Find its eigenvalues and corresponding eigenvectors and show that Theorem 9.2.4 is satisfied.

....

(9) Prove that the set of self-adjoint operators on a Hilbert space X is a real vector space which is a closed subset of X'.

(10) Prove that, for any operator A from a Hilbert space X into itself,

$$\|AA^*\| = \|A^*A\| = \|A\|^2.$$

(Hint: Show that $\|Ax\|^2 \leqslant \|A^*A\|\,\|x\|^2$ for any $x \in X$ and recall that $\|A\| = \sup\{\|Ax\| : x \in X, \|x\| = 1\}$.)

9.6 Complete orthonormal sets; generalised Fourier series

Whenever we have spoken of a linear combination of vectors in a vector space, we have quite explicitly been referring only to finite linear combinations; that is, linear combinations of only a finite number of vectors. To do otherwise immediately means that we are dealing with infinite series of vectors and this has certainly not been the case in this context. The possibility of infinite linear combinations cannot even arise within the theory of vector spaces alone since we cannot talk of infinite series without some concept of convergence, and this requires that a norm or something similar be defined for the space.

This has nothing to do with whether or not the vector space is finite-dimensional. We have already extended to infinite sets the notions of linear independence and span (see Definition 8.2.2), so it is easy now to extend also the definition of a *basis* (Definition 1.11.3(d)): an infinite set that is linearly independent and spans a vector space may be called a basis for that space. It is still only finite linear combinations that are involved. For example, the set of functions $\{1, x, x^2, \dots\}$ is then a basis for the vector space of all polynomial functions. Such a function is a linear combination of only finitely many of 1, x, x^2, ...

However, for other infinite-dimensional vector spaces, such as $C[a, b]$ or l_2, the situation may not be as clear. In the case of l_2, any further discussion would appear to be prompted by the fact that for $\mathbf{C}^n$ the set of n-tuples $\{(1, 0, \dots, 0), (0, 1, \dots, 0), \dots, (0, \dots, 0, 1)\}$ is a basis. This suggests that we ask if the set

$$E = \{(1, 0, 0, \dots), (0, 1, 0, \dots), (0, 0, 1, 0, \dots), \dots\}$$

is a basis for l_2. We quickly see that this is not the case, because a vector such as $(1, \frac{1}{2}, \frac{1}{3}, \frac{1}{4}, \dots)$, which is in l_2 and has infinitely many nonzero components, could not be given as a finite linear combination of vectors in E. It hardly seems likely that we could easily find any other set that would be a basis for l_2.

If we write e_n for the element of E with 1 in the nth place, and allow the notion of infinite linear combinations, then it is easily checked that with the norm for l_2 the point $(1, \frac{1}{2}, \frac{1}{3}, \frac{1}{4}, \dots)$ can be expressed as $\sum_{k=1}^{\infty}(1/k)e_k$. Similarly, any point $(x_1, x_2, \dots)$ in l_2 can be written as $\sum_{k=1}^{\infty} x_k e_k$. A proof of this is called for in Exercise 9.8(1).

It is the aim of this section to generalise this idea. We will obtain necessary and sufficient conditions for the existence of this kind of 'basis', which involves infinite linear combinations, when the vector space is a separable inner product space. We stress that this will not be a basis in the original sense, since that term implies reference to finite linear combinations only. Along with that, we will obtain various properties of such a 'basis' and additional properties available when the space is complete.

We will not be able to give a corresponding development here for the space $C[a, b]$, since it is not an inner product space. It should be mentioned that for $C[a, b]$, and for any infinite-dimensional vector space, it can be proved that a basis (in this context, known as a *Hamel base*) does exist, just as $1, x, x^2, \ldots$ give a basis for the space of polynomial functions. Such bases, even if they could be exhibited, would be of little use since they would surely be too complicated for practical purposes.

Although we cannot handle $C[a, b]$ here, we can find a 'basis', allowing infinite linear combinations, when we consider continuous functions on $[a, b]$ as belonging to $C_2[a, b]$. We will carry this out soon, for the case $a = -1$, $b = 1$, when we have described our aim more precisely. It is apparent that these 'bases' we are talking of depend on the inner product for the space. The 'basis' that we will find for $C_2[a, b]$ will not work for the other spaces $C^w[a, b]$ of the preceding chapter. These are defined on the same vector space but have different inner products, dependent on the weight function w. And certainly it cannot be considered as a 'basis' with the norm of $C[a, b]$.

The following definition is adopted as our starting point.

Definition 9.6.1 Let T be an orthonormal set in an inner product space X. If $\operatorname{Sp} T$ is dense in X, that is, if $\overline{\operatorname{Sp} T} = X$, then the set T is said to be *complete*.

Of course, this use of the word 'complete' is quite distinct from its earlier use. A complete orthonormal set is sometimes referred to as a *complete orthonormal system*, or as an *orthonormal basis*. We will avoid the latter term since it suggests an ordinary basis that happens to be orthonormal. However, as we now show, this is in fact precisely the case when T is a finite set.

Theorem 9.6.2 *A finite complete orthonormal set in an inner product space is a basis for the space.*

To prove this, let T be a finite complete orthonormal set in an inner product space X. Then $\operatorname{Sp} T$ is a finite-dimensional subspace of X, so $\operatorname{Sp} T$ is complete (Theorem 6.5.4), the norm for X being generated by the inner product in the usual way. Hence $\operatorname{Sp} T$ is closed (Theorem 2.7.3), so $\operatorname{Sp} T = \overline{\operatorname{Sp} T}$. Since the orthonormal set T is complete, this means $\operatorname{Sp} T = X$. Finally, by Theorem 8.2.3, it follows that T is a linearly independent set, and so T is a basis for X, as we said.　　　□

This case, when T is a finite set, is not very interesting since it involves us in nothing new. When T is not a finite set, but is countable say, then indeed we must take into account the infinite linear combinations mentioned above. That is, infinite series must be considered. For suppose $T = \{x_1, x_2, \ldots\}$ and $\alpha_1, \alpha_2, \ldots$ is any given sequence of scalars. Then $y_n = \sum_{k=1}^{n} \alpha_k x_k$ belongs to $\operatorname{Sp} T$ for any $n \in \mathbf{N}$. Because we are looking at the closure of $\operatorname{Sp} T$, we must consider the limits of all sequences in $\operatorname{Sp} T$, and in the case of the sequence $\{y_n\}$ this is just an infinite series.

The set $E = \{e_1, e_2, \ldots\}$, considered above, is an orthonormal set in l_2. The fact that for any point $x = (x_1, x_2, \ldots) \in l_2$ we may write $x = \sum_{k=1}^{\infty} x_k e_k$ means that E is a complete orthonormal set in l_2.

Now we can consider in more detail the space $C_2[-1, 1]$. Let f be a given function in $C_2[-1, 1]$. By the Weierstrass approximation theorem (Theorem 6.8.3), we know that, given any $n \in \mathbf{N}$, we can find a positive integer m (depending on n) and numbers $a_{0n}, a_{1n}, \ldots, a_{mn}$ such that

$$|f(t) - (a_{0n} + a_{1n}t + \cdots + a_{mn}t^m)| < \frac{1}{n}$$

for all $t \in [-1, 1]$. Squaring both sides of this inequality and then integrating from -1 to 1 gives

$$\int_{-1}^{1} (f(t) - (a_{0n} + a_{1n}t + \cdots + a_{mn}t^m))^2 \, dt < \frac{2}{n^2}.$$

It is clear from the definition of the Legendre polynomials, in Section 8.2, that the powers t^k here can be expressed as linear combinations of $P_0(t)$, $P_1(t)$, $\ldots$, $P_k(t)$ for $k = 1, 2, \ldots, m$. If P denotes the set of Legendre polynomials, then we have shown that for each $n \in \mathbf{N}$ there exists a polynomial function $Q_n \in \operatorname{Sp} P$ such that

$$\|f - Q_n\| = \sqrt{\int_{-1}^{1} (f(t) - Q_n(t))^2 \, dt} < \frac{\sqrt{2}}{n}.$$

The sequence $\{Q_n\}$ thus converges in $C_2[-1, 1]$, and $\lim Q_n = f$. Hence, $f \in \overline{\operatorname{Sp} P}$ and so $\overline{\operatorname{Sp} P} = C_2[-1, 1]$. Since P is an orthonormal set in $C_2[-1, 1]$, it is therefore a complete orthonormal set.

The set P of Legendre polynomials is certainly a countable set, so, as one consequence of the *Fourier series theorem* to be given soon, there exist real numbers $\alpha_0, \alpha_1, \ldots$ such that $f = \sum_{k=0}^{\infty} \alpha_k P_k$. That theorem will also show that the numbers α_k are unique and easily computable. Thus we will have a situation in $C_2[-1, 1]$ analogous to that in l_2, where we can write $x = \sum_{k=1}^{\infty} x_k e_k$ for any $x = (x_1, x_2, \ldots) \in l_2$. We see now more explicitly the point of this section. Once we have a complete countable orthonormal set in an inner product space, it will turn out to be a simple matter to express any point in the space as an infinite linear combination of the vectors in the orthonormal set. The partial sums of such series provide handy approximations of those points.

We have just found a complete orthonormal set in $C_2[-1, 1]$. The existence of such a set in general is indicated in the following theorem.

Theorem 9.6.3 *An inner product space is separable if and only if it contains a complete orthonormal set that is countable.*

We prove the necessity of the condition first.

Suppose the inner product space X is separable. Then, by definition, X contains a countable dense subset, S say. There exists a subset S_0 of S such that S_0 is linearly independent and such that $\mathrm{Sp}\, S_0 = \mathrm{Sp}\, S$. Certainly, S_0 is countable. The Gram–Schmidt process (Theorem 8.2.4) assures us that there is a countable orthonormal set T in X for which $\mathrm{Sp}\, T = \mathrm{Sp}\, S_0$. To say S is dense in X means $\overline{S} = X$. Now, $S \subseteq \mathrm{Sp}\, S$ and $\mathrm{Sp}\, S \subseteq X$, so $X = \overline{S} \subseteq \overline{\mathrm{Sp}\, S} \subseteq \overline{X} = X$. Thus $\overline{\mathrm{Sp}\, S} = X$ and hence $\overline{\mathrm{Sp}\, T} = X$. As required, X contains a complete orthonormal set, namely T, that is countable.

Now, for the sufficiency, suppose $T = \{x_1, x_2, \ldots\}$ is a countable complete orthonormal set in X. Let S be the set of all finite linear combinations $\sum_{k=1}^{n} \alpha_k x_k$ of vectors in T for which $\mathrm{Re}\, \alpha_k$ and $\mathrm{Im}\, \alpha_k$ are rational numbers for all $k = 1, 2, \ldots, n$. Then S is countable because $\mathbf{Q}$ is. We will show that S is dense in X, thus proving that X is separable. Let w be any vector in X and let $\epsilon > 0$ be arbitrary. Since T is a complete orthonormal set in X, $X = \overline{\mathrm{Sp}\, T}$ and so for some positive integer m and some scalars $\beta_1, \beta_2, \ldots, \beta_m$ there is a vector $x = \sum_{k=1}^{m} \beta_k x_k \in \mathrm{Sp}\, T$ such that $\|w - x\| < \epsilon/2$. Fix this m, and choose, as we may do, complex numbers $\alpha_1, \alpha_2, \ldots, \alpha_m$ with rational real and imaginary parts such that

$$|\beta_k - \alpha_k| < \frac{\epsilon}{2m} \quad \text{for } k = 1, 2, \ldots, m.$$

Then $y = \sum_{k=1}^{m} \alpha_k x_k$ is a vector in S and

$$\|w - y\| \leqslant \|w - x\| + \|x - y\| = \|w - x\| + \left\| \sum_{k=1}^{m} (\beta_k - \alpha_k) x_k \right\|$$

$$< \frac{\epsilon}{2} + \sum_{k=1}^{m} |\beta_k - \alpha_k| \, \|x_k\| < \epsilon,$$

since $\|x_k\| = 1$ for all k. This indeed proves that S is a dense subset of X, and our proof is finished. $\qquad\square$

Thus we see that the notions of separability and countability of a complete orthonormal set are equivalent concepts in an inner product space, and thus in a separable inner product space it makes sense at least formally to talk of an infinite linear combination of the vectors of a complete orthonormal set. What we want to show is that any vector in the space can be expressed as such an infinite linear combination.

As discussed above, this will be one consequence of the Fourier series theorem. The connection with the more familiar theory of Fourier series, and the fact that we are giving a generalisation, becomes apparent when it is realised that in the older theory we take a function f, with domain $\mathbf{R}$ and periodic with period 2π, and try to express it as an infinite linear combination of the functions in the set $T = \{1, \sin t, \cos t, \sin 2t, \cos 2t, \dots\}$ by an equation of the form

$$f(t) = \frac{1}{2} a_0 + \sum_{k=1}^{\infty} (a_k \cos kt + b_k \sin kt).$$

The set T, with its functions restricted to $[-\pi, \pi]$, has been shown before to be orthogonal in $C_2[-\pi, \pi]$. Just as we did for the Legendre polynomials in $C_2[-1, 1]$, we can show that it is a complete orthonormal set in $C_2[-\pi, \pi]$, once its elements are normalised. To do this requires a trigonometric version of the Weierstrass approximation theorem, and we will not give the details. When the above Fourier series representation exists, we recall that the coefficients a_k and b_k are given by the formulas

$$a_k = \frac{1}{\pi} \int_{-\pi}^{\pi} f(t) \cos kt \, dt, \quad k = 0,\ 1,\ 2,\ \dots,$$

$$b_k = \frac{1}{\pi} \int_{-\pi}^{\pi} f(t) \sin kt \, dt, \quad k = 1,\ 2,\ \dots.$$

Ignoring the factor $1/\pi$, which is the normalisation factor, these integrals are the inner products in $C_2[-\pi, \pi]$ of the given function f with the functions of T. Such will be precisely the coefficients in our infinite linear combinations of the general theorem, which now follows.

Theorem 9.6.4 (Generalised Fourier Series Theorem) *Let X be a separable inner product space. Suppose $T = \{x_1, x_2, \dots\}$ is a complete orthonormal set in X. Then each of the following is true.*

(a) *For any point $u \in X$, we have*

$$u = \sum_{k=1}^{\infty} \langle u, x_k \rangle \, x_k.$$

(b) *For any points $u, v \in X$, we have*

$$\langle u, v \rangle = \sum_{k=1}^{\infty} \langle u, x_k \rangle \langle x_k, v \rangle.$$

(c) *For any point $u \in X$, we have*

$$\|u\|^2 = \sum_{k=1}^{\infty} |\langle u, x_k \rangle|^2.$$

(d) *If a point $u \in X$ is such that $\langle u, x_n \rangle = 0$ for all $n \in \mathbf{N}$, then $u = \theta$.*

(e) *If points $u, v \in X$ are such that $\langle u, x_n \rangle = \langle v, x_n \rangle$ for all $n \in \mathbf{N}$, then $u = v$.*

Conversely, if any of the statements (a), (b) *or* (c) *is true for some orthonormal set $T = \{x_1, x_2, \dots\}$ in X, then T is complete.*

Furthermore, if X is a Hilbert space then either of the statements (d) *and* (e) *also implies that the orthonormal set T is complete.*

Notice that the existence of such a set as T here is implied by Theorem 9.6.3. The series on the right in (a) is called the *Fourier series* for the point u, and the numbers $\langle u, x_n \rangle$ are called *Fourier coefficients* of u. Compare this with the classical trigonometric example, just described. The equations in (b) and (c) are known as *Parseval's identities*. In (c), we see how Bessel's inequality (Theorem 8.3.3) may be strengthened with the additional hypothesis. An orthonormal set $S = \{x_1, x_2, \dots\}$ in X satisfying the condition in (d) is often called *total*, in that no further nonzero vectors can be added to S so that the new set remains orthogonal. It is in this sense that we have called such a set 'complete'. In (e), it is seen that the Fourier coefficients of a vector uniquely determine that vector.

Now for the proof of the theorem. We are supposing initially that T is a complete orthonormal set in X. Then $\overline{\operatorname{Sp} T} = X$. Let ϵ be any positive number. Given the point $u \in X$, we know there is a sequence $\{w_n\}$ in $\operatorname{Sp} T$ such that $w_n \to u$. We may write, for each $n \in \mathbf{N}$,

$$w_n = \sum_{k=1}^{t_n} \alpha_{nk} x_k$$

for some scalars α_{nk} and some $t_n \in \mathbf{N}$ (the coefficients in the sum and its number of terms varying from term to term of the sequence). There exists a positive integer N such that $\|u - w_n\| < \epsilon$ when $n > N$. But then, by Theorem 8.3.2,

$$\left\| u - \sum_{k=1}^{t_n} \langle u, x_k \rangle\, x_k \right\| \leqslant \left\| u - \sum_{k=1}^{t_n} \alpha_{nk} x_k \right\| < \epsilon$$

for $n > N$. We may assume $t_1 < t_2 < \cdots$ (if necessary by including extra α's all equal to 0), so $\left\{ \sum_{k=1}^{t_n} \langle u, x_k \rangle\, x_k \right\}_{n=1}^{\infty}$ is a convergent subsequence, with limit u, of the sequence $\{ \sum_{k=1}^{n} \langle u, x_k \rangle\, x_k \}$. The latter is a Cauchy sequence, by Solved Problem 8.5(2), and hence, by Theorem 4.1.1, is itself convergent with limit u. Thus

$$u = \sum_{k=1}^{\infty} \langle u, x_k \rangle\, x_k,$$

and statement (a) is true.

Then, to verify statement (b), we may write $u = \lim u_n$ and $v = \lim v_n$, where

$$u_n = \sum_{k=1}^{n} \langle u, x_k \rangle\, x_k, \qquad v_n = \sum_{j=1}^{n} \langle v, x_j \rangle\, x_j,$$

for $n \in \mathbf{N}$. Using the fact that T is an orthonormal set in X, we have

$$\langle u_n, v_n \rangle = \sum_{k=1}^{n} \sum_{j=1}^{n} \langle u, x_k \rangle\, \overline{\langle v, x_j \rangle}\, \langle x_k, x_j \rangle$$

$$= \sum_{k=1}^{n} \langle u, x_k \rangle\, \overline{\langle v, x_k \rangle} = \sum_{k=1}^{n} \langle u, x_k \rangle\, \langle x_k, v \rangle .$$

By Theorem 8.1.4, $\langle u_n, v_n \rangle \to \langle u, v \rangle$ and hence (b) is true.

Knowing that, we see immediately that (c) is true by putting v equal to u in (b). And then, if in (c) we put $\langle u, x_k \rangle = 0$ for all $k \in \mathbf{N}$, we must have $\|u\| = 0$, so $u = \theta$ and (d) is true. In turn, under the hypothesis of (e) we have $\langle u - v, x_n \rangle = 0$ for all $n \in \mathbf{N}$ and so $u - v = \theta$, and $u = v$. Thus (e) must be true.

Moving to the converse, we show first that (c) $\Rightarrow$ (a), where as usual we read $\Rightarrow$ as 'implies'. We showed above that (a) $\Rightarrow$ (b), and that (b) $\Rightarrow$ (c), so this will mean that all three of these statements are equivalent, any one implying the others. Then we need only show that one of them implies that T is a complete orthonormal set to finish the proof of that part of the theorem.

So suppose that $T = \{x_1, x_2, \dots\}$ is an orthonormal set in X, and that (c) is true. As in the proof of Theorem 8.3.3, we have

$$0 \leqslant \left\| u - \sum_{k=1}^{n} \langle u, x_k \rangle \, x_k \right\|^2 \leqslant \|u\|^2 - \sum_{k=1}^{n} |\langle u, x_k \rangle|^2.$$

By assumption, the final expression here may be made as small as we please by choosing n large enough, so $\sum_{k=1}^{n} \langle u, x_k \rangle \, x_k \to u$. Thus, (a) is true.

Now we show that the truth of (a) implies that T is a complete orthonormal set in X. Assuming (a), if u is any point in X then we may write $u = \lim u_n$, where

$$u_n = \sum_{k=1}^{n} \langle u, x_k \rangle \, x_k.$$

Then $u_n \in \mathrm{Sp}\,T$ for all n, so $u \in \overline{\mathrm{Sp}\,T}$. This means $X \subseteq \overline{\mathrm{Sp}\,T}$. Since clearly $\overline{\mathrm{Sp}\,T} \subseteq X$, then $\overline{\mathrm{Sp}\,T} = X$, as required.

Finally, we assume further that X is a Hilbert space, and suppose again that $T = \{x_1, x_2, \dots\}$ is an orthonormal set in X. We will show, assuming (d) to be true, that T must be complete. Let $v \in X$ be any point and consider the sequence $\{u_n\}$, where

$$u_n = \sum_{k=1}^{n} \langle v, x_k \rangle \, x_k, \qquad n \in \mathbf{N}.$$

By Solved Problem 8.5(2), $\{u_n\}$ is a Cauchy sequence in X. But X is now assumed to be complete, so the sequence converges, with limit w, say. Using Theorem 8.1.4, for each $j \in \mathbf{N}$,

$$\langle v - w, x_j \rangle = \lim_{n \to \infty} \langle v - u_n, x_j \rangle = \langle v, x_j \rangle - \lim_{n \to \infty} \left\langle \sum_{k=1}^{n} \langle v, x_k \rangle \, x_k, x_j \right\rangle.$$

But for each j, if $n \geqslant j$ then

$$\left\langle \sum_{k=1}^{n} \langle v, x_k \rangle \, x_k, x_j \right\rangle = \sum_{k=1}^{n} \langle v, x_k \rangle \, \langle x_k, x_j \rangle = \langle v, x_j \rangle,$$

by the orthonormality of T. Hence $\langle v - w, x_j \rangle = 0$ for each j. Since we are assuming that (d) is true, we thus have $v - w = \theta$. Hence $v = w$, or

$$v = \sum_{k=1}^{\infty} \langle v, x_k \rangle \, x_k.$$

This shows (a) to be a consequence of (d), and in the preceding paragraph we saw that when (a) is true, the set T is complete.

The proof of the generalised Fourier series theorem is finished when we show that (d) is true if (e) is. For this, suppose that (e) is true and that (d) is not. Then there is a point $z \neq \theta$ in X such that $\langle z, x_n \rangle = 0$ for all $n \in \mathbf{N}$. In that case, we have $\langle u, x_n \rangle = \langle u + z, x_n \rangle$ for all n and any point $u \in X$. By (e), this means $u = u + z$, or $z = \theta$, and this is a contradiction. Hence (d) is true when (e) is, so that (e) must also imply, via (d), that T is complete. $\square$

Notice from the statement of this theorem and its proof that the essence of the theorem can be given by the scheme:

$$T \text{ complete} \iff \text{(a)} \iff \text{(b)} \iff \text{(c)} \implies \text{(d)} \iff \text{(e)}$$

for any inner product space X, with

$$\text{(c)} \impliedby \text{(d)}$$

when X is a Hilbert space. (The arrowheads indicate the direction of implication.)

Let us stress again that the convergence of a Fourier series is dependent upon the norm generated by the inner product for the space in which we are working. Reverting to the classical trigonometric case, if we write, for $-\pi \leqslant t \leqslant \pi$,

$$x_1(t) = \frac{1}{\sqrt{2\pi}},$$

$$x_2(t) = \frac{1}{\sqrt{\pi}} \sin t, \quad x_3(t) = \frac{1}{\sqrt{\pi}} \cos t,$$

$$x_4(t) = \frac{1}{\sqrt{\pi}} \sin 2t, \quad x_5(t) = \frac{1}{\sqrt{\pi}} \cos 2t,$$

and so on, so $\{x_1, x_2, \ldots\}$ is a complete orthonormal set in $C_2[-\pi, \pi]$, then we have shown that, for any continuous function f on $[-\pi, \pi]$,

$$\lim_{n \to \infty} \int_{-\pi}^{\pi} \left(f(t) - \sum_{k=1}^{n} \left(\int_{-\pi}^{\pi} f(t) x_k(t)\, dt \right) x_k(t) \right)^2 dt = 0.$$

This is often described by saying that the classical Fourier series for f converges *in mean square* to f, and says nothing about the uniform convergence, say, of the series.

If $\{x_1, x_2, \ldots\}$ is a complete orthonormal set in an inner product space X, then, since $\sum_{k=1}^{\infty} |\langle u, x_k \rangle|^2$ converges for any $u \in X$ by (c) of the Fourier series theorem, we must have $\langle u, x_n \rangle \to 0$ (Theorem 1.8.3). For the trigonometric case above, this means

$$\int_{-\pi}^{\pi} f(t) x_n(t)\, dt \to 0,$$

from which we conclude that

$$\int_{-\pi}^{\pi} f(t) \cos nt\, dt \to 0 \quad \text{and} \quad \int_{-\pi}^{\pi} f(t) \cos nt\, dt \to 0,$$

for any function f, continuous on $[-\pi, \pi]$. This is a version of a result known as the *Riemann–Lebesgue lemma*.

We have shown that the Legendre polynomials P_0, P_1, ... form a complete orthonormal set in $C_2[-1, 1]$. Thus any function f, continuous on $[-1, 1]$, has a Fourier series $\sum_{k=0}^{\infty} \langle f, P_k \rangle P_k$ which converges (in mean square) to f. The other orthonormal sets of polynomials listed at the end of Section 8.2 can also be shown to be complete in their respective inner product spaces. As in the preceding paragraph, this implies, for Chebyshev polynomials for example, that

$$\int_{-1}^{1} \frac{f(t) T_n(t)}{\sqrt{1 - t^2}}\, dt \to 0$$

for any function f, continuous on $[-1, 1]$.

9.7 Hilbert space isomorphism

At the beginning of this chapter, it was stated that in a certain sense l_2 is the only infinite-dimensional separable Hilbert space. We now clarify that statement.

What we intend to show is that any infinite-dimensional separable Hilbert space X is isomorphic to l_2. To do this, we must exhibit a certain kind of mapping, called an *isomorphism*, from X onto l_2. The notion of isomorphism is an essential tool of modern algebra, with important applications in modern analysis. Isomorphisms may be defined between elements of various classes of sets all of which, within any class, have the same algebraic structure. Thus we may speak for example of vector space isomorphisms (as we did in Section 1.11) or inner product space isomorphisms (or, in algebra, of group or field isomorphisms). The definition of an isomorphism may vary from class to class but in all cases an isomorphism between two sets of some class is a one-to-one correspondence (or bijection) between those sets which is such that the algebraic operations in one of the sets is precisely reflected in the other.

In a vector space isomorphism, for example, we require that the sum of two vectors in one space equal the sum of their images in the other space, and similarly for multiplication by scalars. In an inner product space isomorphism, we require further that the value of the inner product for two vectors in one space equal the value of the inner product for their images in the other space. Separable Hilbert spaces have no further algebraic structure beyond that of inner product spaces so that we will only need to define, more precisely than this, what we mean by an inner product space isomorphism. Because we need a preliminary result, important in its own right, we will delay briefly giving that definition.

Theorem 9.7.1 *If $(\alpha_1, \alpha_2, \dots)$ is any point in l_2 and X is an infinite-dimensional separable Hilbert space, then there exists a point $w \in X$ for which $\alpha_1, \alpha_2, \dots$ are the Fourier coefficients with respect to a given complete orthonormal set $\{x_1, x_2, \dots\}$ in X. Moreover, $\|w\|^2 = \sum_{k=1}^{\infty} |\alpha_k|^2$.*

For the proof, we introduce the sequence $\{w_n\}$ in X given by

$$w_n = \sum_{k=1}^{n} \alpha_k x_k, \quad n \in \mathbf{N}.$$

If $n > m$,

$$\|w_n - w_m\|^2 = \langle w_n - w_m, w_n - w_m \rangle$$

$$= \left\langle \sum_{k=m+1}^{n} \alpha_k x_k, \sum_{j=m+1}^{n} \alpha_j x_j \right\rangle$$

$$= \sum_{k=m+1}^{n} \sum_{j=m+1}^{n} \alpha_k \overline{\alpha}_j \langle x_k, x_j \rangle = \sum_{k=m+1}^{n} |\alpha_k|^2,$$

because the set $\{x_1, x_2, \dots\}$ is orthonormal. As $(\alpha_1, \alpha_2, \dots)$ belongs to l_2, the last sum here tends to 0 as $m \to \infty$. Hence $\{w_n\}$ is a Cauchy sequence. As X is a Hilbert space, this sequence converges to w, say, and of course $w \in X$. To show that $\alpha_1, \alpha_2, \ldots$ are the Fourier coefficients of w, we note that

$$\langle w_n, x_k \rangle = \left\langle \sum_{j=1}^{n} \alpha_j x_j, x_k \right\rangle = \sum_{j=1}^{n} \alpha_j \langle x_j, x_k \rangle = \alpha_k,$$

for each $k \in \mathbf{N}$, provided $n \geqslant k$. We then have, using Theorem 8.1.4,

$$\langle w, x_k \rangle = \lim_{n \to \infty} \langle w_n, x_k \rangle = \alpha_k$$

for $k \in \mathbf{N}$, and this is what we had to show.

It is left as an exercise to show further that, for the vector w obtained above, we have

$$\|w\|^2 = \sum_{k=1}^{\infty} |\alpha_k|^2. \qquad \square$$

Theorem 9.7.1 is known as the *Riesz–Fischer theorem*. Now we give the definition discussed above.

Definition 9.7.2 An inner product space X is said to be *isomorphic* to an inner product space Y if there exists a one-to-one mapping A of X onto Y which is linear and which 'preserves inner products', in that, for any vectors $x_1, x_2 \in X$,

$$\langle x_1, x_2 \rangle = \langle Ax_1, Ax_2 \rangle.$$

The mapping A is called an *isomorphism* of X onto Y.

Notice that here we are using the same notation for the inner products for both X and Y.

Since the mapping A is linear, it also has the desired properties of preserving sums and scalar multiples. We know (see Section 7.6) that in this situation the inverse mapping A^{-1} exists and is linear. Furthermore, if for any vectors $y_1, y_2 \in Y$ we have $A^{-1}y_1 = x_1$ and $A^{-1}y_2 = x_2$, then

$$\langle y_1, y_2 \rangle = \langle Ax_1, Ax_2 \rangle = \langle x_1, x_2 \rangle = \langle A^{-1}y_1, A^{-1}y_2 \rangle.$$

Hence A^{-1} is an isomorphism of Y onto X, so that also Y is isomorphic to X. We thus say simply that X and Y are isomorphic inner product spaces when such a mapping A exists.

It is not difficult to show that if X, Y and Z are inner product spaces, with X and Y isomorphic and Y and Z isomorphic, then also X and Z are isomorphic. The details are left as an exercise.

We have been leading up to the following important theorem.

Theorem 9.7.3 *Any infinite-dimensional separable Hilbert space is isomorphic to l_2.*

Of course, we are referring here to inner product space isomorphisms. From the comment just above, it follows that all infinite-dimensional separable Hilbert spaces are mutually isomorphic. The importance of this result in quantum mechanics was mentioned at the beginning of this chapter.

Let X be an infinite-dimensional separable Hilbert space. The theorem will be proved when we have exhibited an isomorphism from X onto l_2. Since X is separable, it contains a countable complete orthonormal set $\{x_1, x_2, \ldots\}$, say. By (a) of the Fourier series theorem (Theorem 9.6.4), it follows that for any point $u \in X$ we may write

$$u = \sum_{k=1}^{\infty} \alpha_k x_k, \qquad \alpha_k = \langle u, x_k \rangle, \ k \in \mathbf{N},$$

and, by (c) of the same theorem, we know that the series $\sum_{k=1}^{\infty} |\alpha_k|^2$ converges. Thus with any point $u \in X$ we may associate the point $\xi = (\alpha_1, \alpha_2, \ldots)$ in l_2, where $\alpha_k = \langle u, x_k \rangle$ for $k \in \mathbf{N}$. Let A be the mapping from X into l_2 such that $Au = \xi$. We will show that A is the desired isomorphism.

Certainly, A is an onto mapping, for this is precisely what Theorem 9.7.1 tells us: for any point $\xi \in l_2$, there is a point $w \in X$ such that $Aw = \xi$.

To show that A is one-to-one, suppose that $Au_1 = \xi_1$ and $Au_2 = \xi_2$, where $u_1, u_2 \in X$ and $u_1 \neq u_2$. Then there is at least one index k for which $\langle u_1, x_k \rangle \neq \langle u_2, x_k \rangle$, by (e) of the Fourier series theorem. For this k, ξ_1 and ξ_2 differ in their kth components and so cannot be equal. This indeed shows that A is one-to-one.

For any points $u_1, u_2 \in X$ and any scalars β_1, β_2, we have

$$\langle \beta_1 u_1 + \beta_2 u_2, x_k \rangle = \beta_1 \langle u_1, x_k \rangle + \beta_2 \langle u_2, x_k \rangle$$

for each $k \in \mathbf{N}$. Hence $A(\beta_1 u_1 + \beta_2 u_2) = \beta_1 A u_1 + \beta_2 A u_2$. That is, A is a linear mapping.

It remains to show that A preserves inner products. For this, we use (b) of the Fourier series theorem: for any points $u_1, u_2 \in X$, we have

$$\langle u_1, u_2 \rangle = \sum_{k=1}^{\infty} \langle u_1, x_k \rangle \overline{\langle u_2, x_k \rangle}.$$

But, by definition of the inner product for l_2, this says precisely that $\langle u_1, u_2 \rangle = \langle Au_1, Au_2 \rangle$, as required, and this proves the theorem. $\quad\square$

It can be shown in a similar fashion that any complex inner product space, of dimension n, is isomorphic to $\mathbf{C}^n$, and this is left as an exercise.

9.8 Exercises

(1)   (a) Consider l_2 as a normed space and let e_k be the point in l_2 with all components 0 except for the kth, which is 1. Show that, if $x = (x_1, x_2, \dots) \in l_2$, then $x = \sum_{k=1}^{\infty} x_k e_k$.
       (b) Give an example in which this series for x is not absolutely convergent.

(2) Complete the proof of Theorem 9.7.1 by proving that

$$\|w\|^2 = \sum_{k=1}^{\infty} |\alpha_k|^2.$$

(3) Let X, Y, Z be inner product spaces and suppose the mappings $A\colon X \to Y$ and $B\colon Y \to Z$ are isomorphisms. Prove that the mapping BA is an isomorphism of X onto Z.

(4) Prove that any (complex) inner product space of dimension n is isomorphic to $\mathbf{C}^n$.

(5) Let $\{x_1, x_2, \dots\}$ be the usual complete orthonormal set (of trigonometric functions) in $C_2[-\pi, \pi]$ and let f and g be continuous functions on $[-\pi, \pi]$. Define functions F_n, G_n $(n \in \mathbf{N})$ and G for $-\pi \leqslant u \leqslant \pi$ by

$$F_n(u) = \sum_{k=1}^{n} \langle f, x_k \rangle \, x_k,$$

$$G_n(u) = \int_{-\pi}^{u} \sum_{k=1}^{n} \langle f, x_k \rangle \, x_k(t) g(t) \, dt,$$

$$G(u) = \int_{-\pi}^{u} f(t) g(t) \, dt.$$

By the Fourier series theorem, $\{F_n\}$ converges in mean square to f. Prove that $\{G_n\}$ converges uniformly on $[-\pi, \pi]$ to G. Prove also that the same is true for any function g for which $|g|^2$ is integrable on $[-\pi, \pi]$.

. . . .

(6) A linear mapping U from a Hilbert space X onto itself is called *unitary* if $\langle Ux, Uy \rangle = \langle x, y \rangle$ for all $x, y \in X$.

    (a) Show that U is an operator and that $\|U\| = 1$.

    (b) Show that $UU^* = U^*U = I$ (the identity operator on X), where U^* is the adjoint of U.

    (c) Show that $\{Ux_1, Ux_2, \dots\}$ is a complete orthonormal set in X, whenever $\{x_1, x_2, \dots\}$ is, if X is separable.

(7) A bounded sequence $\{y_n\}$ in a Hilbert space X is said to be *weakly convergent* if the sequence $\{\langle z, y_n \rangle\}$ in $\mathbf{C}$ converges for every $z \in X$. Use the Riesz representation theorem (Theorem 9.2.1) to show that there exists $y \in X$ such that $\lim \langle z, y_n \rangle = \langle z, y \rangle$ for every $z \in X$.

(8) Continuing, the sequence $\{y_n\}$ is then said to converge weakly to y, and we write $y_n \overset{\text{w}}{\to} y$. In contrast, if $y_n \to y$ (in norm, as usual) we may say the sequence $\{y_n\}$ converges *strongly*.

    (a) Show that if $y_n \to y$ then $y_n \overset{\text{w}}{\to} y$.

    (b) Show that if $y_n \overset{\text{w}}{\to} y$ and $\|y_n\| \to \|y\|$ then $y_n \to y$. (See also Exercise 8.6(15).)

(9) Continuing, show that any complete orthonormal set in l_2 forms a sequence which converges weakly, but not strongly, to $\theta \in l_2$.

Bibliography

The books in the list that follows are generally easier introductions to their content matter than are available in a host of other texts. The general principle has been to include texts written in a similar spirit to this book. After the list, there is a detailed description of particular books relevant to the different chapters.

The books (often in earlier editions) that have been of particular benefit to the writing of sections of this one are given special mention.

1. I. T. Adamson, *A General Topology Workbook*, Birkhäuser, Boston (1996).
2. G. Bachman and L. Narici, *Functional Analysis*, Academic Press, New York (1966). Available as a Dover Publications reprint (2000).
3. R. G. Bartle and D. R. Sherbert, *Introduction to Real Analysis* (third edition), Wiley, New York (1999).
4. S. K. Berberian, *Introduction to Hilbert Space* (second edition), Oxford, New York (1999).
5. E. W. Cheney, *Introduction to Approximation Theory* (second edition), AMS Chelsea Publications, Providence, RI (1999).
6. L. Collatz, *Functional Analysis and Numerical Mathematics*, Academic Press, New York (1966).
7. R. Courant, *Differential and Integral Calculus* (Volume 1), Wiley, New York (1988).
8. P. J. Davis, *Interpolation and Approximation*, Blaisdell, New York (1963). Available as a Dover Publications reprint (1975).
9. L. Debnath and P. Mikusiński, *Introduction to Hilbert Spaces with Applications* (second edition), Academic Press, Boston (1998).

10. F. Deutsch, *Best Approximation in Inner Product Spaces*, Springer, New York (2001).
11. R. R. Goldberg, *Methods of Real Analysis* (second edition), Wiley, New York (1976).
12. J. R. Kirkwood, *An Introduction to Analysis* (second edition), PWS, Boston (1995).
13. A. N. Kolmogorov and S. V. Fomin, *Introductory Real Analysis*, Prentice–Hall, Englewood Cliffs, NJ (1970). Available as a Dover Publications reprint (1975).
14. E. Kreyszig, *Introductory Functional Analysis with Applications*, Wiley, New York (1978).
15. R. G. Kuller, *Topics in Modern Analysis*, Prentice–Hall, Englewood Cliffs, NJ (1969).
16. B. L. Moiseiwitsch, *Integral Equations*, Longman, London (1977).
17. A. F. Monna, *Functional Analysis in Historical Perspective*, Wiley, New York (1973).
18. A. W. Naylor and G. R. Sell, *Linear Operator Theory in Engineering and Science*, Holt, Rinehart and Winston, New York (1971). Reprinted by Springer–Verlag (2000).
19. J. T. Oden and L. F. Demkowicz, *Applied Functional Analysis*, CRC Press, Boca Ratan (1996).
20. J. M. Ortega, *Numerical Analysis, A Second Course*, SIAM, Philadelphia (1990).
21. G. M. Phillips and P. J. Taylor, *Theory and Applications of Numerical Analysis* (second edition), Academic Press, London (1996).
22. C. G. C. Pitts, *Introduction to Metric Spaces*, Oliver and Boyd, Edinburgh (1972).
23. D. Roseman, *Elementary Topology*, Prentice–Hall, New Jersey (1999).
24. H. L. Royden, *Real Analysis* (third edition), Macmillan, New York (1988).
25. W. Rudin, *Functional Analysis* (second edition), McGraw–Hill, New York (1991).
26. M. Schechter, *Principles of Functional Analysis* (second edition), American Mathematical Society, Providence, RI (2001).
27. M. A. Snyder, *Chebyshev Methods in Numerical Approximation*, Prentice–Hall, Englewood Cliffs (1966).
28. D. L. Stancl and M. L. Stancl, *Real Analysis with Point-Set Topology*, Marcel Dekker, New York (1987).

29. S. Vickers, *Topology via Logic*, Cambridge University Press, Cambridge (1990).

30. B. Z. Vulikh, *Introduction to Functional Analysis for Scientists and Technologists*, Pergamon, Oxford (1963).

For further general work in functional analysis, see any of 2, 13–15, 18, 19, 24–26 and 30. (The last of these was especially useful for the writing of this book, but will now be difficult to find.) In particular, 14, 18, 19 and 30 stress applications of functional analysis. For background reading in functional analysis, see 17.

More detailed treatments of the work of Sections 1.2–1.10 are given in 3, 7, 11, 12 and 28. In particular, 3, 12 and 28 cover the work of Sections 1.4 and 1.6. Any introductory text on linear algebra will include the work of Section 1.11.

On the work of Chapters 2–4 in general, see 22. Further work on integral equations, as in Section 3.3, is given in 16. The work of Section 3.4 was adapted from 6. For more on approximation theory, see 5, 8 and 10.

The reference 28 covers much of the work of Chapter 5, as do 1 and 23 much more comprehensively. See 29 for an unusual but readable account of topology that emphasises its applications in computer science.

The approximation theory of Chapter 6 is contained in 5 and 8. For more on Chebyshev polynomials, see 27.

In Chapter 7, again see 16 for the work on integral equations. For the applications to numerical analysis, see 6, 20 and 21. Section 7.10, with the application to quantum mechanics, was adapted from 6; see also 9.

The references 8 and 10 contain the further applications to approximation theory in Chapter 8, and 4 and 9 are general references to the work of Chapter 9.

Selected Solutions

Exercises 2.4

(1) Take any $x, y, z, u \in X$. Using (M3), we have

$$d(x, z) \leqslant d(x, y) + d(y, z) \leqslant d(x, y) + d(y, u) + d(u, z),$$
$$d(y, u) \leqslant d(y, x) + d(x, u) \leqslant d(y, x) + d(x, z) + d(z, u),$$

Using (M2), these imply, respectively,

$$d(x, z) - d(y, u) \leqslant d(x, y) + d(z, u),$$
$$d(x, z) - d(y, u) \geqslant -d(y, x) - d(z, u) = -(d(x, y) + d(z, u)).$$

Then $|d(x, z) - d(y, u)| \leqslant d(x, y) + d(z, u)$, as required.

(3) We will verify (M3) for d_4, using the fact that it is true for d_1 and d_2. Take any $x, y, z \in X$. Then

$$
\begin{aligned}
d_1(x, y) &\leqslant d_1(x, z) + d_1(z, y) \\
&\leqslant \max\{d_1(x, z), d_2(x, z)\} + \max\{d_1(z, y), d_2(z, y)\} \\
&= d_4(x, z) + d_4(z, y).
\end{aligned}
$$

In exactly the same way, also $d_2(x, y) \leqslant d_4(x, z) + d_4(z, y)$, and it follows then that $d_4(x, y) = \max\{d_1(x, y), d_2(x, y)\} \leqslant d_4(x, z) + d_4(z, y)$.

(7) Let x, y be elements of X and assume that the function x is zero outside the the interval I and the function y is zero outside the interval J. (Note that I and J may be disjoint.) Clearly, $x - y$ is zero outside (that is, in the complement of) $I \cup J$. Let the left and right endpoints of I be a and b, respectively, and let those of J be c and d, respectively. Then $I \cup J \subseteq [\min\{a, c\}, \max\{b, d\}]$. Define a function f on this closed interval by $f(t) = |x - y|(t) = |x(t) - y(t)|$. Then f is continuous since x and y are, and hence f attains its maximum value (by Theorem 1.9.6).

312

Since $(x - y)(t) = 0$ for t outside the interval, so $|x - y|$ also attains its maximum value. This shows that $d(x, y)$ is well defined, and in fact it is clear that $d(x, y) = \max_{t \in I \cup J} |x(t) - y(t)|$.

It is also clear from this formulation that d satisfies requirements (M1) and (M2) of a metric. To verify (M3), let x, y, z be elements of X; that is, x, y, z are continuous functions defined on $\mathbf{R}$, which are zero outside the intervals I, J, K, say. We have

$$d(x, z) = \max_{t \in \mathbf{R}} |x(t) - z(t)| = \max_{t \in I \cup K} |x(t) - z(t)| = \max_{t \in I \cup J \cup K} |x(t) - z(t)|,$$

since $I \cup K \subseteq I \cup J \cup K$ and $x(t) = z(t) = 0$ for t outside $I \cup K$. Then, with similar reasoning for $x - y$ and $y - z$,

$$
\begin{aligned}
d(x, z) &= \max_{t \in I \cup J \cup K} |x(t) - z(t)| \\
&\leqslant \max_{t \in I \cup J \cup K} |x(t) - y(t)| + \max_{t \in I \cup J \cup K} |y(t) - z(t)| \\
&= \max_{t \in I \cup J} |x(t) - y(t)| + \max_{t \in J \cup K} |y(t) - z(t)| \\
&= d(x, y) + d(y, z).
\end{aligned}
$$

This shows that (M3) is satisfied.

Exercises 2.9

(1) Take any $\epsilon > 0$. Since $\lim x_n = x$ and $\lim y_n = y$, there exist positive integers N_1 and N_2 such that

$$d(x_n, x) < \frac{\epsilon}{2} \quad \text{for } n > N_1 \quad \text{and} \quad d(y_n, y) < \frac{\epsilon}{2} \quad \text{for } n > N_2.$$

Let $N = \max\{N_1, N_2\}$. Then, using the inequality in Exercise 2.4(1), we have

$$|d(x_n, y_n) - d(x, y)| \leqslant d(x_n, x) + d(y_n, y) < \frac{\epsilon}{2} + \frac{\epsilon}{2} = \epsilon,$$

for $n > N$. This shows that $d(x_n, y_n) \to d(x, y)$.

(5) Let $\{x_n\}$ be a Cauchy sequence in (X, d), where d is the discrete metric. There exists a positive integer N such that $d(x_n, x_m) < 1$ when $m, n > N$. Then $x_n = x_{N+1}$ when $n > N$, since d is the discrete metric. Now take any $\epsilon > 0$. Then $d(x_n, x_{N+1}) = 0 < \epsilon$ when $n > N$, and hence the Cauchy sequence converges (to x_{N+1}). So (X, d) is complete.

(10) The triangle inequality in $\mathbf{C}$ implies that $||u| - |v|| \leqslant |u - v|$ for

any $u, v \in \mathbf{C}$. Take any $\epsilon > 0$. Since $z_n \to z$, we have

$$0 \leqslant ||z_n| - |z|| \leqslant |z_n - z| < \epsilon,$$

for all n large enough, implying that $|z_n| \to |z|$.

Now let $S = \{w \in \mathbf{C} : |w| \leqslant c\}$, where c is any positive number, and let $\{z_n\}$ be a sequence in S which converges in $\mathbf{C}$. Then $|z_n| \leqslant c$ for all $n \in \mathbf{N}$. Put $z = \lim z_n$. We will show that $|z| \leqslant c$, so that $z \in S$ and this will prove that S is closed. If this is not so then $|z| > c$; let $\epsilon = |z| - c > 0$. From the earlier result, we have $|z_n| \to |z|$, so there exists a positive integer N such that $||z_n| - |z|| < \epsilon = |z| - c$, when $n > N$. In particular, for such n, $|z_n| - |z| > -(|z| - c)$, or $|z_n| > c$. This contradicts the statement that $|z_n| \leqslant c$ for all $n \in \mathbf{N}$. Therefore we must have $|z| \leqslant c$.

(12) When $a = 0$, the sequence $\{x_n\}$ in Example 2.6(6) is given by

$$x_n(t) = \begin{cases} nt, & 0 \leqslant t \leqslant 1/n, \\ 1, & 1/n \leqslant t \leqslant b, \end{cases}$$

for $n \in \mathbf{N}$. This is still a Cauchy sequence. However, we will show that it is a convergent sequence in $C_1[0, b]$, its limit being the (continuous) function h, given by $h(t) = 1$ for $0 \leqslant t \leqslant b$. For this,

$$d(x_n, h) = \int_0^b |x_n(t) - h(t)|\, dt$$
$$= \int_0^{1/n} |nt - 1|\, dt = \int_0^{1/n} (1 - nt)\, dt = \frac{1}{2n},$$

so $x_n \to h$, as stated. Hence this sequence cannot serve to show that $C_1[a, b]$ is not complete.

Exercises 3.5

(2)(b) The given equation is equivalent to $\frac{1}{8} \sin x + \frac{1}{4} \sinh x + \frac{1}{4} = x$. Consider the function f defined by $f(x) = \frac{1}{8} \sin x + \frac{1}{4} \sinh x + \frac{1}{4}$, for $0 \leqslant x \leqslant \frac{1}{2}\pi$. We have $0 \leqslant \sin x \leqslant 1$ and $0 \leqslant \sinh x \leqslant \sinh(\frac{1}{2}\pi) < 2.4$, so that

$$\frac{1}{4} \leqslant f(x) \leqslant \frac{1}{8} + \frac{1}{4} \sinh \frac{\pi}{2} + \frac{1}{4} < 1.$$

Thus the range of f is a subset of its domain $[0, \frac{1}{2}\pi]$. Also,

$$|f'(x)| = \left| \frac{1}{8}\cos x + \frac{1}{4}\cosh x \right|$$

$$\leqslant \frac{1}{8}|\cos x| + \frac{1}{4}\cosh x \leqslant \frac{1}{8} + \frac{1}{4}\cosh\frac{\pi}{2} < 1.$$

Therefore, $f(x) = x$ (and hence the given equation) has a unique root in $[0, \frac{1}{2}\pi]$. With $x_0 = 0$, the next iterate to this root, using the method of successive approximations, is $x_1 = f(x_0) = 0.25$, and the next few are $x_2 = f(x_1) \doteq 0.3440$, $x_3 = f(x_2) \doteq 0.3799$, $x_4 = f(x_3) \doteq 0.3936$. (The actual root is 0.4022, to four decimal places.)

(4)(b) The given system is equivalent to

$$\begin{aligned}
\tfrac{3}{4}x + \tfrac{1}{6}y - \tfrac{2}{5}z &= 3, \\
-\tfrac{1}{2}x + \tfrac{1}{3}y + \tfrac{1}{4}z &= -1, \\
\tfrac{1}{8}x \quad\quad + \tfrac{5}{4}z &= -5.
\end{aligned}$$

Let A be the matrix of coefficients from the left-hand sides, and put

$$C = I - A = \begin{pmatrix} \tfrac{1}{4} & -\tfrac{1}{6} & \tfrac{2}{5} \\ \tfrac{1}{2} & \tfrac{2}{3} & -\tfrac{1}{4} \\ -\tfrac{1}{8} & 0 & -\tfrac{1}{4} \end{pmatrix}.$$

The sums of the absolute values of the elements in the three columns of C are all less than 1. Thus the condition developed in Exercise 3.5(3) for the existence of a unique solution to a system of linear equations is satisfied (but the two earlier such conditions are not, in this case).

(7)(b) This is a Fredholm integral equation. In the notation of the text, we have $|k(s,t)| = 1$ for all $s,t \in [0,2]$, so that, taking $M = 1$, $\lambda = \frac{1}{3} < \frac{1}{2} = 1/M(b-a)$. So the equation has a unique solution.

We solve the integral equation first by integrating both sides with respect to s over $[0,2]$:

$$\int_0^2 x(s)\,ds = \int_0^2 \left(\frac{1}{3}\int_0^2 x(t)\,dt\right) ds + \int_0^2 s^2\,ds = \frac{2}{3}\int_0^2 x(t)\,dt + \frac{8}{3}.$$

Since $\int_0^2 x(s)\,ds = \int_0^2 x(t)\,dt$, so $\int_0^2 x(t)\,dt = 8$. Hence we obtain the solution $x(s) = \frac{8}{3} + s^2$.

Alternatively, we observe that $\int_0^2 x(t)\,dt = c$, for some number c. Then

$x(s) = \frac{1}{3}c + s^2$. Substituting this into the integral equation, we have

$$\frac{c}{3} + s^2 = \frac{1}{3} \int_0^2 \left(\frac{c}{3} + t^2\right) dt + s^2 = \frac{1}{3}\left(\frac{2c}{3} + \frac{8}{3}\right) + s^2,$$

from which $c = 8$, so that $x(s) = \frac{8}{3} + s^2$.

Exercises 4.5

(2)(b) Let $S = \{x, x_1, x_2, \ldots\}$. Let σ be a sequence in S that contains
infinitely many distinct elements of S but does not include x as a term.
Form a subsequence of σ by choosing elements from σ in increasing
order of their subscripts. This will be a subsequence of $\{x_n\}$, and hence
will be convergent to x by the result of Exercise 4.5(1). Any sequence
in S that contains infinitely many distinct elements of S, including x,
has a subsequence σ in which x is omitted, and this may be treated
as above. Any sequence in S that contains only finitely many distinct
elements of S will have a constant subsequence, and constant sequences
are convergent. Hence any sequence in S has a convergent subsequence,
so S is compact.

(4) We prove first that a union of finitely many compact subsets of a
metric space is compact. It is sufficient to show that if S_1 and S_2 are
compact subsets of a metric space X then $S = S_1 \cup S_2$ is also a compact
subset of X. The more general result will follow by induction. Any
sequence $\{x_n\}$ in S must have subsequence $\{x_{n_k}\}$ all of whose terms are
in S_1, or all of whose terms are in S_2. Since S_1 and S_2 are compact,
$\{x_{n_k}\}$ itself has a convergent subsequence which will thus be a convergent
subsequence of $\{x_n\}$. This shows that S is compact.

We note however that the union of an infinite number of compact
subsets of a metric space need not be compact. For example, consider
the subsets $S_n = [n, n+1]$, for $n \in \mathbf{Z}$, of the metric space $\mathbf{R}$. For each n,
S_n is closed and bounded and hence is a compact subset of $\mathbf{R}$. But
$\bigcup_{n=-\infty}^{\infty} S_n = \mathbf{R}$ is not compact.

Now we prove that the intersection of any number of compact subsets
of a metric space is compact. Let T be any nonempty set, finite or
infinite, perhaps uncountable, and suppose, for each $t \in T$, that S_t is
a compact subset of some metric space. Put $S = \bigcap_{t \in T} S_t$. Assume
$S \neq \varnothing$ (else certainly S is compact). If $\{x_n\}$ is a sequence in S then
also $\{x_n\}$ is a sequence in S_t, for $t \in T$. Since S_t is compact, $\{x_n\}$ has a
convergent subsequence with limit $x \in S_t$. Since this is true for all $t \in T$,

then $x \in S$, so $\{x_n\}$, as a sequence in S, has a convergent subsequence. Hence S is compact.

(9) Let G be the set of all functions g, where $g(x) = \int_a^x f(t)\, dt$, for $f \in F$, $a \leqslant x \leqslant b$. Since F is a bounded subset of $C[a,b]$, so F is a uniformly bounded family (by the result of Exercise 4.5(7)). Then there exists $M > 0$ such that $|f(x)| \leqslant M$ for all $f \in F$, $a \leqslant x \leqslant b$. Thus, for all $g \in G$ and $a \leqslant x \leqslant b$,

$$|g(x)| = \left| \int_a^x f(t)\, dt \right| \leqslant \int_a^x |f(t)|\, dt \leqslant M(x-a) \leqslant M(b-a).$$

Hence G is uniformly bounded. Also, given $\epsilon > 0$, take $\delta = \epsilon/M$. Then for all $g \in G$ we have

$$|g(x') - g(x'')| = \left| \int_a^{x'} f(t)\, dt - \int_a^{x''} f(t)\, dt \right| = \left| \int_{x'}^{x''} f(t)\, dt \right|$$

$$\leqslant \left| \int_{x'}^{x''} |f(t)|\, dt \right| \leqslant M|x' - x''| < M\delta < \epsilon,$$

whenever $x', x'' \in [a,b]$ and $|x' - x''| < \delta$. This shows that G is equicontinuous.

Exercises 5.7

(2) Let S be any subset of X. In $(X, \mathscr{T}_{\max})$, $\operatorname{int} S = \overline{S} = S$. In $(X, \mathscr{T}_{\min})$, $\operatorname{int} S = \varnothing$ except that $\operatorname{int} X = X$, and $\overline{S} = X$ except that $\overline{\varnothing} = \varnothing$.

(4) (a) Let $b(x_0, r)$ be an open ball in a metric space (X, d), and take any $x \in b(x_0, r)$. Put $\epsilon = r - d(x_0, x)$. We will show that $b(x, \epsilon) \subseteq b(x_0, r)$, and this will imply that $b(x_0, r)$ is an open set. So take any $y \in b(x, \epsilon)$. Since $d(x, y) < \epsilon$, we have

$$d(x_0, y) \leqslant d(x_0, x) + d(x, y) < (r - \epsilon) + \epsilon = r.$$

Thus $y \in b(x_0, r)$, and this shows that $b(x, \epsilon) \subseteq b(x_0, r)$, as required.

(b) We must verify (T1), (T2) and (T3) for the metric topology $\mathscr{T}_d$ on a metric space (X, d).

We have $\varnothing \in \mathscr{T}_d$, by definition, and clearly the whole space X is open, so $X \in \mathscr{T}_d$. This confirms (T1).

Let $\mathscr{S}$ be any subcollection of $\mathscr{T}_d$ and consider $\bigcup_{T \in \mathscr{S}} T$. Take any $x \in \bigcup_{T \in \mathscr{S}} T$; then $x \in T$ for some $T \in \mathscr{S}$. Since $T \in \mathscr{T}_d$ (that is, T is an open set), there is an open ball $b(x, r)$ such that $b(x, r) \subseteq T \subseteq \bigcup_{T \in \mathscr{S}} T$. This shows that $\bigcup_{T \in \mathscr{S}} T \in \mathscr{T}_d$, confirming (T2).

Let $T_1, T_2 \in \mathscr{T}_d$. Take any $x \in T_1 \cap T_2$. Then $x \in T_1$ and, since T_1 is open, there is an open ball $b(x, r_1) \subseteq T_1$. Similarly, there is an open ball $b(x, r_2) \subseteq T_2$. Let $r = \min\{r_1, r_2\}$. Then $b(x, r) \subseteq b(x, r_i) \subseteq T_i$ for $i = 1$ and 2, so that $b(x, r) \subseteq T_1 \cap T_2$. This shows that $T_1 \cap T_2 \in \mathscr{T}_d$, confirming (T3).

(7) To prove that $\{x\}$ is a closed subset of a Hausdorff space X, for any $x \in X$, we will prove that $\{x\}$ contains its cluster points. Let y be any cluster point of $\{x\}$ and assume that $y \notin \{x\}$. Then we have $y \neq x$. Since X is a Hausdorff space, there exist neighbourhoods U_x of x and U_y of y such that $U_x \cap U_y = \varnothing$. Hence the neighbourhood U_y of y does not contain any point of $\{x\}$. This contradicts the fact that y is a cluster point of x. Hence we must have $y \in \{x\}$.

(10) (a) Let $\{x_n\}$ be a sequence in $(X, \mathscr{T}_{\min})$ and let x be any point in this space. Since $\mathscr{T}_{\min} = \{\varnothing, X\}$, so X is the only neighbourhood of x. Furthermore, x_n is a point in this neighbourhood for all $n \geqslant 1$. Hence, $x_n \to x$.

(b) Let $\{x_n\}$ be a convergent sequence in a Hausdorff space X, and suppose that $\{x_n\}$ has two distinct limits, x and y. Then $x, y \in X$ and $x \neq y$, so there exist disjoint neighbourhoods U_x of x and U_y of y. Since $x_n \to x$ and $x_n \to y$, there exist positive integers N_1 and N_2 such that $x_n \in U_x$ for $n > N_1$, and $x_n \in U_y$ for $n > N_2$. Then, if $N = \max\{N_1, N_2\}$, we have $x_n \in U_x \cap U_y$ for $n > N$, and this contradicts the fact that $U_x \cap U_y = \varnothing$. Hence a convergent sequence in a Hausdorff space cannot have distinct limits: the limit must be unique.

(11) We use the fact that a mapping $A \colon (X, \mathscr{T}_1) \to (X, \mathscr{T}_2)$ is continuous if and only if $A^{-1}(T) \in \mathscr{T}_1$ whenever $T \in \mathscr{T}_2$ (from Theorem 5.4.4). Suppose the identity map $I \colon (X, \mathscr{T}_1) \to (X, \mathscr{T}_2)$ is continuous, and let $T \in \mathscr{T}_2$. Then $T = I^{-1}(T) \in \mathscr{T}_1$. Hence $\mathscr{T}_2 \subseteq \mathscr{T}_1$. Conversely, suppose $\mathscr{T}_2 \subseteq \mathscr{T}_1$ and let $T \in \mathscr{T}_2$. Then $I^{-1}(T) = T \in \mathscr{T}_1$, so I is continuous. We have shown that I is continuous if and only if $\mathscr{T}_2 \subseteq \mathscr{T}_1$, that is, if and only if $\mathscr{T}_1$ is stronger than $\mathscr{T}_2$.

Exercises 6.4

(3)(b) Noticing that

$$\alpha_n x_n - \alpha x = \alpha(x_n - x) + (\alpha_n - \alpha)x + (\alpha_n - \alpha)(x_n - x)$$

makes the following proof not so obscure.

Take any $\epsilon > 0$. Then let $\eta > 0$ be such that $\eta|\alpha| < \frac{1}{3}\epsilon$ (which allows for the possibility that $\alpha = 0$), $\eta\|x\| < \frac{1}{3}\epsilon$ (which allows for the possibility that x is the zero vector) and $\eta^2 < \frac{1}{3}\epsilon$. Since $x_n \to x$ and $\alpha_n \to \alpha$, we can find positive integers N_1, N_2 such that $\|x_n - x\| < \eta$ when $n > N_1$ and $|\alpha_n - \alpha| < \eta$ when $n > N_2$. Then, provided n is greater than both N_1 and N_2, we have

$$\|\alpha_n x_n - \alpha x\| \leqslant |\alpha|\,\|x_n - x\| + |\alpha_n - \alpha|\,\|x\| + |\alpha_n - \alpha|\,\|x_n - x\|$$
$$< \eta|\alpha| + \eta\|x\| + \eta^2 < \epsilon.$$

Hence $\alpha_n x_n \to \alpha x$.

(5) Suppose $\|x\| \leqslant M$ for some $M > 0$ and all $x \in S$ and let d be the metric induced by the norm $\|\ \|$. Take any points $x, y \in S$. Then $\|x\| \leqslant M$ and $\|y\| \leqslant M$, and

$$d(x,y) = \|x - y\| \leqslant \|x\| + \|-y\| = \|x\| + \|y\| \leqslant 2M.$$

Hence $\sup\{d(x,y) : x, y \in S\} \leqslant 2M$, so S has a finite diameter. Thus S is bounded.

Conversely, suppose S is bounded, with diameter Δ. Let x_0 be some particular point of S. Then, for any $x \in S$,

$$\|x\| = \|(x - x_0) + x_0\| \leqslant \|x - x_0\| + \|x_0\| \leqslant \Delta + \|x_0\|.$$

That is, taking $M = \Delta + \|x_0\|$, we have $\|x\| \leqslant M$ for all $x \in S$, as required to complete the proof.

(6) We will give the verification of (N3) here, for the normed space P. Let p, q, where $p(t) = a_0 + a_1 t + \cdots + a_n t^n$, $q(t) = b_0 + b_1 t + \cdots + b_m t^m$ be elements of P and assume that $n \geqslant m$. Then

$$(p + q)(t) = (a_0 + b_0) + (a_1 + b_1)t + \cdots + (a_m + b_m)t^m$$
$$+ a_{m+1}t^{m+1} + \cdots + a_n t^n,$$

so that

$$\|p + q\| = |a_0 + b_0| + |a_1 + b_1| + \cdots + |a_m + b_m| + |a_{m+1}| + \cdots + |a_n|$$
$$\leqslant (|a_0| + |a_1| + \cdots + |a_n|) + (|b_0| + |b_1| + \cdots + |b_m|)$$
$$= \|p\| + \|q\|.$$

This confirms (N3).

However, P is not a Banach space. To see this, consider (among many possible examples) the sequence $\{p_n\}$ in P given by

$$p_n(t) = \sum_{k=0}^{n} \frac{t^k}{k!}, \qquad n \in \mathbf{N}.$$

If $n > m$, then we have $p_n(t) - p_m(t) = \sum_{k=m+1}^{n} t^k/k!$, so

$$\|p_n - p_m\| = \frac{1}{(m+1)!} + \cdots + \frac{1}{n!}.$$

Since $\sum_{k=0}^{\infty} 1/k!$ is a convergent series, this is arbitrarily small for m sufficiently large (by the Cauchy convergence criterion), so $\{p_n\}$ is a Cauchy sequence in P. But the only candidate for $\lim p_n$ must be the function p given by $p(t) = e^t$. Since the exponential function is not a polynomial function, $\{p_n\}$ is not convergent in P.

Exercises 6.10

(6) For $-1 \leqslant x \leqslant 1$,

$$T_6(x) = 32x^6 - 48x^4 + 18x^2 - 1,$$
$$T_7(x) = 64x^7 - 112x^5 + 56x^3 - 7x,$$
$$T_8(x) = 128x^8 - 256x^6 + 160x^4 - 32x^2 + 1;$$
$$x^6 = \tfrac{1}{32}T_6(x) + \tfrac{3}{16}T_4(x) + \tfrac{15}{32}T_2(x) + \tfrac{5}{16}T_0(x),$$
$$x^7 = \tfrac{1}{64}T_7(x) + \tfrac{7}{64}T_5(x) + \tfrac{21}{64}T_3(x) + \tfrac{35}{64}T_1(x),$$
$$x^8 = \tfrac{1}{128}T_8(x) + \tfrac{1}{16}T_6(x) + \tfrac{7}{32}T_4(x) + \tfrac{7}{16}T_2(x) + \tfrac{35}{128}T_0(x).$$

(10) Answer: (a) $\tfrac{1}{8} + x$, (b) $\tfrac{17}{24} + \tfrac{1}{3}x$.

(11) If the required function is $a + bx$, $0 \leqslant x \leqslant 1$, for some a, b, then the error function E is given by

$$E(x) = a + bx - \frac{1}{1+x}, \qquad 0 \leqslant x \leqslant 1.$$

Let the maximum absolute error be E_M, occurring when $x = \xi$. A sketch of the graph of $1/(1+x)$ on $[0, 1]$ will confirm that to find a and b (and E_M and ξ) we must solve the equations

$$a - 1 = -E_M, \quad a + b\xi - \frac{1}{1+\xi} = E_M,$$
$$a + b - \frac{1}{2} = -E_M, \quad b + \frac{1}{(1+\xi)^2} = 0.$$

We find that $b = -\frac{1}{2}$ and $a = (2\sqrt{2}+1)/4$, so the linear function that is the best uniform approximation of $1/(1+x)$ on $[0,1]$ is $0.957 - 0.5x$, using three decimal places for a.

(12) Let $A: X \to Y$ be a uniformly continuous mapping, for normed spaces X, Y. Then, given $\epsilon > 0$, there exists a number $\delta > 0$ such that $\|Ax' - Ax''\| < \epsilon$ whenever $x', x'' \in X$ and $\|x' - x''\| < \delta$. Now let $\{x_n\}$ be a Cauchy sequence in X. Then for this value of δ there exists a positive integer N such that $\|x_n - x_m\| < \delta$ whenever $m, n > N$. But then $\|Ax_n - Ax_m\| < \epsilon$ whenever $m, n > N$, so $\{Ax_n\}$ is a Cauchy sequence in Y.

(16) Answer: (a) $\frac{1}{2}\sqrt{2} + \frac{1}{4} - \frac{1}{2}x^2$, (b) $\frac{1}{8} + x^2$.

Exercises 7.5

(2) First, consider A as a mapping from $C_1[a,b]$ into itself. Then, for any $x \in C_1[a,b]$, we have $\|x\| = \int_a^b |x(t)|\,dt$. Thus

$$\|Ax\| = \|y\| = \int_a^b |y(s)|\,ds = \int_a^b \left|\lambda \int_a^b k(s,t)x(t)\,dt\right| ds$$

$$\leqslant |\lambda| \int_a^b \int_a^b |k(s,t)|\,|x(t)|\,dt\,ds \leqslant |\lambda|M \int_a^b \left(\int_a^b |x(t)|\,dt\right) ds$$

$$= |\lambda|M(b-a)\|x\|,$$

for all $x \in C_1[a,b]$. Hence A is bounded and $\|A\| \leqslant |\lambda|M(b-a)$.

Now consider A as a mapping from $C_2[a,b]$ into itself. Then, for any $x \in C_2[a,b]$, we have $\|x\| = \sqrt{\int_a^b (x(t))^2\,dt}$. Using the integral form of the Cauchy–Schwarz inequality we have, for $a \leqslant s \leqslant b$,

$$(y(s))^2 = \lambda^2 \left(\int_a^b k(s,t)x(t)\,dt\right)^2$$

$$\leqslant \lambda^2 \left(\int_a^b (k(s,t))^2\,dt\right)\left(\int_a^b (x(t))^2\,dt\right) \leqslant \lambda^2 M^2(b-a)\|x\|^2.$$

Then, for any $x \in C_2[a,b]$, we have

$$\|Ax\| = \|y\| = \sqrt{\int_a^b (y(s))^2\,ds} \leqslant \sqrt{\int_a^b \lambda^2 M^2(b-a)\|x\|^2\,ds}$$

$$= |\lambda|M\sqrt{b-a}\|x\|\sqrt{\int_a^b ds} = |\lambda|M(b-a)\|x\|.$$

Hence A is bounded and $\|A\| \leqslant |\lambda| M(b-a)$.

(7) Using the definition of the norm for l_2, we have

$$|f(x)| = |x_j| \leqslant \sqrt{\sum_{k=1}^{\infty} |x_k|^2} = \|x\|,$$

so f is bounded and $\|f\| \leqslant 1$. Consider now $x_0 = (0, \ldots, 0, 1, 0, \ldots) \in l_2$, in which the jth component is 1 and the others are all 0. Then $\|x_0\| = 1$ and $f(x_0) = 1$. If $\|f\| < 1$, then $1 = |f(x_0)| \leqslant \|f\| \, \|x_0\| < 1 \cdot 1 = 1$, a clear contradiction. Therefore, $\|f\| = 1$.

(10) Consider the normed space $X \times Y$, which we will assume initially to be normed by $\|(x,y)\| = \|x\| + \|y\|$ ($x \in X$, $y \in Y$). Let $\{(x_n, y_n)\}$ be a Cauchy sequence in this space. Then, given $\epsilon > 0$, there is a positive integer N such that

$$\|(x_n, y_n) - (x_m, y_m)\| = \|(x_n - x_m, y_n - y_m)\|$$
$$= \|x_n - x_m\| + \|y_n - y_m\| < \epsilon$$

whenever $m, n > N$. Thus $\|x_n - x_m\| < \epsilon$ and $\|y_n - y_m\| < \epsilon$ whenever $m, n > N$ and so $\{x_n\}$ is a Cauchy sequence in X and $\{y_n\}$ is a Cauchy sequence in Y. Since X and Y are Banach spaces, so $\{x_n\}$ and $\{y_n\}$ converge in X and Y, respectively. Put $x = \lim x_n$ and $y = \lim y_n$, so $x \in X$, $y \in Y$, and therefore $(x,y) \in X \times Y$. We will show that $(x_n, y_n) \to (x,y)$. Let K_1, K_2 be positive integers such that

$$\|x_n - x\| < \frac{\epsilon}{2} \text{ for } n > K_1 \quad \text{and} \quad \|y_n - y\| < \frac{\epsilon}{2} \text{ for } n > K_2.$$

If we take $K = \max\{K_1, K_2\}$, then

$$\|(x_n, y_n) - (x, y)\| = \|(x_n - x, y_n - y)\|$$
$$= \|x_n - x\| + \|y_n - y\| < \frac{\epsilon}{2} + \frac{\epsilon}{2} = \epsilon$$

when $n > K$. Hence $\{(x_n, y_n)\} \to (x,y)$, so $X \times Y$ is a Banach space when X and Y are Banach spaces.

It follows that the same is true when $X \times Y$ is normed alternatively by $\|(x,y)\|' = \max\{\|x\|, \|y\|\}$, since the norms $\|\ \|'$ and $\|\ \|$ are equivalent (Exercise 7.5(9)).

(11) Let X, Y be normed spaces and $A \colon X \to Y$ be an operator, so that the mapping A is linear and bounded. Let $\{x_n\}$ be a sequence in X which converges to x, say, and which is such that the sequence $\{Ax_n\}$

in Y is also convergent, to y, say. To show that A is closed, we must show that $Ax = y$. For this, we have

$$
\begin{aligned}
0 \leqslant \|Ax - y\| &= \|(Ax - Ax_n) + (Ax_n - y)\| \\
&\leqslant \|Ax - Ax_n\| + \|Ax_n - y\| \\
&= \|A(x - x_n)\| + \|Ax_n - y\| \\
&\leqslant \|A\| \, \|x - x_n\| + \|Ax_n - y\|.
\end{aligned}
$$

Since $\|x_n - x\| \to 0$ and $\|Ax_n - y\| \to 0$, we must have $\|Ax - y\| = 0$, or $Ax = y$, as required.

Exercises 7.9

(1) (a) Introduce the operator $K \colon C[a,b] \to C[a,b]$ by $Kx = y$, where $y(s) = \lambda \int_a^b k(s,t)x(t)\,dt$, for $x \in C[a,b]$ and $a \leqslant s \leqslant b$. Then the given Fredholm equation may be given as $f(s) = x(s) - (Kx)(s)$, for $a \leqslant s \leqslant b$, or simply $f = (I - K)x$, where I is the identity operator on $C[a,b]$.

(b) It is known (see Exercise 7.5(2)) that $\|K\| \leqslant |\lambda| M(b-a)$, so $\|K\| < 1$ if $|\lambda| < 1/M(b-a)$. Then it is a known result (Theorem 7.6.5) that $(I - K)^{-1}$ exists and $(I - K)^{-1}f = \sum_{j=0}^\infty K^j f$. But from $f = (I - K)x$, we have $x = (I - K)^{-1}f$, and the result follows.

(2) (a) We use mathematical induction. The result is true when $n = 1$, since $k_1 = k$. Assume the result is true when $n = m$, where $m \geqslant 1$, and suppose that $y = K^{m+1}x$. Then

$$
\begin{aligned}
y(s) = (K^{m+1}x)(s) &= (K(K^m x))(s) \\
&= \lambda \int_a^b k(s,u)\left(\lambda^m \int_a^b k_m(u,t)x(t)\,dt\right)du \\
&= \lambda^{m+1} \int_a^b \int_a^b k(s,u)k_m(u,t)x(t)\,dt\,du \\
&= \lambda^{m+1} \int_a^b \left(\int_a^b k(s,u)k_m(u,t)\,du\right)x(t)\,dt \\
&= \lambda^{m+1} \int_a^b k_{m+1}(s,t)x(t)\,dt.
\end{aligned}
$$

This shows that the result is then also true when $n = m+1$, and hence it is true for all $n \in \mathbf{N}$.

(b) Again, use induction. The result is clearly true when $n = 1$.

Assume the result is true when $n = m$, where $m \geqslant 1$. Then

$$|k_{m+1}(s,t)| = \left| \int_a^b k(s,u)k_m(u,t)\,du \right| \leqslant \int_a^b |k(s,u)|\,|k_m(u,t)|\,du$$

$$\leqslant M \cdot M^m (b-a)^{m-1} \int_a^b du = M^{m+1}(b-a)^m,$$

so the result is true also when $n = m+1$. Hence it is true for all $n \in \mathbf{N}$.

(3) Putting the preceding results together, we have, if $|\lambda| < 1/M(b-a)$,

$$x(s) = \sum_{j=0}^{\infty} (K^j f)(s) = f(s) + \sum_{j=1}^{\infty} \lambda^j \int_a^b k_j(s,t)f(t)\,dt$$

$$= f(s) + \int_a^b f(t) \sum_{j=1}^{\infty} \lambda^j k_j(s,t)\,dt.$$

(4) Answer: (a) $x(s) = s$, (b) $x(s) = \sin s + 24s/(96 - \pi^3)$, (c) $x(s) = f(s) + (s/3) \int_1^2 (f(t)/t)\,dt$.

(5) Answer: (a) $x(s) = \sin s$, (b) $x(s) = s^2 e^s$, (c) $x(s) = se^{s-1}$.

(8) (a) Observe first that if $(\gamma A)x = \theta$ then $\gamma(Ax) = \theta$, so $Ax = \theta$ since $\gamma \neq 0$. But then $x = \theta$ since A^{-1} exists, and this implies that $(\gamma A)^{-1}$ exists. (We have made two applications of Theorem 7.6.2.) Therefore, if $y = (\gamma A)x$, where $x \in X$, then $x = (\gamma A)^{-1}y$. On the other hand, we have $y = \gamma(Ax)$, so that $Ax = \gamma^{-1}y$. Then $x = A^{-1}(\gamma^{-1}y) = \gamma^{-1}A^{-1}y$. Hence $(\gamma A)^{-1} = \gamma^{-1}A^{-1}$.

(b) Let $\gamma = 1 + \alpha \neq 0$. We have $A + E = A + \alpha A = (1+\alpha)A = \gamma A$, so, by (a), $(A+E)^{-1}$ exists and, furthermore,

$$(A+E)^{-1} = (\gamma A)^{-1} = \gamma^{-1}A^{-1} = \frac{1}{1+\alpha}A^{-1}.$$

(c) We have $x = A^{-1}v$ and, from (b),

$$y = (A+E)^{-1}v = \frac{1}{1+\alpha}A^{-1}v.$$

Then

$$\|x - y\| = \left\| A^{-1}v - \frac{1}{1+\alpha}A^{-1}v \right\| = \left\| \frac{\alpha}{1+\alpha}A^{-1}v \right\|$$

$$= \left\| \frac{\alpha}{1+\alpha}x \right\| = \frac{|\alpha|}{|1+\alpha|}\|x\|,$$

and the result follows.

(11) Since B is approximated by $\widetilde{B}$, so there is an operator E on X such that $\widetilde{B} = B + E$. Hence $\widetilde{B} - B = E$ and, since $AB = C$, so $A\widetilde{B} - C = A(B+E) - AB = AE$. Using the result of Exercise 7.9(6)(a), we have both $\|C\| = \|AB\| \leqslant \|A\|\,\|B\|$, so $\|B\| \geqslant \|C\|/\|A\|$, and

$$\|E\| = \|A^{-1}AE\| \leqslant \|A^{-1}\|\,\|AE\|.$$

Thus

$$\frac{\|\widetilde{B} - B\|}{\|B\|} = \frac{\|E\|}{\|B\|} \leqslant \frac{\|A^{-1}\|\,\|AE\|}{\|C\|/\|A\|} = k(A)\frac{\|A\widetilde{B} - C\|}{\|C\|},$$

as required, since $\|A\|\,\|A^{-1}\|$ is the condition number $k(A)$.

Exercises 8.6

(4) Let V be a finite-dimensional vector space and let $\{v_1, \ldots, v_n\}$ be a basis for V. For vectors $x = \sum_{k=1}^{n} \alpha_k v_k$ and $y = \sum_{k=1}^{n} \beta_k v_k$ in V define a mapping $\langle\ ,\ \rangle : V \times V \to \mathbf{C}$ by $\langle x, y \rangle = \sum_{k=1}^{n} \alpha_k \overline{\beta}_k$. It is straightforward to verify that this defines an inner product for V.

(5) We prove that $\{\langle x_n, y_n \rangle\}$ is a Cauchy sequence in $\mathbf{C}$. Then the sequence will be convergent, since $\mathbf{C}$ is complete. Notice first that since $\{x_n\}$ and $\{y_n\}$ are Cauchy sequences, they are bounded so there exist positive constants K, L such that $\|x_n\| \leqslant K$, $\|y_n\| \leqslant L$ for all $n \in \mathbf{N}$. Let $\epsilon > 0$ be given and let $N_1, N_2 \in \mathbf{N}$ be such that $\|x_n - x_m\| < \epsilon/(2L)$ when $n > N_1$ and $\|y_n - y_m\| < \epsilon/(2K)$ when $n > N_2$. We then have, provided $m, n > \max\{N_1, N_2\}$,

$$
\begin{aligned}
|\langle x_n, y_n \rangle - \langle x_m, y_m \rangle| &= |\langle x_n, y_n \rangle - \langle x_n, y_m \rangle + \langle x_n, y_m \rangle - \langle x_m, y_m \rangle| \\
&= |\langle x_n, y_n - y_m \rangle + \langle x_n - x_m, y_m \rangle| \\
&\leqslant |\langle x_n, y_n - y_m \rangle| + |\langle x_n - x_m, y_m \rangle| \\
&\leqslant \|x_n\|\,\|y_n - y_m\| + \|x_n - x_m\|\,\|y_m\| \\
&< K\frac{\epsilon}{2K} + L\frac{\epsilon}{2L} = \epsilon,
\end{aligned}
$$

using the general Cauchy–Schwarz inequality. Hence $\{\langle x_n, y_n \rangle\}$ is a Cauchy sequence, as required.

(10) Since $x \perp y$, we have $\langle x, y \rangle = \langle y, x \rangle = 0$. Also, $\langle x, x \rangle = \|x\|^2$ and $\langle y, y \rangle = \|y\|^2$. Then

$$
\begin{aligned}
\|x + y\|^2 &= \langle x + y, x + y \rangle = \langle x, x \rangle + \langle x, y \rangle + \langle y, x \rangle + \langle y, y \rangle \\
&= \|x\|^2 + \|y\|^2,
\end{aligned}
$$

as required. More generally, if $\{x_1, \ldots, x_n\}$ is an orthogonal set in X, we have

$$\left\| \sum_{k=1}^{n} x_k \right\|^2 = \left\langle \sum_{k=1}^{n} x_k, \sum_{j=1}^{n} x_j \right\rangle = \sum_{k=1}^{n} \sum_{j=1}^{n} \langle x_k, x_j \rangle.$$

But $\langle x_k, x_j \rangle = 0$ for $k \neq j$, so the terms on the right are all zero except for those with $k = j$. Since $\langle x_k, x_k \rangle = \|x_k\|^2$ for $k = 1, \ldots, n$, it follows that $\| \sum_{k=1}^{n} x_k \|^2 = \sum_{k=1}^{n} \|x_k\|^2$, as required.

(11) (a) Answer:

$$\left\{ \left(\frac{1}{\sqrt{2}}, \frac{1}{\sqrt{2}}, 0 \right), \left(\frac{1}{\sqrt{6}}, \frac{-1}{\sqrt{6}}, \frac{2}{\sqrt{6}} \right), \left(\frac{-1}{\sqrt{3}}, \frac{1}{\sqrt{3}}, \frac{1}{\sqrt{3}} \right) \right\},$$

using the vectors in the given order.

(b) Answer: Obtain

$$\left\{ \left(\frac{1}{\sqrt{3}}, \frac{1}{\sqrt{3}}, \frac{i}{\sqrt{3}} \right), \left(\frac{2-i}{2\sqrt{3}}, \frac{-1+2i}{2\sqrt{3}}, \frac{1-i}{2\sqrt{3}} \right), \left(\frac{i}{2}, \frac{1}{2}, \frac{1-i}{2} \right) \right\}$$

using the vectors in the given order, or $\{(i, 0, 0), (0, i, 0), (0, 0, i)\}$ using the vectors in the reverse order.

(c) Answer: $\{u_1, u_2, u_3\}$, where

$$u_1 = \left(\frac{2}{\sqrt{5}}, 0, \frac{1}{\sqrt{5}}, 0 \right),$$

$$u_2 = \left(\frac{-2}{3\sqrt{5}}, 0, \frac{4}{3\sqrt{5}}, \frac{5}{3\sqrt{5}} \right),$$

$$u_3 = \left(\frac{-2}{3\sqrt{13}}, \frac{3}{\sqrt{13}}, \frac{4}{3\sqrt{13}}, \frac{-4}{3\sqrt{13}} \right),$$

using the vectors in the given order.

(13) Answer: (a) $\frac{4}{15} + \frac{4}{5}x$, (b) $\frac{20}{27} + \frac{44}{135}x$.

Exercises 9.5

(2) Let A be an operator on a Hilbert space X.

By definition of the adjoint A^*, and using the fact that $A^{**} = A$, we have, for all $x, y \in X$, $\langle A^*Ax, y \rangle = \langle Ax, A^{**}y \rangle = \langle Ax, Ay \rangle = \langle x, A^*Ay \rangle$. Hence $(A^*A)^* = A^*A$, so A^*A is self-adjoint.

Next, for all $x, y \in X$,

$$\langle (A + A^*)x, y \rangle = \langle Ax + A^*x, y \rangle = \langle Ax, y \rangle + \langle A^*x, y \rangle$$
$$= \langle x, A^*y \rangle + \langle x, Ay \rangle = \langle x, A^*y + Ay \rangle = \langle x, (A^* + A)y \rangle.$$

Hence $(A + A^*)^* = A^* + A = A + A^*$, so $A + A^*$ is self-adjoint. The proof that $i(A - A^*)$ is self-adjoint is similar.

(4) Answer: The eigenvalues and corresponding eigenvectors are -1, $(\sqrt{3}, 3i, -i)^T$; $2 + \frac{1}{3}\sqrt{3}$, $(2i, \sqrt{3}+1, \sqrt{3}+3)^T$; and $2 - \frac{1}{3}\sqrt{3}$, $(2i, \sqrt{3}-1, \sqrt{3}-3)^T$.

(8) (a) In order that $Ax = \lambda x$ for some number λ and some nonzero point $x = (x_1, x_2, \dots) \in l_2$, we must have

$$(x_2, x_3, x_4, \dots) = (\lambda x_1, \lambda x_2, \lambda x_3, \dots).$$

Taking $x_1 = 1$, this implies $x_k = \lambda^{k-1}$ for $k = 1, 2, \dots$. Notice that, by definition of l_2, $x = (1, \lambda, \lambda^2, \dots) \in l_2$ provided $\sum_{k=0}^\infty |\lambda|^{2k}$ is convergent. This is the case when $|\lambda| < 1$. Thus any such number λ is an eigenvalue of A and $(1, \lambda, \lambda^2, \dots)$ is a corresponding eigenvector.

(b) Take any points $x = (x_1, x_2, \dots) \in l_2$ and $y = (y_1, y_2, \dots) \in l_2$, and suppose $A^*y = z = (z_1, z_2, \dots) \in l_2$. Using the definition of the inner product in l_2, we have

$$\langle Ax, y \rangle = \langle (x_2, x_3, \dots), (y_1, y_2, \dots) \rangle = x_2 \bar{y}_1 + x_3 \bar{y}_2 + x_4 \bar{y}_3 + \cdots,$$
$$\langle x, A^*y \rangle = \langle (x_1, x_2, \dots), (z_1, z_2, \dots) \rangle = x_1 \bar{z}_1 + x_2 \bar{z}_2 + x_3 \bar{z}_3 + \cdots.$$

Since $\langle Ax, y \rangle = \langle x, A^*y \rangle$ for any $x, y \in l_2$, we must have $z_1 = 0$, $z_2 = y_1$, $z_3 = y_2$, $\dots$, so the adjoint A^* is given by

$$A^*(y_1, y_2, y_3, \dots) = (0, y_1, y_2, \dots).$$

If λ is an eigenvalue of A^*, then $A^*y = \lambda y = (\lambda y_1, \lambda y_2, \lambda y_3, \dots)$, implying that $\lambda y_1 = 0$ and $\lambda y_k = y_{k-1}$ for $k = 2, 3, \dots$. It may be checked that whether $\lambda = 0$ or $\lambda \neq 0$, we obtain $y = (0, 0, 0, \dots) = \theta$, the zero of l_2; but the zero vector cannot be an eigenvector. Hence A^* has no eigenvalues.

(c) Let B^* be the adjoint of B. Take any points $x = (x_1, x_2, \dots) \in l_2$ and $y = (y_1, y_2, \dots) \in l_2$. Suppose $B^*y = z = (z_1, z_2, \dots) \in l_2$. By definition of the inner product in l_2, we have

$$\langle Bx, y \rangle = \langle (x_1, \tfrac{1}{2}x_2, \dots), (y_1, y_2, \dots) \rangle = x_1 \bar{y}_1 + \tfrac{1}{2}x_2 \bar{y}_2 + \tfrac{1}{3}x_3 \bar{y}_3 + \cdots,$$
$$\langle x, B^*y \rangle = \langle (x_1, x_2, \dots), (z_1, z_2, \dots) \rangle = x_1 \bar{z}_1 + x_2 \bar{z}_2 + x_3 \bar{z}_3 + \cdots.$$

For these to be equal for all x and y, we must have $z_1 = y_1$, $z_2 = \tfrac{1}{2}y_2$, $z_3 = \tfrac{1}{3}y_3$, $\dots$, so that $B^*y = By$. Hence $B^* = B$, so B is self-adjoint. If λ is an eigenvalue of B, and x is a corresponding eigenvector, then the equation $Bx = \lambda x$ implies $\lambda x_k = x_k/k$, for each $k = 1, 2, \dots$. Since

$x \neq \theta$, there is some k such that $x_k \neq 0$ and thus $\lambda = 1/k$ for this k. Then, for all $j = 1, 2, \ldots$ with $j \neq k$, we have $x_j/k = x_j/j$ so that $x_j = 0$. Taking $x_k = 1$, we may thus give the eigenvector corresponding to the eigenvalue $1/k$ as $(0, \ldots, 0, 1, 0, \ldots)$, where the 1 is in the kth place, for $k = 1, 2, \ldots$. We observe that these eigenvalues are real and the eigenvectors corresponding to distinct eigenvalues are orthogonal, in accord with Theorem 9.2.4.

Exercises 9.8

(1) (a) Take any $\epsilon > 0$. Using the definition of the norm in l_2, we have

$$\left\| x - \sum_{k=1}^{n} x_k e_k \right\| = \left\| \sum_{k=n+1}^{\infty} x_k e_k \right\| = \| (0, \ldots, 0, x_{n+1}, x_{n+2}, \ldots) \|$$

$$= \sqrt{\sum_{k=n+1}^{\infty} |x_k|^2} < \epsilon,$$

provided n is large enough, since $\sum |x_k|^2$ converges. We have shown that $\sum_{k=1}^{n} x_k e_k \to x$, that is, $x = \sum_{k=1}^{\infty} x_k e_k$.

(b) The series for x is absolutely convergent if the series $\sum_{k=1}^{\infty} \| x_k e_k \|$ of real numbers is convergent. Take $x_k = 1/k$ for $k \in \mathbf{N}$. Then

$$\| x_k e_k \| = \left\| \left(0, \ldots, 0, \frac{1}{k}, 0, \ldots \right) \right\|$$

$$= \sqrt{0^2 + \cdots + 0^2 + \frac{1}{k^2} + 0^2 + \cdots} = \frac{1}{k},$$

and $\sum 1/k$ diverges. Hence this is an example in which the series for x is not absolutely convergent.

(3) We must show that the mapping $BA \colon X \to Z$ is a linear bijection that preserves inner products. For this, we make use of the corresponding properties of the mappings $A \colon X \to Y$ and $B \colon Y \to Z$.

To show that BA is onto, take any $z \in Z$. Let $y \in Y$ be such that $By = z$ and let $x \in X$ be such that $Ax = y$. Then $(BA)x = B(Ax) = By = z$, so BA is onto.

To show that BA is one-to-one, suppose that $(BA)x_1 = (BA)x_2$ for $x_1, x_2 \in X$. Then $B(Ax_1) = B(Ax_2)$ so $Ax_1 = Ax_2$, and then $x_1 = x_2$, so BA is one-to-one.

To show that BA is linear, take any $x_1, x_2 \in X$ and any scalars α_1, α_2.

Then

$$(BA)(\alpha_1 x_1 + \alpha_2 x_2) = B(A(\alpha_1 x_1 + \alpha_2 x_2))$$
$$= B(\alpha_1 A x_1 + \alpha_2 A x_2)$$
$$= \alpha_1 B(A x_1) + \alpha_2 B(A x_2)$$
$$= \alpha_1 (BA) x_1 + \alpha_2 (BA) x_2,$$

so BA is linear.

Finally, to show that BA preserves inner products, take $x_1, x_2 \in X$. Then

$$\langle (BA) x_1, (BA) x_2 \rangle = \langle B(A x_1), B(A x_2) \rangle = \langle A x_1, A x_2 \rangle = \langle x_1, x_2 \rangle,$$

so BA preserves inner products.

(5) Take any $\epsilon > 0$. We need to find a positive integer N such that $|G_n(u) - G(u)| < \epsilon$ for all $n > N$ and all $u \in [-\pi, \pi]$. Suppose $|g|^2$ is integrable on $[-\pi, \pi]$ (which will be the case if g is continuous on $[-\pi, \pi]$), and put $H = \int_{-\pi}^{\pi} |g(t)|^2 \, dt$. (We may assume $H > 0$—the result is obvious otherwise.) Since $\{F_n\}$ converges in mean square to f, there exists a positive integer N such that

$$\int_{-\pi}^{\pi} \left(\sum_{k=1}^{n} \langle f, x_k \rangle x_k(t) - f(t) \right)^2 dt < \frac{\epsilon^2}{H}$$

for $n > N$. We now have, for any $u \in [-\pi, \pi]$,

$$|G_n(u) - G(u)| = \left| \int_{-\pi}^{u} \left(\sum_{k=1}^{n} \langle f, x_k \rangle x_k(t) - f(t) \right) g(t) \, dt \right|$$
$$\leq \int_{-\pi}^{u} \left| \sum_{k=1}^{n} \langle f, x_k \rangle x_k(t) - f(t) \right| |g(t)| \, dt$$
$$\leq \sqrt{\int_{-\pi}^{u} \left(\sum_{k=1}^{n} \langle f, x_k \rangle x_k(t) - f(t) \right)^2 dt} \sqrt{\int_{-\pi}^{u} (g(t))^2 \, dt}$$
$$< \sqrt{\frac{\epsilon^2}{H}} \sqrt{H} = \epsilon,$$

using the Cauchy–Schwarz inequality for integrals. The proof is finished.

Index

$\aleph_0$, 19
absolute value, 4, 54
adjoint, 285
annihilator, 279
approximation theory, 149, 192, 280, 283
Arzelá, 148
Ascoli's theorem, 146

$B(X, Y)$, 215
Banach space, 178
basis, 75, 294
  orthonormal, 295
Bernstein polynomials, 202
Bessel's inequality, 270, 299
best approximation, 150, 194, 268
bijection, 11
binomial coefficient, 51
binomial theorem, 51, 201
Bolzano–Weierstrass property, 162
Bolzano–Weierstrass theorem, 23
  for sequences, 40
bounds, lower and upper, 21, 35

c, 19
c, 73
C, 5
$\mathbf{C}^n$, 8, 72, 80, 90, 177, 254
$C[a, b]$, 59, 73, 93, 177
$C_1[a, b]$, 94, 178
$C_2[a, b]$, 94, 178, 255
$C^{(1)}[a, b]$, 74
$C'[a, b]$, 246
$C^w[a, b]$, 265
cardinal number, 19
cartesian product, 7
Cauchy convergence criterion, 41, 45
Cauchy sequence, 101, 178
Cauchy–Schwarz inequality, 88, 96, 256, 257

Chebyshev
  norm, 195
  polynomials, 198, 266
    of the second kind, 267
  theory, 195
closed disc, 108
closed set, 28, 107, 156
closure, 157
cluster point, 22, 35, 159
  greatest, least, 25, 36
collection, 29
compactness, 29, 141
  countable, 162
  relative, 141
comparison test, 48
complement, 7
complete orthonormal set, 295
completeness, 4, 25, 42, 47, 57, 102, 179
complex number, 5
condition number, 239
conjugate, 5, 80
connectedness, 169
continuum hypothesis, 19
contraction constant, 116
contraction mapping, 116
convergence
  absolute, 46, 179
  in mean square, 303
  pointwise, 60
  of a sequence, 36, 43, 98, 164, 178
  of a series, 45, 179
  strong, 308
  uniform, 59
  weak, 308
countability, 14

δ-neighbourhood, 20
de Morgan's laws, 9
definite integral, 58, 65
derivative, 57

Printed in the United States
By Bookmasters